LITHIUM BATTERIES

THE ELECTROCHEMICAL SOCIETY SERIES

ECS-The Electrochemical Society
65 South Main Street
Pennington, NJ 08534-2839
http://www.electrochem.org

A complete list of the titles in this series appears at the end of this volume.

LITHIUM BATTERIES
Advanced Technologies and Applications

Edited by

BRUNO SCROSATI
K. M. ABRAHAM
WALTER VAN SCHALKWIJK
JUSEF HASSOUN

Published by John Wiley & Sons, Inc., Hoboken, New Jersey.
Published simultaneously in Canada.

Library of Congress Cataloging-in-Publication Data:

Lithium batteries : advanced technologies and applications / edited by Bruno Scrosati,
K. M. Abraham, Walter van Schalkwijk, Jusef Hassoun.
    pages cm.
  Includes index.
  ISBN 978-1-118-18365-6 (hardback)
 1. Lithium cells.   I. Scrosati, Bruno.
  TK2945.L58L553 2013
  621.31′2424–dc23

                                                              2012047246

Printed in the United States of America

10 9 8 7 6 5 4 3 2 1

CONTENTS

CONTRIBUTORS

K. M. Abraham, Northeastern University Center of Renewable Energy Technology, Boston, Massachusetts

Khalil Amine, Chemical Sciences and Engineering Division, Argonne National Laboratory, Lemont, Illinois

Catia Arbizzani, Dipartimento di Scienza dei Metalli, Elettrochimica e Tecniche Chimiche, University of Bologna, Bologna, Italy

D. Aurbach, Bar-Ilan University, Ramat-Gan, Israel

P. Barpanda, Department of Chemical System Engineering, University of Tokyo, Tokyo, Japan

I. Belharouak, Chemical Sciences and Engineering Division, Argonne National Laboratory, Lemont, Illinois

Zonghai Chen, Chemical Sciences and Engineering Division, Argonne National Laboratory, Lemont, Illinois

Owen Crowther, MaxPower, Inc., Harleysville, Pennsylvania

Swapnil Dalavi, Department of Chemistry, University of Rhode Island, Kingston, Rhode Island

Libero Damen, Dipartimento di Scienza dei Metalli, Elettrochimica e Tecniche Chimiche, University of Bologna, Bologna, Italy

Hubert Gasteiger, Chemistry Department, Technische Universität München, Munich, Germany

Y. Gofer, Bar Ilan University, Ramat-Gan, Israel

A. Guerfi, Institut de Recherche d'Hydro-Québec, Varennes, Québec, Canada

Juan Herranz, Chemistry Department, Technische Universität München, Munich, Germany

Nobuyuki Imanishi, Mie University, Tsu, Japan

Per Jacobsson, Department of Applied Physics, Chalmers University of Technology, Goteborg, Sweden

Patrik Johansson, Department of Applied Physics, Chalmers University of Technology, Goteborg, Sweden

K. Kanamura, Tokyo Metropolitan University, Tokyo, Japan

Amine Khalil, Electrochemical Technology Program, Chemical Sciences and Engineering Division, Argonne National Laboratory, Argonne, Illinois

Katharina Krischer, Physics Department, Technische Universität München, Munich, Germany

Mariachiara Lazzari, Dipartimento di Scienza dei Metalli, Elettrochimica e Tecniche Chimiche, University of Bologna, Bologna, Italy

Brett L. Lucht, Department of Chemistry, University of Rhode Island, Kingston, Rhode Island

Marina Mastragostino, Dipartimento di Scienza dei Metalli, Elettrochimica e Tecniche Chimiche, University of Bologna, Bologna, Italy

Seung-Taek Myung, Department of Nano Engineering, Sejong University, Seoul, South Korea

Michele Piana, Chemistry Department, Technische Universität München, Munich, Germany

N. Pour, Bar-Ilan University, Ramat-Gan, Israel

Mark Salomon, MaxPower, Inc., Harleysville, Pennsylvania

Bruno Scrosati, Department of Chemistry, University of Rome, Sapienza, Italy

Francesca Soavi, Dipartimento di Scienza dei Metalli, Elettrochimica e Tecniche Chimiche, University of Bologna, Bologna, Italy

Yang-Kook Sun, Department of WCU Energy Engineering and Department of Chemical Engineering, Sejong University, Seoul, South Korea

J.-M. Tarascon, Laboratoire de Reáctivité et Chimie des Solides, Université de Picardie Jules Verne, Amiens, France

Nikolaos Tsiouvaras, Chemistry Department, Technische Universität München, Munich, Germany

A. Vijh, Institut de Recherche d'Hydro-Québec, Varennes, Québec, Canada

Susanne Wilken, Department of Applied Physics, Chalmers University of Technology, Goteborg, Sweden

Mengqing Xu, Department of Chemistry, University of Rhode Island, Kingston, Rhode Island

Osamu Yamamoto, Mie University, Tsu, Japan

K. Zaghib, Institut de Recherche d'Hydro-Québec, Varennes, Québec, Canada

PREFACE

Lithium-ion batteries are indispensable for everyday life as power sources for laptop and tablet computers, cellular telephones, e-book readers, digital cameras, power tools, electric vehicles, and numerous other portable devices. The exponential evolution of these batteries from being a laboratory curiosity only three decades ago to multibillion-dollar consumer products today has been nothing short of spectacular. This success has come from the contributions of many scientists and engineers from research laboratories around the world on electrode materials, nonaqueous electrolytes, membrane separators, and engineering and manufacturing of cells and battery packs. Early research on rechargeable lithium batteries focused on systems based on lithium metal anodes (negative electrodes) and lithium intercalation cathodes (positive electrodes). Progress to develop a lithium metal anode-based practical rechargeable battery was slow, due to the less than satisfactory rechargeability of the lithium metal anode coupled with its safety hazards. While it was recognized early on that many of the problems of the rechargeable lithium metal anode could be solved by replacing it with a lithium intercalation anode, a practically attractive solution had to wait for the discovery that lithiated carbon could be charged and discharged in an appropriate organic electrolyte solution that produced a stable surface film, known as the solid electrolyte interphase, on the graphite electrode. Thus, lithium-ion batteries emerged with graphite anodes (negative electrodes) and lithitated metal dioxide cathodes in which complementary lithium intercalation (insertion) and deintercalation (extraction) processes occur in the anode and cathode during charge–discharge cycling.

Rapid progress in the development of new electrode and electrolyte materials followed, with a concomitant increase in the energy density of commercial lithium-ion cells, which has more than doubled in the last two decades. Commercial 18650 cells today have gravimetric energy densities of about 250 Wh/kg and volumetric energy densities approaching 650 Wh/L. Lithium-ion battery cells and packs are now manufactured and sold with a variety of cathode materials tailored to myriad applications. Commercial lithium-ion batteries are available with three classes of cathode materials: lithiated layered transition metal dioxides, Li_xMO_2, where M = Co, Ni, Mn, or their mixtures; transition metal spinel oxides, LiM_2O_4, in which M = Mn or mixtures of Mn, Co, and Ni; and transition metal phosphates, $LiMPO_4$, where M = Fe. A variety of other cathode materials, which are variations of these or altogether new materials, aimed at higher capacity, longer cycle life, and improved cell safety are being developed, although they are not yet available in commercial cells. The anode material in all commercial lithium-ion cells today is graphite with different manufacturers using different types of graphite for proprietary advantages. Progress is being made in developing higher-capacity anode materials, such as silicon, germanium, and other metal alloys of lithium, as higher-capacity anodes. There is also active research and development of improved electrolytes for longer cycle and shelf life, and better low-temperature performance and safety in lithium-ion batteries.

It is now recognized that despite the spectacular progress in the last two decades in lithium-ion battery materials, engineering, and manufacturing, the energy density of today's lithium-ion batteries are inadequate to meet the energy and power demands of many present and future power-hungry applications of consumer communication devices, power tools, and electric vehicles. Electrode materials and battery chemistries having a step change in energy density and performance must be identified and developed to meet these demands. The goal of this book is to bring attention to this need, with a focus on identifying battery chemistry and electrode and electrolyte materials for future high-energy-density rechargeable batteries. A group of recognized leaders in the various aspects of advanced battery chemistry and materials have contributed to this book, which is directed to university students and to researchers, engineers, and decision makers in academia and in industry. Such a book is not currently available.

Chapter 2 provides a brief account of the history of rechargeable lithium batteries and sets the stage for subsequent chapters. Its evolution from the early lithium metal anode systems to today's lithium-ion batteries is outlined and the key materials and developments in chemistry that have made lithium-ion batteries a household word are identified. To significantly increase the energy density of lithium-ion batteries, new electrode materials, particularly cathode materials, with significantly higher specific capacities are required. Presently, lithium-on battery cathode materials are approaching capacity limits equivalent to the transfer of one electron per transition metal atom or about 250 mAh/g, which is expected to yield 18650 cells with nearly 4 Ah or approximately 300 Wh/kg. As shown in Chapter 8, rechargeable batteries with twice this energy density are needed for electric vehicles capable of a 300-mile driving range on a single charge. Clearly, a paradigm shift in battery chemistry and materials is required to achieve this step change in energy density. Work on advanced cathode materials for lithium-ion batteries is summarized in Chapter 5. Discussed in Chapter 7 is the research being carried out on lithium intercalation electrodes with cathode materials such as transition metal fluorosulfates capable of multielectron transfer per transition metal atom to achieve a potential doubling of the energy density of lithium-ion batteries. However, such lithium intercalation/deintercalation reactions are often fraught with thermodynamic and kinetic difficulties that limit electrode capacity. These limitations must be understood to realize the full capabilities of lithium intercalation electrodes capable of multielectron reactions. Very high-energy-density lithium-ion batteries will ultimately be based not only on these new high-energy-density cathodes but will also utilize high-capacity anodes such as lithium alloys of tin and silicon, as discussed in Chapter 6.

The search for ultrahigh-energy-density rechargeable batteries is focused beyond intercalation cathodes to materials that exhibit displacement-type reactions, such as sulfur and oxygen. Indeed, the $Li-O_2$ battery, commonly called the lithium–air battery, is perhaps the highest-energy-density rechargeable practical battery that could be envisioned. There is a worldwide effort to develop various types of rechargeable lithium–air batteries, as discussed in Chapters 8, 9, 10, and 11. The anode in the lithium–air cell and its close relative with a lower energy density, the lithium–sulfur cell, is lithium metal, which is characterized by recognized shortcomings of cycle life and safety that must be understood and solved.

The discharge–charge rates, rechargeability, and cycle and calendar life of batteries are strongly influenced by electrolytes. Advanced organic and ionic liquid electrolytes are described in Chapters 3, 4 and 14. Utilization of such electrolytes for battery applications is discussed in Chapters 12 and 14.

Finally, alternative anode rechargeable batteries are sought for new, lower-cost battery technologies. Two types of such batteries are the magnesium and sodium anode systems. A review of their state of the art and advantages and limitations are the topics of Chapters 15 and 16.

Although there are other books dealing with the chemistry and materials for various types of lithium-ion batteries, this is the first book devoted exclusively to future rechargeable battery technologies. We expect this book to serve both as a textbook for graduate students and as a general reference book for the wider battery community.

Bruno Scrosati
K. M. Abraham
Walter van Schalkwijk
Jusef Hassoun

CHAPTER 1

ELECTROCHEMICAL CELLS: BASICS

Hubert Gasteiger, Katharina Krischer, and Bruno Scrosati

1 ELECTROCHEMICAL CELLS AND ION TRANSPORT

An electrochemical cell is a device with which electrical energy is converted into chemical energy, or vice versa. We can consider two types: *electrolytic cells*, in which electric energy is converted into chemical energy (corresponding to the charging of a battery), and *galvanic cells*, in which chemical energy is converted into electric energy (corresponding to a battery in discharge). In its most basic structure, an electrochemical cell is formed by two electrodes, one positive and one negative, separated by an ionically conductive and electronically insulating *electrolyte*, which may be a liquid, a liquid imbibed into a porous matrix, an ionomeric polymer, or a solid. At the negative electrode, an oxidation or anodic reaction occurs during discharge (e.g., the release of electrons and lithium ions from a graphite electrode: $LiC_6 \rightarrow C_6 + Li^+ + e^-$), while the process is reversed during charge, when a reduction or cathodic reaction occurs at the negative electrode (e.g., $C_6 + Li^+ + e^- \rightarrow LiC_6$). Even though the negative electrode is in principle an anode during discharge and a

Lithium Batteries: Advanced Technologies and Applications, First Edition.
Edited by Bruno Scrosati, K. M. Abraham, Walter van Schalkwijk, and Jusef Hassoun.
© 2013 John Wiley & Sons, Inc. Published 2013 by John Wiley & Sons, Inc.

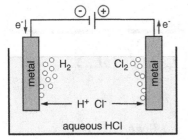

FIGURE 1 Electrolytic cell, illustrating the decomposition of aqueous hydrochloric acid into hydrogen and chlorine. Here, aqueous HCl serves as an ionically conducting and electronically insulating electrolyte, facilitating the overall reaction: $2HCl \rightarrow Cl_2 + H_2$.

reduction (cathode)	oxidation (anode)
$2H^+ + 2e^- \rightarrow H_2$	$2Cl^- \rightarrow Cl_2 + 2e^-$

cathode during charge, the *negative electrode is commonly referred to as an anode* in the battery community (i.e., the discharge process is taken as the nominal defining process). Similarly, a reduction or cathodic reaction occurs at the positive electrode during discharge (e.g., the uptake of lithium ions and electrons by iron phosphate: $FePO_4 + Li^+ + e^- \rightarrow LiFePO_4$), and thus the *positive electrode is commonly referred to as a cathode*, even though, of course, an anodic process occurs on the positive electrode during charge. Since this convention can be somewhat confusing, referring to the electrodes as a negative or positive electrode would elminate the ambiguity introduced by using the terms *anode* and *cathode*.

Figure 1 schematizes an HCl/H_2/Cl_2 *electrolytic cell*. The electrochemical processes are the cathodic reduction of hydrogen ions (protons) at the negative electrode ($2H^+ + 2e^- \rightarrow H_2$) and the anodic oxidation of the chloride ions at the positive electrode ($2Cl^- \rightarrow Cl_2 + 2e^-$). These two *half-cell reactions* can be added up to the overall reaction of this electrolytic cell: namely, the evolution of chlorine and hydrogen from hydrochloric acid (used industrially to recycle waste HCl in chemical plants): $2HCl \rightarrow H_2 + Cl_2$. As illustrated in Figure 1, an electrochemical reaction leads to a flow of electrons in the external circuit which is balanced by the migration of positive ions (cations) to the cathode and of negative ions (anions) to the anode; the principle of electroneutrality demands that the external electronic current must be matched by an internal ionic current (i.e., by the sum of the cation flow to the cathode and the anion flow to the anode).

The reversible cell voltage, $E_{cell,rev}$, also referred to as the *electromotoric force* (emf), can be obtained from the Gibbs free energy change of the reaction, ΔG_R:

$$E_{cell,rev} = -\frac{\Delta G_R}{nF} \tag{1}$$

where n is the number of electrons involved in the electrochemical reaction (for Fig. 1, $n = 2$) and F is the Faraday constant, equal to 96,485 As/mol. The standard Gibbs free energy of reaction, ΔG_R^0, can readily be obtained from the standard Gibbs free energies of formation, ΔG_f^0, as is shown for the HCl electrolysis process above:

$$\Delta G_R^0 = \Delta G_{f(H_2)}^0 + \Delta G_{f(Cl_2)}^0 - 2 \cdot \Delta G_{f(HCl)}^0 \tag{2}$$

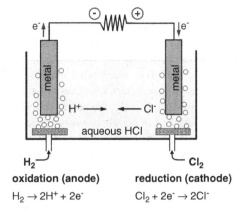

H$_2$
oxidation (anode)
H$_2 \rightarrow$ 2H$^+$ + 2e$^-$

Cl$_2$
reduction (cathode)
Cl$_2$ + 2e$^-$ $\rightarrow$ 2Cl$^-$

FIGURE 2 Galvanic cell, illustrating the reaction between hydrogen and chlorine to yield aqueous hydrochloric acid: Cl$_2$ + H$_2$ $\rightarrow$ 2HCl.

Considering that $\Delta G^0_{f(\mathrm{H}_2)}$ and $\Delta G^0_{f(\mathrm{Cl}_2)}$ are defined to be zero under standard conditions (conventionally defined as 25°C and gas partial pressures of 100 kPa) and taking a value of -131.1 kJ/mol for $\Delta G^0_{f(\mathrm{HCl})}$ at the standard condition for species dissolved in water (conventionally, 1 M solution at 25°C), one can determine the standard Gibbs free energy of reaction of 262.2 kJ/mol. If the reaction is run under these standard conditions, the reversible cell voltage can be calculated from ΔG^0_R using equation (1):

$$E^0_{\mathrm{cell,rev}} = -\frac{262.2 \text{ kJ/mol}}{2 \times 96{,}485 \text{ As/mol}} = -1.36 \text{ V} \qquad (3)$$

Since ΔG_R is greater than 0, energy is required to decompose HCl into H$_2$ and Cl$_2$; a negative emf value (i.e., $E_{\mathrm{cell,rev}}$ is < 0) therefore means that electric energy is needed to drive the electrochemical reaction, as in fact is expected for an electrolytic cell, which also corresponds to a battery cell during charge.

Figure 2 schematizes an HCl/H$_2$/Cl$_2$ *galvanic cell*. The electrochemical half-cell reactions are the anodic oxidation of hydrogen molecules at the negative electrode (H$_2$ $\rightarrow$ 2H$^+$ + 2e$^-$) and the cathodic reduction of chlorine molecules at the positive electrode (Cl$_2$ + 2e$^-$ $\rightarrow$ 2Cl$^-$), resulting in the overall formation of HCl from hydrogen and chlorine (H$_2$ + Cl$_2$ $\rightarrow$ 2HCl), just the opposite reaction to that of an electrolytic cell. A comparison of Figures 1 and 2 once more illustrates that, for example, the hydrogen electrode is always the negative electrode and that irrespective of whether the cell is operated as an electrolytic or a galvanic cell, the half-cell reaction at the hydrogen electrode is either anodic (Fig. 2) or cathodic (Fig. 1). As discussed for the electrolytic cell (Fig. 1), the flow of electrons through the external circuit has to be exactly balanced by the flow of cations to the cathode and anions to the anode.

The standard Gibbs free energy of reaction for the galvanic cell in Figure 2 is given by

$$\Delta G^0_R = 2 \cdot \Delta G^0_{f(\mathrm{HCl})} - (\Delta G^0_{f(\mathrm{H}_2)} + \Delta G^0_{f(\mathrm{Cl}_2)}) = -262.2 \text{ kJ/mol} \qquad (4)$$

equating to a reversible cell voltage of $E_{\mathrm{cell,rev}} = +1.36$ V [see equation (1)] under standard conditions. The fact that ΔG_R is < 0 and $E_{\mathrm{cell,rev}}$ is > 0 signifies that the reaction proceeds spontaneously by converting chemical energy into electrical energy, as in fact is expected for a galvanic cell or for a battery during discharge.

The rate of energy conversion in galvanic or electrolytic cells is typically expressed as current (in units of amperes, A) or as current density (in units of A/cm^2). It depends on the kinetics of the half-cell reactions as well as on many other materials (e.g., ionic conductivity of the electrolyte) and design parameters (thickness of the electrolyte-gap between positive and negative electrodes). Clearly, the actual size of a galvanic or electrolytic device for a required energy or materials conversion rate decreases with increasing current density, so that the maximum power density of galvanic (e.g., fuel cells) and electrolytic (e.g., chlorine–alkaline electrolyzer) cells is an important figure of merit:

$$P_{electric}(W/cm^2) = E_{cell}\ (V) \times i(A/cm^2) \tag{5}$$

Power densities vary from ≈ 0.1 W/cm^2 for high-power lithium-ion batteries at discharge C rates of $\approx$10 h^{-1} [the *C rate* is defined as the number of times the full capacity of the battery is (dis)charged per hour] to $\approx$1 W/cm^2 for proton-exchange membrane (PEM) fuel cells. Obviously, the higher the achievable power density, the lower the necessary electrode area and, generally, the smaller the device. The efficiency of galvanic and electrolytic cells is often given in terms of the cell voltage efficiency, η_{cell}, which relates the actual cell voltage to the reversible cell voltage of a galvanic or electrolytic cell:

$$\eta_{cell,galvanic} = \frac{E_{cell}}{E_{cell,rev}} \quad \text{and} \quad \eta_{cell,electrolytic} = \frac{E_{cell,rev}}{E_{cell}} \tag{6}$$

For all galvanic and electrolytic cells, the deviation between the actual cell voltage and the reversible cell voltage increases with increasing current density, which means that η_{cell} increases with current and power density.

The ionic conductivity, κ, of an electrolyte solution derives from the movement of anions and cations in the electrolyte solution caused by an electric field, that is, by a gradient of the electrostatic potential in the electrolyte solution phase, $\nabla\phi_s$. The flow of ions produces an electric current which can be expressed in terms of the ionic mobility u, the ion concentration c, and the charge number z:

$$i = -F(z_+c_+u_+ + |z_-|c_-u_-)\nabla\phi_s = -\kappa\nabla\phi_s \tag{7}$$

where the subscripts $+$ and $-$ refer to the cations and anions, respectively, and F is the Faraday constant; the ionic mobility quantifies the terminal velocity of an ion in an electric field and has units of (cm/s)/(V/cm). The ionic conductivity is most commonly expressed in units of S/cm (a siemens, S, is a reciprocal ohm). The conductivity of typical battery electrolytes is on the order of 1 to 10 mS/cm (see Table 1), while the conductivity of aqueous or ionomeric electrolytes used in fuel cells and electrolyzers is on the order of 100 mS/cm.

TABLE 1   Conductivities of Typical Lithium Ion Battery Electrolytes at 25°C and 1 M Salt Concentration

Solvent	Mixing ratio (g/g)	Salt	Conductivity, κ (mS/cm)
EC/DMC	1:1	$LiAsF_6$	11
EC/DMC	1:1	$LiPF_6$	11
EC/DMC/DEC	1:1:1	$LiPF_6$	10
EC/PC	3:2	$LiBF_4$	2.7

EC, ethylene carbonate; DMC, dimethyl carbonate; DEC, diethyl carbonate; PC, propylene carbonate.

2   CHEMICAL AND ELECTROCHEMICAL POTENTIAL

2.1   Temperature Dependence of the Reversible Cell Voltage

The temperature dependence of the Gibbs free energy change of a reaction under isobaric conditions is proportional to the entropy change of reaction, ΔS_R:

$$\left(\frac{\partial \Delta G_R}{\partial T}\right)_{p,n_i} = -\Delta S_R \tag{8}$$

where ΔS_R is the entropy change of the reaction. Combining equation (8) with equation (1) yields the temperature dependence of the reversible cell voltage:

$$\left(\frac{\partial E_{\text{cell,rev}}}{\partial T}\right)_{p,n_i} = \frac{\Delta S_R}{nF} \tag{9}$$

2.2   Chemical Potential

Let us consider a generic chemical reaction: $v_A A + v_B B \rightarrow v_C C + v_D D$. The Gibbs free energy change of reaction, ΔG_R, is given by

$$\Delta G_R = v_C \mu_C + v_D \mu_D - (v_A \mu_A + v_B \mu_B) \tag{10}$$

where μ_i is the chemical potential of species i. The chemical potential defines the change of the Gibbs free energy when an infinitesimal number of moles of species i is added to a mixture, with all other components remaining constant:

$$\mu_i = \left(\frac{\partial G}{\partial n_i}\right)_{p,T,n_{j \neq i}} \tag{11}$$

The concentration dependence of the chemical potential is

$$\mu_i = \mu_i^0 + RT \ln(a_i) \tag{12}$$

where μ_i^0 is the standard chemical potential (conventionally defined at 25°C and 100 kPa$_{\text{abs}}$), R is the gas constant (8.314 J/(mol·K)), and T is the temperature in Kelvin. In the case of chemical activity, a_i, one has to distinguish between the activity of pure solid substances ($a_i = 1$), of gases ($a_i = f_i p_i / p^0$, where f_i is the fugacity coefficient, p_i is the partial pressure, and p^0 is the standard pressure); for ideal gases, $f_i = 1$), of dissolved species ($a_i = \gamma_i c_i / c^0$, where γ_i is the activity coefficient, c_i is the concentration, and c^0 is the standard concentration of 1 mol/L; for ideal solutions, $\gamma_i = 1$), and of pure solvents ($a_i = 1$; e.g., water in aqueous electrolytes).

If two phases (I and II) with a common species i are brought into contact, i is exchanged between the two phases until

$$\mu_i(I) = \mu_i(II) \tag{13}$$

Thus, equation (13) is the thermodynamic condition for phase equilibrium in the absence of charge separation at the interface between the two phases.

2.3 Electrochemical Potential

When a metal (Me) electrode (e.g., Cu) is brought into contact with a solution of its Me^{z+} ions (e.g., an aqueous solution of Cu^{2+}), the following reactions will occur until equilibrium is attained:

$$Cu^0(M) \rightleftharpoons Cu^{2+}(s) + 2e^-(M) \tag{14}$$

This reaction leads to a charge separation between the electrolyte phase (s) and the metal phase (M), which occurs spontaneously when equilibrium is being established between the two phases. As can be seen in Figure 3, two different processes are conceivable when pure copper (Cu^0) is brought into contact with a dissolved copper salt (Cu^{2+}): on the left-hand side, partial dissolution of copper, leading to the accumulation of negative charges in the metal at the metal–solution interface which are counterbalanced by positive ionic charges on the solution side of the interface; on the right-hand side, partial plating of copper, leading to the accumulation of positive charges in the metal at the metal–solution interface which are counterbalanced by negative ionic charges on the solution side of the interface. The accumulation of charges on each side of the metal–solution interface can be described in terms

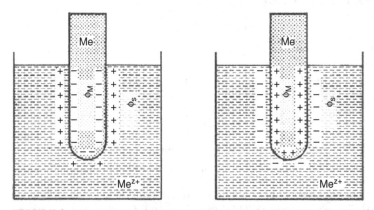

FIGURE 3   Charge separation between the electrode phase (Me phase) and the electrolyte solution phase (Me_z^{+} phase) for cases of spontaneous metal dissolution (left-hand side) and of spontaneous metal deposition (right-hand side). In each case, a potential difference between the metal phase potential, ϕ_M, and the solution phase potential, ϕ_s, is being created. (Adapted from [1].)

of the electrostatic potential in both the metal, ϕ_M, and the electrolyte solution, ϕ_s (see Fig. 3).

In contrast to the simple chemical phase equilibrium described by equation (13), which considers only the chemical equilibrium between species in the two different phases in the absence of charge separation at the interface, description of the electrochemical phase equilibrium for electrochemical reactions where a charge separation across the interface is established requires that the energy of formation of the charge accumulation or depletion at the interface be included in the chemical potential. Therefore, a complete description of the *electrochemical equilibrium* has to consider the electrostatic potential ϕ in each phase, in addition to the chemical potentials of species i in the two phases. With the definition of the electrochemical potential $\tilde{\mu}_i$,

$$\tilde{\mu}_i = \mu_i^0 + RT \ln(a_i) + z_i F \phi \tag{15}$$

the electrochemical phase-equilibrium condition reads

$$\tilde{\mu}_i(\mathrm{I}) = \tilde{\mu}_i(\mathrm{II}) \tag{16}$$

The electrochemical potential difference between the solid and the solution phase is illustrated in Figure 3 for the copper dissolution/plating example: (1) if $\tilde{\mu}_{Cu}^{2+}(M) = \tilde{\mu}_{Cu}(M) - 2\tilde{\mu}_{e^-}(M) > \tilde{\mu}_{Cu^{2+}}(s)$ equilibrium will be established via copper dissolution (left-hand side); and (2) if $\tilde{\mu}_{Cu}(M) < \tilde{\mu}_{Cu^{2+}}(s) + 2\tilde{\mu}_{e^-}(M)$, equilibrium is reached via copper plating (right-hand side). Establishing equilibrium gives rise to the charge separation illustrated in Figure 3 for both copper plating and copper dissolution, creating an electrical potential difference between the two phases which is referred to as the *Galvani potential difference*: $\Delta\phi = \phi_M - \phi_s$. Consequently, the electrochemical equilibrium condition based on equation (16) and using equation (15) for each phase can be written as

$$\begin{aligned} \mu_{Cu}(M) &= \mu_{Cu^{2+}}(s) + 2F\phi_s + 2\,\mu_{e^-}(M) - 2F\phi_M \\ &= \mu_{Cu^{2+}}(s) + 2\mu_{e^-}(M) - 2F\Delta\phi \end{aligned} \tag{17}$$

Inserting the activity dependence of the chemical potential, μ_i, from equation (15), equation (17) can be expanded to

$$\begin{aligned} \mu_{Cu}^0(M) + RT \ln[(a_{Cu}(M)] &= \mu_{Cu^{2+}}^0(s) + RT \ln[(a_{Cu^{2+}}(s)] + 2\mu_{e^-}^0(M) \\ &+ RT \ln[(a_{e^-}(M)] - 2F\Delta\phi \end{aligned} \tag{18}$$

In summary, equation (18) represents a detailed description of the electrochemical phase equilibrium between a copper metal electrode and copper ions in solution, including the Galvani potential difference which is produced during the phase equilibration process.

2.4   The Nernst Equation

Equation (18) can be simplified further by considering that the activity of a pure solid phase is 1 [i.e., $a_{Cu}(M) \equiv 1$] and that the activity of electrons in the metal phase, $a_{e^-}(M)$, will not be affected by establishing the electrochemical phase equilibrium due to the high and essentially unperturbed electron concentration in the metal, so that we can define $a_{e^-}(M) \equiv 1$. Under these assumptions, equation (18) can be rewritten as

$$\Delta\phi - \frac{\mu_{e^-}^0(M)}{F} = \frac{\mu_{Cu^{2+}}^0(s) - \mu_{Cu}^0(M)}{2F} + \frac{RT}{2F} \ln[a_{Cu^{2+}}(s)] \tag{19}$$

By defining the electrode potential E and the standard potential E^0 for the case when the activity of Cu^{2+} ions is 1 (i.e., under standard conditions), equation (19) becomes the well-known *Nernst equation* applied to the Cu/Cu^{2+} redox couple:

$$E \equiv \Delta\phi - \frac{\mu_{e^-}^0(M)}{F} = E^0 + \frac{RT}{2F} \ln[a_{Cu^{2+}}(s)] \tag{20}$$

The specific Nernst equation derived above for the Cu/Cu^{2+} redox couple can also be generalized for a generic half-cell reaction:

$$\nu_A A + \nu_B B + \cdots + ne^- \rightarrow \nu_C C + \nu_D D + \cdots \tag{21}$$

yielding the general Nernst equation:

$$E = E^0 - \frac{RT}{nF} \ln \frac{(a_C)^{\nu_C}(a_D)^{\nu_D} \cdots}{(a_A)^{\nu_A}(a_B)^{\nu_B} \cdots} = E^0 + \frac{RT}{nF} \ln \frac{\prod\limits_{\text{oxidized}} (a_i)^{\nu_i}}{\prod\limits_{\text{reduced}} (a_i)^{\nu_i}} \tag{22}$$

whereby the species on the reduced side of the half-cell reaction (i.e., the side where the electrons are written) are placed in the denominator in the logarithic term and the species on the oxidized side of the half-cell reaction are placed in the numerator. The activity a_i of nonionic species is related to their concentration c_i by the activity coefficient γ_i and the standard concentration c^0 according to $a_i = \gamma_i c_i / c^0$. For ionic species, one often uses the mean ionic activity coefficient, $\gamma_\pm$, instead of γ ($\gamma_\pm$ is defined for neutral cation–anion pairs, since the activity coefficient of single ions cannot be measured). The activity of a solvent is given in terms of the mole fraction x_i and the activity coefficient, which is usually also denoted by γ_i (i.e., $a_i = \gamma_i x_i$). The activity of pure solvents (e.g., of water) is always equal to 1. Finally, for gaseous species the activity is linked to the partial pressure p_i of the species through the fugacity coefficient f_i and the standard pressure p^0 (100 kPa), $a_i = f_i p_i / p^0$. For ideal gases the fugacity becomes the partial pressure of the gas p_i. E^0 is, as described above, the *standard half-cell potential* of the reaction, that is, the potential at which each of the species involved is present at an activity equal to 1.

It is common to list standard half-cell potentials, E^0, as to *standard reduction potentials*, that is, for half-cell reactions written as reduction reactions, as done in equation (21). A series of half-cell reactions and their standard reduction potentials are listed in Table 2.

For example, in the case of the hydrogen electrode reaction with the reduction reaction written on the left side of the electrochemical equation,

$$2H^+(s) + 2e^-(M) \rightleftharpoons H_2(g) \tag{23}$$

TABLE 2   Examples of the Electrochemical Standard Reduction Potentials of Some Common Half-Cell Reactions

Electrode reaction	E^0 (V)	Electrode reaction	E^0 (V)
$Li^+ + e \rightleftharpoons Li$	−3.01	$Tl^+ + e \rightleftharpoons Tl$	−0.34
$Rb^+ + e \rightleftharpoons Rb$	−2.98	$Co^{2+} + 2e \rightleftharpoons Co$	−0.27
$Cs^+ + e \rightleftharpoons Cs$	−2.92	$Ni^{2+} + 2e \rightleftharpoons Ni$	−0.23
$K^+ + e \rightleftharpoons K$	−2.92	$Sn^{2+} + 2e \rightleftharpoons Sn$	−0.14
$Ba^{2+} + 2e \rightleftharpoons Ba$	−2.92	$Pb^{2+} + 2e \rightleftharpoons Pb$	−0.13
$Ca^{2+} + 2e \rightleftharpoons Ca$	−2.84	$H^+ + e \rightleftharpoons 1/2H_2$	0.000
$Na^+ + e \rightleftharpoons Na$	−2.71	$Cu^{2+} + 2e \rightleftharpoons Cu$	0.34
$Mg^{2+} + 2e \rightleftharpoons Mg$	−2.38	$\frac{1}{2}O_2 + H_2O + 2e \rightleftharpoons 2OH^-$	0.40
$Ti^{2+} + 2e \rightleftharpoons Ti$	−1.75	$Cu^+ + e \rightleftharpoons Cu$	0.52
$Be^{2+} + 2e \rightleftharpoons Be$	−1.70	$Hg^{2+} + 2e \rightleftharpoons Hg$	0.80
$Al^{3+} + 3e \rightleftharpoons Al$	−1.66	$Ag^+ + e \rightleftharpoons Ag$	0.80
$Mn^{2+} + 2e \rightleftharpoons Mn$	−1.05	$Pd^{2+} + 2e \rightleftharpoons Pd$	0.83
$Zn^{2+} + 2e \rightleftharpoons Zn$	−0.76	$Ir^{3+} + 3e \rightleftharpoons Ir$	1.00
$Ga^{3+} + 3e \rightleftharpoons Ga$	−0.52	$Br_2 + 2e \rightleftharpoons 2Br^-$	1.07
$Fe^{2+} + 2e \rightleftharpoons Fe$	−0.44	$O_2 + 4H^+ + 4e \rightleftharpoons 2H_2O$	1.23
$Cd^{2+} + 2e \rightleftharpoons Cd$	−0.40	$Cl_2 + 2e \rightleftharpoons 2Cl^-$	1.36
$In^{3+} + 3e \rightleftharpoons In$	−0.34	$F_2 + 2e \rightleftharpoons F^-$	2.87

the Nernst equation would be written

$$E_{H_2/H^+} = E^0_{H_2/H^+} + \frac{RT}{2F} \ln \frac{(a_{H^+})^2}{a_{H_2}} = E^0_{H_2/H^+} + \frac{RT}{F} \ln \frac{a_{H^+}}{(a_{H_2})^{0.5}} \qquad (24)$$

where the H_2 fugacity coefficient is often assumed to be 1 (ideal gas behavior, i.e., $a_{H_2} = p_{H_2}/p^0$) and where the proton activity is often written in terms of pH [i.e., $pH \equiv -\log(a_{H^+})$]. In the example above of the copper electrode, the corresponding half-cell reaction reads

$$Cu^{2+}(aq) + 2e^-(M) \rightleftharpoons Cu(s) \qquad (25)$$

and the Nernst equation can be written

$$E_{Cu/Cu^{2+}} = E^0_{Cu/Cu^{2+}} + \frac{RT}{2F} \ln(a_{Cu^{2+}}) = E^0_{Cu/Cu^{2+}} + \frac{RT}{2F} \ln\left(\frac{\gamma_\pm c_{Cu^{2+}}}{c^0_{Cu^{2+}}}\right) \qquad (26)$$

E cannot be determined experimentally since measurement requires the use of a suitable instrument (e.g., a voltmeter) that necessarily has a second terminal to which a second electrode has to be connected. Commonly, the material of the second electrode will differ from that of the first one. In Figure 4 we assume that the two electrodes are made of metal M and M_1, respectively, and for simplicity we assume that the electrical connections in our voltmeter are made from M_1. Then the voltage measured is equal to the potential drop across three interfaces, the two metal solution interfaces M/S and M_1/S with potential drops $\Delta\phi(M)$ and $\Delta\phi(M_1)$, respectively, and the metal–metal interface M/M'_1, where the prime indicates that the electrostatic potential of M_1, ϕ_M, at the two terminals is, in general, different.

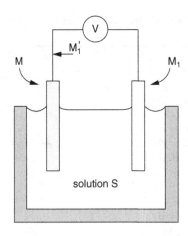

FIGURE 4   Various metal–metal and metal–solution interfaces in an electrochemical cell.

The potential read at the voltmeter is thus given by

$$V = \Delta\phi(M_1) - \Delta\phi(M) + \left(\phi_M - \phi_{M_1'}\right) \tag{27}$$

At the metal–metal interface, the electrochemical potentials adjust, and hence

$$\tilde{\mu}_{e^-}(M_1') = \mu_{e^-}(M_1) - F\phi_{M_1'} = \mu_{e^-}(M) - F\phi_M = \tilde{\mu}_{e^-}(M) \tag{28}$$

Combining equations (27) and (28), we obtain

$$V = \left[\Delta\phi(M_1) + \frac{\mu_{e^-}(M_1)}{F}\right] - \left[\Delta\phi_M(M) - \frac{\mu_{e^-}(M)}{F}\right] \tag{29}$$

Furthermore, when comparing equation (28) with Nernst equation (21), it is obvious that the measured voltage is equal to the difference between the Nernst potentials of the two electrodes:

$$V = E(M_1) - E(M) \tag{30}$$

Since we cannot measure individual electrode potentials, we can compare electrode potentials of two different electrodes only when they have been measured with respect to the same reference electrode (i.e., electrode M_1 in Fig. 4 would be replaced by a reference electrode).

The zero of the electrode potential scale has been chosen arbitrarily as the electrode potential of the standard hydrogen electrode (SHE), with the potential being determined by the H_2/H^+ reaction equation (23) at standard conditions. In practice, this typically involves the use of a Pt electrode immersed in a solution of unit activity of protons in equilibrium with H_2 gas bubbling at a pressure of 100 kPa (see Fig. 5). Electrode potentials given with respect to this zero point are reported "versus SHE."

In addition to the SHE, other reference electrodes can be used. One of the most common is the saturated calomel electrode (SCE) formed by mercury in contact with an insoluble Hg_2Cl_2 paste and a saturated KCl solution (see Fig. 6). Here the electrode potential is determined by the reaction

$$2Hg \rightleftharpoons Hg_2^{2+} + 2e^- \tag{31}$$

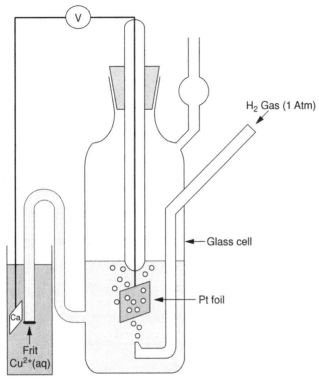

FIGURE 5   Cell for the measurement of an electrode potential (e.g., the Cu/Cu^{2+} electrode) vs. the SHE reference electrode. (From [1].)

The Hg$_2^{2+}$ ions are in equilibrium with insoluble calomel, Hg$_2$Cl$_2$, which, to a small extent, is dissociated in Hg$_2^{2+}$ and Cl$^-$ ions. The electrode potential can, therefore, be expressed through the Cl$^-$ concentration:

$$E = E^0_{Hg/Hg_2Cl_2} - \frac{RT}{F} \ln(a_{Cl}^-) = 0.241 \text{ V vs. SHE} \tag{32}$$

It is always possible to reconvert the electrode potential from the scale based on the chosen reference electrode (e.g., SCE) to that of the SHE:

$$E_{SHE} = E_{SCE} + 0.241 \text{ V} \tag{33}$$

2.5   Electrochemical Double Layer

The interfacial region in which the excess charges on the electrode and in the solution accumulate (see, e.g., Fig. 3) is called the *electrochemical double layer* (DL). On the metal electrode, the excess charge resides in a thin layer (<0.1Å) at the electrode surface. The charge is counterbalanced by the accumulation of ion in the electrolyte close to the electrode (i.e., cations in the case of negative excess charge on the electrode, anions in the case of positive excess charge on the

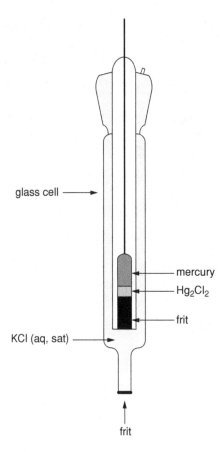

glass cell ⟶

mercury

Hg₂Cl₂

frit

KCl (aq, sat) ⟶

frit

FIGURE 6   Standard calomel electrode. (From [1].)

electrode). The driving forces for the formation of this *space charge layer* in the electrolyte are primarily coulombic forces but are also chemical interactions with the electrode surface and entropic forces which determine the distribution of ions in the DL.

Figure 7 is a model of the double layer. It can be seen that the ions in the electrolyte arrange themselves in a layered structure. The layer closest to the electrode surface is formed by ions that interact strongly with the electrode surface. These are mainly anions (e.g., halide anions) that tend to lose part of their solvation shell to adsorb directly at the electrode surface, even if the surface carries the same charge. The center of charge of these chemisorbed ions coincides approximately with the radius of the adsorbing species. The plane through the center of these specifically adsorbed ions is called the *inner Helmholtz plane* (IHP). Ions that are attracted to the electrode electrostatically (i.e., those with only minor chemical interactions with the electrode surface) keep their hydration shell. Thus, they approach the electrode at most up to a distance that corresponds to the radius of the hydrated ions. The center of charge of these hydrated ions (cations in the example in Fig. 7) is called the *outer Helmholtz plane* (OHP).

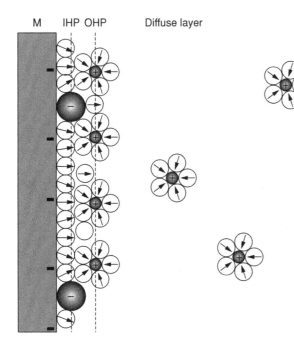

M   IHP OHP   Diffuse layer

FIGURE 7
Electrochemical double
layer. M, metal electrode
(e.g., Cu); IHP, inner
Helmholtz plane; OHP,
outer Helmholtz plane.
The solvation (hydration)
shell on the cations and the
dipole alignment of the
solvent molecules (water)
are also shown. (After [4].)

Counteracting the electrostatic force is an entropic force that entails a continuous decay of the concentration of the excess ions from the OHP into the bulk, with increasing distance to the electrode until the bulk concentration is reached. The extension of this *diffuse layer* depends on the ion concentrations of the electrolyte solution, being negligible for concentrated solutions ($\sim$1 M) but reaching several tens of nanometers for dilute electrolytes ($\leq 10^{-3}$ M).

The electrochemical double layer behaves like a capacitor (hence the name), whereby in concentrated solutions the metal surface and the OHP take the role of capacitor plates, with the gap between them filled by water molecules that because they possess a permanent dipole moment behave as a high-dielectric-constant medium. Due to the microscopic distance between the plates (i.e., a few tens of angstroms), the DL capacitance is much higher than that of common electronic devices, typically on the order of 10 to 40 μF/cm^2. As in a capacitor, the DL can be charged or discharged by changing the electrode–electrolyte potential difference, keeping the charge on the electrode surface always equal to that of the space charge layer in the electrolyte solution.

3   OHMIC LOSSES AND ELECTRODE KINETICS

Whereas in Section 2 we considered the thermodynamic description of electrolytic and galvanic electrochemical cells as well as of half-cell reactions, in this section we detail the various potential losses that are caused by the ion transport resistance in the electrolyte and the kinetics of the electrode reactions. Potential losses caused by

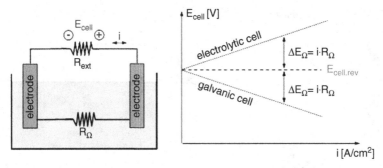

FIGURE 8 Left-hand side: simplified equivalent circuit of a galvanic or electrolytic cell in the absence of kinetic and diffusion resistances, representing the ohmic ion conduction resistance of the electrolyte by a simple resistor, R_Ω; right-hand side: i vs. E curve of an idealized electrolytic and galvanic cell.

concentration gradients in the electrolyte (diffusion overpotentials) which can arise at high current densities are not discussed here.

3.1 Ohmic Potential Losses

When a current is drawn from an electrochemical cell through an external load, represented by an ohmic resistor, R_{ext}, the externally flowing electronic current must be balanced by an ionic current through the electrolyte between the two electrodes. In the simplified case in which kinetic and diffusion resistances are negligible, the behavior of the electrochemical cell can be approximated by an equivalent electronic circuit as shown in Figure 8, where R_Ω represents the ionic electrolytic resistance and is most commonly expressed as areal resistance in units of $\Omega \cdot cm^2$. For an electrolytic cell, where external voltage has to be applied to drive the reaction (see Fig. 1), the external cell voltage that must be applied to the cell, $E_{cell,electrolytic}$, is the sum of the reversible cell voltage $E_{cell,rev}$ and the ohmic potential loss ΔE_Ω across the electrolyte between the electrodes:

$$E_{cell,electrolytic} = E_{cell,rev} + iR_\Omega = E_{cell,rev} + \Delta E_\Omega \qquad (34)$$

where i is the current density (in units of A/cm^2; i.e., the total current devided by the cross-sectional area of the electrodes). For a galvanic cell (see Fig. 2), the cell voltage obtained, $E_{cell,galvanic}$, is reduced by the internal ionic resistance:

$$E_{cell,galvanic} = E_{cell,rev} - iR_\Omega = E_{cell,rev} - \Delta E_\Omega \qquad (35)$$

The resulting current density/potential relationship (i vs. E curve) is illustrated in Figure 8.

Using the approximation above, we can estimate the ohmic potential loss in a lithium-ion battery at high charge–discharge rates. The latter are commonly not given in terms of current density but as a *C rate*, which describes how many times the battery capacity is being charged or discharged per hour. For high-power batteries (e.g., in hybrid electric vehicles), C rates can be as high as $20\ h^{-1}$ (i.e., the battery is

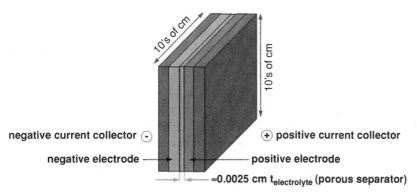

FIGURE 9 Configuration of a typical lithium battery, showing the negative and positive electrodes sandwiched between negative and positive current collectors, separated by a separator, a microporous material (usually, polypropylene and/or polyethylene) into which liquid electrolyte (solvent and salt) is imbibed.

fully charged or discharged in 1/20 of an hour). Considering a typical areal capacity of ≈ 1 mAh/cm^2 for high-power lithium-ion batteries, the corresponding current density would be ≈ 20 mA/cm^2. As illustrated in Figure 9., the electrode area of lithium-ion batteries (x/y-dimension) is very large (tens of centimeters) compared to the thickness of the electrolyte layer between the electrodes, $t_{electrolyte}$ (z-dimension; tens of micrometers), so that the ohmic resistance R_Ω between the two electrodes is well described by a simple one-dimensional relationship:

$$R_\Omega = \frac{t_{electrolyte}}{\kappa \varepsilon / \tau} \cong \frac{t_{electrolyte}}{\kappa \varepsilon^{1.5}} \tag{36}$$

where κ is the electrolyte conductivity, ε is the electrolyte volume fraction in the typically used electrolyte-imbibed porous *separator* materials used to separate the two electrodes, and τ is the tortuosity of the ionic conduction path within the separator. The latter is often approximated by the *Bruggeman relationship* ($\varepsilon/\tau \approx \varepsilon^{1.5}$), as shown in equation (36). For a typical electrolyte conductivity of 10 mS/cm (see Table 1), a separator thickness of 25 μm, and a electrolyte volume fraction of 0.5, equation (36) yields $R_\Omega \approx 0.7$ $\Omega \cdot$ cm^2. This would result in an ohmic potential loss at 20 mA/cm^2 of $\Delta E_\Omega \approx 14$ mV, which is rather modest compared to the average cell voltage of ≈ 4 V for most lithium-ion batteries.

3.2 Kinetic Overpotential

Additional potential losses are usually caused by the fact that the electrode reactions are not infinitely fast, so that an additional driving force is required to sustain a given reaction rate. This required driving force is an additional potential loss. Let us consider first the equilibrium condition of a generic electrochemical reaction:

$$Ox + ne^- \underset{i_{anodic}}{\overset{i_{cathodic}}{\rightleftarrows}} Red \tag{37}$$

The half-cell reaction above represents a dynamic state in which there is a continuous and reversible exchange between the oxidized and the reduced species, referred to as *dynamic equilibrium*. For example, for the equilibrium between copper metal and its copper ions in solution shown in equation (14), equilibrium is reached when the anodic copper dissolution ($Cu \rightarrow Cu^{+2} + 2e^-$) occurs as fast as the cathodic copper deposition ($Cu^{+2} + 2e^- \rightarrow Cu$). In this example, electrochemical equilibrium implies a flow of electrons out of the copper electrode for the copper dissolution reaction (anodic current) which is simultaneously counterbalanced by an equal flow of electrons into the copper electrode from the copper deposition reaction (cathodic current). Thus, in equilibrium, the anodic current, i_{anodic}, and the cathodic current, $i_{cathodic}$, are equal in magnitude and correspond to the *exchange current density*, i_0:

$$i_0 = i_{anodic} = |i_{cathodic}| \tag{38}$$

Most commonly, anodic currents are defined as positive currents and cathodic currents are defined as negative currents (note, however, that in the older literature the opposite sign convention is often used). Since anodic (positive) and cathodic (negative) currents are equal in magnitude and opposite in sign, no externally observable net current is flowing in equilibrium. A net external current is obtained when deviating from equilbrium in either the anodic (positive current) or cathodic (negative current) direction, whereby the equilibrium in equation (37) is shifted to the right if $|i_{cathodic}| > i_{anodic}$ or to the left if $i_{anodic} > |i_{cathodic}|$.

A net current flow across an electrode is accompanied by deviations of the electrode potential from its equilibrium half-cell potential value, E_{rev}, described by a *kinetic overpotential*, η. To sustain a net anodic current, a positive deviation from the equilibrium potential is required, i.e., an *anodic overpotential*, η_{anodic}. On the other hand, to sustain a net cathodic current, a negative deviation from the equilibrium potential is required, i.e., a cathodic overpotential, $\eta_{cathodic}$. In summary, the electrode potential for an anodic net current can be described as $E_{electrode} = E_{rev} + \eta_{anodic}$ and as $E_{electrode} = E_{rev} - |\eta_{cathodic}|$ in the case of a net cathodic current. The implications of overpotentials for an electrochemical cell consisting of two electrodes can now be determined for an electrolytic cell and for a galvanic cell. For an electrolytic cell (see Fig. 1) with a net current flow, a cathodic reaction occurs on the negative electrode, and its potential is thus decreased by $|\eta_{cathodic}|$, while an anodic reaction occurs on the positive electrode, the potential of which increases by η_{anodic}. Therefore, when drawing a net current from an an electrolytic cell, its potential, $E_{cell,electrolytic}$, can be described as

$$E_{cell,electrolytic} = E_{cell,rev} + \eta_{anodic} + |\eta_{cathodic}| + iR_\Omega \tag{39}$$

On the other hand, for a current flow in a galvanic cell (see Fig. 2), an anodic reaction occurs on the negative electrode, and its potential is thus increased by η_{anodic}, while a cathodic reaction occurs on the positive electrode, so that its potential decreases by $|\eta_{cathodic}|$. Therefore, the overall cell voltage, $E_{cell,galvanic}$, of a galvanic cell is

$$E_{cell,galvanic} = E_{cell,rev} - \eta_{anodic} - |\eta_{cathodic}| - iR_\Omega \tag{40}$$

The functionality between the anodic or cathodic overpotential and the current density is discussed below.

3.3   The Butler–Volmer Equation

The relation between current and kinetic overpotential for a generic reaction involving n electrons is provided by the well-known Butler–Volmer equation:

$$
\begin{aligned}
i &= i_0 \left(e^{\frac{\alpha n F}{RT} \cdot \eta} - e^{-\frac{(1-\alpha)n F}{RT} \cdot \eta} \right) \\
&= i_0 \left(10^{\frac{\alpha n F}{2.303 \cdot RT} \cdot \eta} - 10^{-\frac{(1-\alpha)n F}{2.303 \cdot RT} \cdot \eta} \right)
\end{aligned}
\tag{41}
$$

where α is the *transfer coefficient*, determining what fraction of electric energy resulting from the displacement of the potential from equilibrium affects the rate of the electrochemical reaction. Often, the Butler–Volmer equation is written as a function of the base 10 [as on the right-hand side of equation (41)], in which case the terms $2.303\,RT/(\alpha n F)$ and $2.303\,RT/[(1-\alpha)n F]$ in the exponents are referred to as anodic and cathodic *Tafel slopes*, respectively. The latter describe the overpotential (in mV) that is required to increase the anodic or cathodic current by a factor of 10. If $\alpha = 0.5$, the Butler–Volmer curve is inversion symmetric, and for $n = 1$, the Tafel slopes at room temperature have a value of 120 mV/decade. An example plot of the Butler–Volmer equation, including the anodic and cathodic partial currents, is shown in Figure 10.

As one can see from the figure, for small overpotentials, the i vs. η curve is approximately linear. Using a Taylor series expansion, it can be shown that the Butler–Volmer equation simplifies to a linear equation for overpotentials smaller than one-third of the value of the Tafel slope:

$$
i = \frac{i_0 F}{RT}\eta
\tag{42}
$$

This equation resembles Ohm's law, from which one can define a *charge transfer resistance*, R_{ct}:

$$
R_{\text{ct}} \equiv \frac{\eta}{i} = \frac{RT}{i_0 F}
\tag{43}
$$

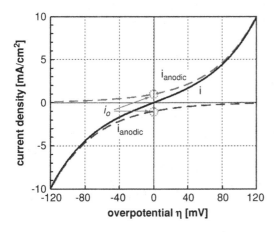

FIGURE 10   Solid curve: Butler–Volmer equation for $\alpha = 0.5$ and $n = 1$ ($\equiv$ 120 mV/decade Tafel slope) and for an exchange current density of $i_0 = 1$ mA/cm². Dashed curves: anodic (top) and cathodic (bottom) partial current densities.

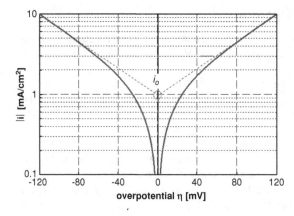

FIGURE 11   Tafel plot for the the same parameters as those used in Figure 10. The dashed lines show the linear fits at high overpotentials, with a y-axis intercept at i_0.

This simplified relationship is frequently used in electrochemical modeling for fast electrode processes in batteries (e.g., lithium anode) and fuel cells (e.g., hydrogen anode).

For anodic (cathodic) overpotentials larger than roughly half of the value of the Tafel slope, the anodic (cathodic) current dominates the net current, i. In this case, the Butler–Volmer equation can be approximated by the *Tafel equation* for either the anodic branch at high positive overpotentials,

$$\log(i) = \log(i_0) + \frac{\alpha n F}{2.303\, RT}\eta \tag{44}$$

or the cathodic branch at high negative overpotentials,

$$\log(|i|) = \log(i_0) + \frac{(1-\alpha)nF}{2.303\, RT}|\eta| \tag{45}$$

Thus, under these conditions, a plot of log i vs. η, known as the *Tafel plot*, yields a straight line (see Fig. 11). A Tafel plot is a very useful diagnostic tool to determine important kinetic parameters such as the value of α from the slope [see equation (44) and (45)] and of i_0 by interpolation of the linear segments of the Tafel plot to the equilibrium potential ($\eta = 0$; see Fig. 11). The Tafel equations above are frequently used to model slow electrochemical reactions in batteries and fuel cells (e.g., the oxygen reduction reaction).

4   CONCLUDING REMARKS

We have tried to introduce the most essential electrochemical concepts that are needed to understand the physicochemical processes controlling the performance of batteries. This includes (1) fundamental thermodynamic relationships such as reversible cell voltage and the half-cell potential and its dependence on the operating parameters (temperature, etc.); (2) the basic properties of electrolytes and the electrode–electrolyte interface; and (3) the fundamental kinetic model and its approximations. Although very restricted in scope, we hope that our brief introduction will provide the general reader with the background necessary to more easily comprehend the chapters that follow. For those who would like to get a deeper insight into the various aspects of electrochemistry, we refer to the monographs listed below.

BIBLIOGRAPHY

Parts of this chapter have been taken from the following texts, to which the reader is referred for more details on the matters presented here.

1. Hamann, C. H., A. Hammet, and W. Vielstich, *Electrochemistry*, Wiley, Hoboken, NJ, 2007.
2. O'M. Bockris, J., and A. K. N. Reddy, *Modern Electrochemistry*, Plenum Press, New York, 1977.
3. Bard, A. J., and L. R. Faulkner, *Electrochemical Methods*, Wiley, New York, 1980.
4. Krischer, K. and E. Savinova, *Handbook of Heterogeneous Catalysis*, 2nd ed., G. Ertl, H. Knözinger, F. Schüth, and J. Weitkamp, Eds., Wiley-VCH, Weinheim, Germany, 2008, Chap. 8.1.1.

CHAPTER 2

LITHIUM BATTERIES: FROM EARLY STAGES TO THE FUTURE

Bruno Scrosati

1 INTRODUCTION

The evolution of any device is influenced by its past history, and this also applies to lithium batteries. In 1800, when Alessandro Volta, professor at the University of Pavia in Italy, went to the court of Napoleon in France to disclose his "electric pile," he certainly could not have imagined that his invention, simply a result of a dispute with his colleague and competitor Luigi Galvani at the University of Bologna (Fig. 1), would, via various progressive steps in technological evolution, open up a route that lead eventually to the development of the electrochemical power source that today dominates the consumer electronics market and that is about to revolutionize our common concept of road transportation. This is the lithium battery. Without its advent we could not rely on multifunctional, year-by-year more sophisticated cellular phones and long-lasting laptops, or enjoy Mp3 music to enliven our exercise routines.

Indeed, Volta's work had a tremendous impact on the progress of electrochemical science by catalyzing the rapid evolution of battery history through the cumulative discoveries of many important electrochemical systems, most of them in the nineteenth century. These include the zinc–manganese oxide cell, invented in 1866 by the French engineer Georges-Lionel Leclanché; the lead–acid rechargeable battery, invented in 1859 by the French scientist Gaston Planté; and the rechargeable nickel–cadmium battery, invented in 1901 by the Swedish engineer Waldmar Jungner [1] . It is interesting to notice that although improved by innovations in construction design,

Lithium Batteries: Advanced Technologies and Applications, First Edition.
Edited by Bruno Scrosati, K. M. Abraham, Walter van Schalkwijk, and Jusef Hassoun.
© 2013 John Wiley & Sons, Inc. Published 2013 by John Wiley & Sons, Inc.

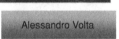

Luigi Galvani

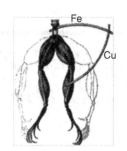

Alessandro Volta

FIGURE 1 In 1781 at the University of Bologna, Italy, Luigi Galvani carried out his classical experiment in which he observed that because a frog's leg twitched when touched by a series of two different metals, the animal's muscle could generate electricity. In 1800, Alessandro Volta at the University of Pavia in Italy claimed the opposite: namely, that the muscle responded by stimulus of the two metals, arguing his point by the demonstration of electricity production from his "voltaic pile," formed by an alternating sequence of two different metals (zinc and silver disks) separated by a cloth soaked in a sodium chloride solution. Actually, the two discoveries are not in total contrast since Galvani's concept may imaginatively be associated with bioenergy and that of Volta with energy storage.

all of the electrochemical power systems cited above are still used for the development of commercial batteries designed for important applications, such as powering electronic devices, portable tools, and car engine ignition. In today's versions, the early batteries have been reengineered from the original concept: for example, in the case of the Leclanche cell, by changing the electrolyte from a liquid to a mixed manganese dioxide–carbon paste, the zinc rod to a core of mixed powdered zinc and electrolyte paste, and the container to a stainless-steel case to form the common alkaline battery. Similar to the improved electrode structure and case design of Planté cells are the lead–acid batteries used widely for car lighting and ignition. Both alkaline and lead–acid batteries are currently produced and commercialized at yearly rates of several billion units and, accordingly, one may wonder whether other batteries are really needed to fulfill the needs of the portable electronic and transportation markets. Effectively, not much innovation took place in the battery market for more than a century after the discovery of alkaline, nickel–cadmium, and lead–acid batteries, since, with minor changes in their chemistry and configuration, these early systems met the requirements of the technology at that time. It was not until the late 1960s that a change in battery technology took place, triggered by a series of innovations in the demand for portable energy, including progress in implantable medical devices, the first oil crisis, and the outbreak of the consumer electronics market.

In this respect, it soon became clear that the most crucial deficiency of conventional batteries was their low energy density, which is associated with their electrode

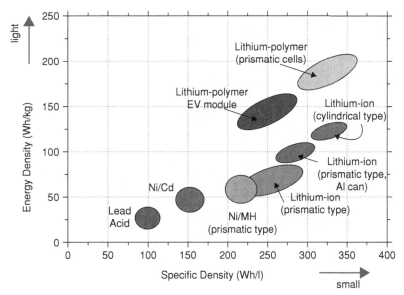

FIGURE 2 Energy density (Wh/kg) vs. specific density (Wh/L) for a series of batteries. A battery directed to the consumer electronics market, especially to the electric vehicle market, should be as light (high energy density) and as small (high specific density) as possible, that is, the ideal battery should lay in the upper right corner of the diagram. Conventional batteries such as lead–acid, Ni/Cd, and Ni/MH are in the opposite corner, being heavy and bulky. Different is the case for lithium batteries, which approach the ideal target. Due to their unique energy content, today lithium batteries are the power source of choice for the portable electronic market and are considered to be the most promising powering system for sustainable electric road transport. (From [2].)

combinations, which could offer only a limited specific capacity value (in terms of ampere-hours per gram, Ah/g), which reflects in a low energy density (i.e., watthours stored by weight, Wh/kg, or by volume, Wh/L). As shown in Figure 2, which illustrates the relationship between gravimetric and volumetric energy density for a series of batteries, it is clear that alkaline–manganese, nickel–cadmium, and lead–acid batteries offer very low energy density. Simply put, these batteries were too heavy and too large to serve the evolving technologies satisfactorily.

Illustrative is the case of implantable medical devices, in particular the cardiac pacemaker. This device, essential to save the lives of patients affected by serious arrhythmia, requires a battery to power the electronic microcircuitry that, on demand, assists the heart to beat. In its early stages, pacemakers had to rely on the only battery available at the time, the primary zinc–mercury battery (Fig. 3). To provide the energy required for proper operation, four Zn–HgO batteries were needed, occupying almost three-fourths of the device's size, so early pacemakers were very heavy and bulky (Fig. 4). In addition, zinc–mercury oxide batteries lost their charge within two years, and their replacement required a new placement operation, with serious stress on patients. Clearly, new batteries, assuring higher energy from the weight/volume ratio, combined with a long operational life, were urgently needed.

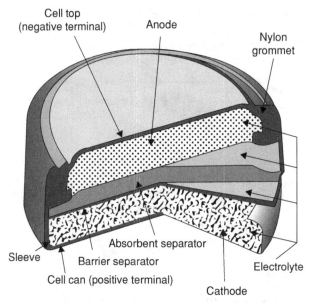

FIGURE 3 Zinc–mercuric oxide battery, button configuration. This battery was used to power cardiac pacemakers in their very earliest stage. To provide the energy required for correct pacemaker operation, four Zn/HgO batteries were needed, which resulted in high weight and volume encumbrance.

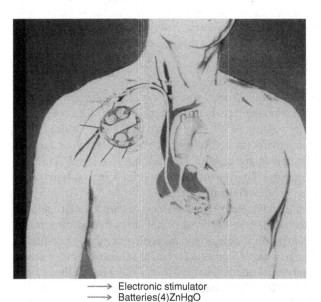

FIGURE 4 At their early implant stage, cardiac pacemakers used four Zn–HgO batteries to power the electronic stimulator. The batteries alone occupied almost three-fourths of the device's size, so the device was very heavy and bulky. In addition, the zinc–mercury oxide batteries lost their charge within two years, and their replacement required a new operation, with serious stress on patients.

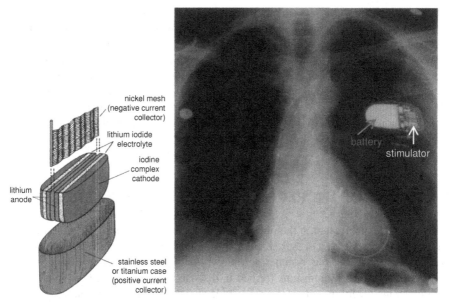

FIGURE 5   The lithium–iodine battery is fabricated with a special shape (on the left) designed to fit the pacemaker case.

A breakthrough arrived with the development of new-concept batteries exploiting lithium as one of the electrode materials. Due to its electrochemical equivalent, the highest among all metals, lithium can provide a much higher specific capacity than that of zinc: 3860 Ah/g vs. 820 Ah/g. Obviously, lithium metal is not compatible with water, and its use required moving from common aqueous electrolytes to more electrochemically stable organic media, generally formed by a solution of a lithium salt in a carbonate organic solvent (e.g., propylene carbonate, ethylene carbonate) or in a mixture of them.

It is easy to imagine that in moving from a zinc-based to a lithium-based battery, a considerable increase in energy density could be obtained. Indeed, a battery combining a lithium metal anode with an iodine-based cathode, the lithium–iodine battery [3], provided a practical energy density of about 250 Wh/kg, almost five times higher than that of a zinc–mercury oxide battery, which is limited to 50 Wh/kg.

The lithium–iodine battery was fabricated with a special shape designed to fit the pacemaker case (Fig. 5). The development of this battery had a tremendous impact on pacemaker efficiency and, consequently, on patient comfort, since its use to replace the zinc–mercury oxide battery resulted in a great reduction in weight and volume (compare Fig. 5 with Fig. 4) as well as an increase in the operational life of the cardiac device, extending it from two years to six or seven years [4]. Indeed, today, almost all implanted pacemakers are powered by a lithium–iodine battery.

The success of the lithium–iodine battery highlighted the validity of lithium as a high-energy electrode material, and it opened a route for the development of a series of new batteries addressed to meet various diversified applications. The lithium battery evolution was accelerated by the explosion in the 1970s of consumer electronics, which brought onto the market a series of popular devices such as electronic watches, toys, and cameras. These devices required batteries capable of combining good powering operation with a small volume size and a contained price. This need promoted extended research and development and a series of batteries exploiting new cathode materials (e.g., manganese dioxide cathode, MnO_2 [5]; carbon fluoride, CF_x [6]; iron sulfide, FeS_2 [7]; copper oxide, CuO [8]) most fabricated in a coin-type shape that could fit easily into a device case, were soon commercialized.

2   ADVENT OF THE RECHARGEABLE LITHIUM BATTERY

All the batteries fabricated in the initial stage of the lithium technology were of the primary type. The success of these batteries stimulated an obvious interest in moving to secondary, rechargeable systems. In theory, there was no apparent difficulty on the anode side since lithium ions formed in discharge were expected to plate back reversibly onto the lithium metal in charge. Indeed, lithium deposition from propylene carbonate–based solutions was demonstrated as early as 1958 [9]. Although the assumption of a facile lithium electrodeposition process involving many lithium plating and stripping (charge–discharge) cycles was not totally true (see below), initially the attention was directed toward the cathode side, with the aim of identifying materials that could withstand a long cycle life. The breakthrough arrived in 1978 with the development of insertion or intercalation electrodes [10]: namely, compounds that can reversibly accept and release lithium ions from their open structure and at the same time can assume various valence states. These conditions are met by transition metal compounds, of which titanium sulfide is a classic example.

The basic structure of a Li–TiS_2 cell is schematized in Figure 6. The overall process involves lithium oxidation at the anode, with the formation of lithium ions that travel through the nonaqueous electrolyte to reach the cathode to finally insert into its layer structure. Because the guest lithium ions keep their charge when intercalated in the TiS_2, to maintain electroneutrality, a modification of the electronic structure of the TiS_2 host also occurs by a variation in the oxidation state of the transition metal, which passes from Ti(IV) to Ti(III). In addition, to allow the electrochemical reaction to go on to multiple cycles, hence to assure the cycle life of the battery, a highly reversible evolution of both the electronic structure (to balance the positive charge of the inserted lithium ions) and the crystal structure (to prevent the lattice to collapse) is required.

Basically, batteries using intercalation electrodes (e.g., TiS_2) can be regarded as concentration cells where the activity of lithium varies from zero (at the TiS_2 pristine state) to 1 (at the fully intercalated $LiTiS_2$ state). The total electrochemical process may in fact be written as

$$x\text{Li} + \text{TiS}_2 \rightleftharpoons \text{Li}_x\text{TiS}_2 \tag{1}$$

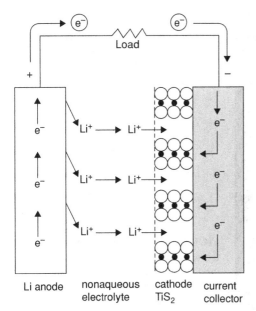

Li anode   nonaqueous   cathode   current   **FIGURE 6**   Basic structure of a
            electrolyte   TiS₂   collector Li–TiS₂ cell.

where x, the intercalation degree, may vary from 0 to 1. Accordingly, the potential of the cell is given by the Nernst equation:

$$E = E^0 - \frac{RT}{xe} ln \frac{a_{Li+}(inLi_x TiS_2)}{a_{Li} x a_{TiS_2}} = E^0 - \frac{RT}{xe} ln \left[a_{Li+}(inLi_x TiS_2)\right] \qquad (2)$$

As the intercalation proceeds, the activity of Li^+ in $Li_x TiS_2$ increases and thus the cell potential E decreases. A voltage decay upon operation is the typical behavior of lithium intercalation batteries. If the intercalation degree is maintained within limited levels (generally, ≤ 1), the electrochemical process is highly reversible and the cell can be cycled many times, provided that the same reversibility conditions are matched at the anode side.

The success of intercalation electrodes stimulated large academic and industrial interest, and accordingly, many announcements of the development of new types of rechargeable lithium batteries appeared in the 1980s. One that stimulated considerable debate was the proposal to use conducting polymers such as poly(acetylene) or poly(pyrrole) as positive electrodes [11]. However, the initial enthusiasm cooled down rapidly when it became clear that these polymer materials had a very poor electrochemical behavior [12].

Nevertheless, despite the limitations, the interest in lithium batteries remained high and the number of laboratories involved in lithium battery R&D grew rapidly, making the time ripe for worldwide discussion and confrontation; in fact, a first international meeting on lithium batteries (IMLB 1) was held in 1992 in Rome, Italy. Although only 80 participants attended, the scientific value and concrete prospects of a fast evolution of lithium batteries were already evident. Indeed, the IMLBs grew into valuable series that now experience the participation of over 1000 delegates.

FIGURE 7 Large Li–TiS$_2$ battery fabricated in 1977. Lithium metal was used as the anode, and a solution of LiB(CH$_3$)$_4$ salt in dioxolane was used as the electrolyte. The operation of this early rechargeable lithium battery was affected by unsafe events associated primarily with the highly reactive anode material.

All these important events focused notable attention on rechargeable lithium batteries. The first commercial prototypes appeared in the late 1970s, one produced by the Exxon Company in the United States using a TiS$_2$ cathode [13] (Fig. 7) and another by the Moli Energy Battery group in Canada using a MoS$_2$ cathode [14], both with liquid organic electrolytes. However, some operational faults, including fire incidents, led to the rapid conclusion that there were some problems that prevented safe, extended operation of these first lithium batteries. It was soon realized that the problems were associated with the anode; due to its very high reactivity, lithium metal reacts readily with the electrolyte, with the formation of a passivation layer on its surface. The layer, usually called the solid electrolyte interphase (SEI) [15], is permeable to lithium ions, thus allowing continuation of the discharge process; however, irregularities at the SEI may lead to uneven lithium deposition upon charge, with consequent dendrite formation that eventually grows to short the cell. In extreme cases, these uncontrolled events give rise to overheating effects with associated thermal runaway and explosions. Clearly, to assure cycle life and safety, two options were possible: (1) a careful choice of the electrolyte system to assure optimized, smooth lithium deposition, or (2) replacement of the lithium metal with a less aggressive anode material.

The feasibility of the first choice was demonstrated as early as the early 1980s by Abraham and co-workers, who showed that by selecting proper organic electrolytes, for example, those based on solutions of lithium hexafluoroarsenate, LiAsF$_6$, in aliphatic ethers such as 2-methyltetrahydrofuran, 2MeTHF, with an additive of 2-methylfuran, lithium batteries using a range of positive electrodes, including TiS$_2$,

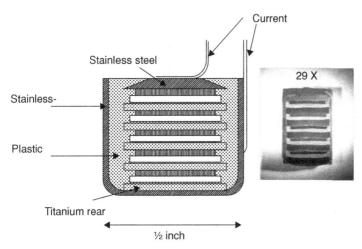

FIGURE 8   Solid-state battery using a solid $RbAg_4I_5$ electrolyte. To obtain an useful voltage of 3 V, five cells in series were used.

MoS_3, and V_6O_{13}, could be developed with operational lives extending several hundred cycles [16,17].

Another popular route for assuring safe lithium cycling was to switch from liquid, reacting solutions to solid, inert electrolytes. Obviously, to prevent serious ohmic polarization, only solid media having a high level of lithium ion conductivity over a wide temperature range could be used. Not many materials fulfill this condition. A known solid having a conductivity comparable to that of liquid solutions is silver rubidium iodide, $RbAg_4I_5$ [18,19]. Ceramic-type materials such as $RbAg_4I_5$ are ideal electrolytes since they allow the construction of very reliable solid-state batteries. This is, in fact, the case with $RbAg_4I_5$-based batteries. In a recent study it was reported that these batteries maintained their charge unchanged even after 33 years of storage [20]. To obtain a useful voltage of 3 V, a battery formed by five cells connected in series was produced (Fig. 8). Unfortunately, this $RbAg_4I_5$ battery operates by transport of silver ions and thus is of no use in lithium batteries.

Much effort has been devoted to developing lithium ion–conducting solids. One example is, of course, the lithium iodide used in the above discussed pacemaker batteries. This material, however, has, in its pure state, a too low an ionic conductivity level to be used for applications which are more energy demanding than a cardiac pacemaker. Upgrading the material by adding aluminum oxide particles was attempted in 1980 by Liang [21]. The idea was to promote preferential conductive paths along the surface of the filler particles to favor lithium transport. Effectively, an increase in conductivity of two orders of magnitude (i.e., 10^{-4} S/cm) was obtained when passing from pure LiI to Al_2O_3–added LiI; however, still a too low level to be of practical interest. Other potential lithium solid conductors were sought in the LISICON family [22] and β-aluminas [23], but generally their input was marginal. This situation may change by virtue of a recently disclosed new solid electrolyte, $Li_{10}GeP_2S_{12}$, which has a conductivity on the order of 10^{-3} S/cm [24],

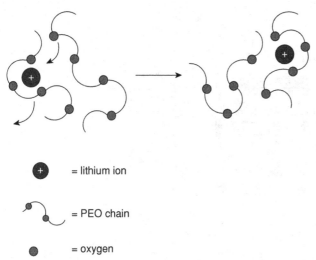

= lithium ion

= PEO chain

= oxygen

FIGURE 9   Lithium-ion transport mechanism in a polymer electrolyte formed by a complex between a lithium salt and a coordinating poly(ethylene oxide) (PEO) polymer. By coulombic attraction the excess of negative charge on the oxygen in PEO chains coordinates the Li^+ ions, thus separating them from the anions. Because of the large size and encumbrance of the chains, Li^+ ion transport requires sufficient chain flexibility so that they can unfold to free the Li^+ ions to move from one chain loop to the other.

although its practical application in an effective lithium solid battery is yet to be fully demonstrated.

A breakthrough arrived in the late 1970s with the discovery by Michel Armand of a solid-state polymer electrolyte formed by an aggregate between a lithium salt and a coordinating polymer [e.g., lithium triflate and poly(ethylene oxide) (PEO)] [25]. The excess of negative charge on the oxygen in the PEO chains coordinates the Li^+ ions through coulombic attraction, thus separating them from the anions. By this process, the lithium salt is "dissolved" in the PEO matrix, in analogy with the process of salt dissolution in liquid solvents. The main difference is that whereas in liquids the ions can move within their salvation shell, this is not possible in PEO complexes, due to the large size and encumbrance of the polymer chains. Therefore, ion transport in the polymer electrolytes requires flexibility in the PEO chains so that they can unfold to free the Li^+ ions to move from one chain loop to the other (Fig. 9) [26].

Figure 10, which illustrates the general configuration of a PEO–lithium salt (here LiBr) polymer electrolyte, demonstrates the complex structure of the system, where in addition to Li^+ ions trapped in the PEO chain cages and the separated Br^- anions, various cation–anion ion-pair interactions and elaborate chain entangling occur. Due to this configuration, fast ion transport may occur only when the polymer is in the amorphous state, which is reached above 70°C [26], a condition that obviously limits the range of application of PEO-based polymer batteries. However, the unique advantages of polymer electrolytes, which include chemical stability, compatibility with the lithium metal electrode, and moderate cost, still make these

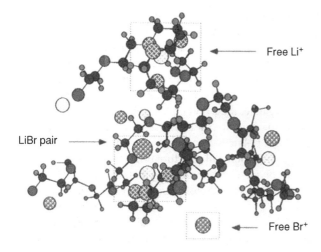

Free Li⁺

LiBr pair

Free Br⁺

FIGURE 10   General configuration of a PEO–lithium salt (here, LiBr) polymer electrolyte.

media quite suitable and interesting for the development of lithium batteries designed for applications where temperature is not a limiting factor (e.g., uses associated with electric vehicles). The polymer version of the lithium battery is the one that assures the highest energy content (Fig. 2).

PEO-based polymer electrolytes were exploited in the late 1990s for the fabrication of large laminated battery modules based on cells formed by a lithium foil anode, a PEO-based electrolyte layer, and a vanadium oxide cathode (Fig. 11),

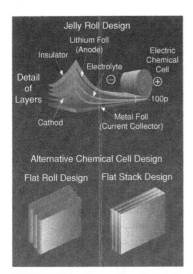

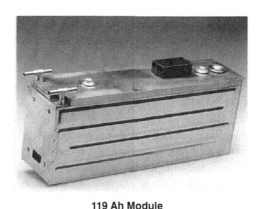

119 Ah Module

FIGURE 11   Concept and module of the all-solid advanced electric vehicle battery using a dry polymer electrolyte and a metallic lithium anode developed jointly in 1993 by 3M and Hydro Québec under U.S. Advanced Battery Consortium contracts. The goal was to produce a series of full-sized lithium polymer battery packs for extensive platform and road testing. Unfortunately, despite initial success, the project was abandoned.

developed jointly by Hydro Québec in Canada and 3M in the United States. The battery module had very good performance in terms of energy density (155 Wh/kg) and cycle life (600 cycles at 80% depth of discharge), and it was proposed as a power source for electric vehicles [27], a very futuristic application in 1996. However, despite this and other successful demonstration projects, the lithium polymer battery project was abandoned and only very recently reconsidered by a French battery manufacturer for use in an electric vehicle, the Bolloré blue car.

The recognition of the low conductivity of PEO-based solid polymer electrolytes led to the development of gel polymer electrolytes, in which a liquid electrolyte is introduced to plasticize the polymer, which in turn produces highly conductive solid polymer electrolytes with conductivity approaching that of liquid electrolyte [28]. These gel-type electrolytes, with modifications, form the engineering basis of today's lithium batteries used in cell phones, iPads, and other high-tech consumer products. Clearly, the route for the development of a successful rechargeable lithium battery had to pass through the second option: replacement of the lithium metal with another, more reliable anode. The winning approach relied on a totally new concept, based on a combination of two insertion electrodes, one capable of accepting lithium ions, operating as the anode, and the other, capable of releasing lithium ions, operating as the cathode. During charge the negative intercalation electrode acts as a "lithium sink" and the positive electrode as a "lithium source," and the total electrochemical process of the cell involves the transfer of x equivalents of lithium ions between the two intercalation electrodes; the process is then reversed upon discharge and repeated cyclically. These systems are actually concentration cells in which lithium ions "rock" across the electrodes, giving birth to a new type of system, the *lithium rocking chair battery* (Fig. 12).

It is interesting to note that the concept of a rocking chair battery dates back to the late 1970s [29–31] and was demonstrated practically as early as the 1980s [32–34]. However, more than 10 years had to pass before the concept could reach a practical application with a battery introduced by the Japanese manufacturer Sony in 1991 [35]. The key feature of the Sony battery, which was called "lithium-ion battery", was in the definition of proper electrode materials, identified in graphite as the lithium sink anode and in lithium cobalt oxide as the lithium source cathode (Fig. 12).

The use of a reversible graphite electrode for the anode can be considered the key advance that enabled commercial lithium-ion batteries [36, 37]. Effectively, the choice of graphite as anode is rather surprising, since its electrochemical process evolves outside the stability window of most common electrolytes. The electrolyte, in fact, decomposes at the graphite surface; however, the formation of an electronically blocking but ionically conducting layer that stops the decomposition, still allows the electrochemical process to continue. Therefore, graphite is thermodynamically unstable but kinetically protected. The investigation and clarification of the electrode interface, promoted particularly by the pioneering work of Peled, who named the passivating layer the (solid electrolyte interphase), SEI [15], and later by Jeong et al., who demonstrated its formation [38], have been of fundamental importance to an understanding of the lithium-ion battery process and for correct use of the graphite anode.

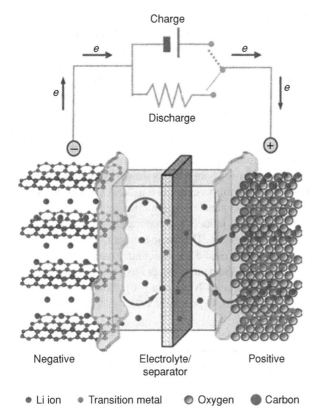

FIGURE 12 The lithium-ion battery. In its most conventional configuration the battery uses a graphite lithium-sink negative electrode, a transition metal oxide (e.g., lithium cobalt oxide) lithium-source positive electrode, and a liquid organic solution (e.g., $LiPF_6$) in carbonate solvent (e.g., ethylene carbonate–dimethyl carbonate electrolyte). During charge, lithium ions, removed from the $LiCoO_2$ positive electrode travel through the electrolyte and reach the negative electrode, where they are intercalated into the graphite structure. To compensate for the transfer of the ionic charges, electrons are also exchanged between the two electrodes.

The work of Sony triggered interest worldwide, and many battery manufacturers, located mainly in Asia, are now producing lithium ion batteries. The success of these batteries has been outstanding. Due to their specific properties, primarily in terms of their energy density, which generally exceeds that of conventional nickel–cadmium and younger systems, such as the nickel–metal hydride battery (Fig. 2), lithium-ion batteries are today the power sources of choice for very popular portable devices, such as cellular phones, notebooks, camcorders, and Mp3 players. Accordingly, the production of these batteries amounts to several billions of units per year.

3   A LOOK INTO THE FUTURE

Although graphite is generally the anode of choice in lithium-ion batteries, other materials are under consideration as valid alternatives. One example is lithium titanium oxide, $Li_4Ti_5O_{12}$ (LTO). This material, disclosed in the middle 1990s by Ohzuku [39], has a defective spinel-framework structure and is characterized by a two-phase electrochemical process evolving with a flat voltage profile. The theoretical specific capacity is lower, and the voltage level higher, than those of conventional graphite.

Both differences may result in a lower specific energy for the battery; however, the interest in LTO remains high because of its specific properties, including negligible changes in the lattice structure upon accepting and releasing lithium ions and an operation process evolving within the stability domain of the electrolyte. Indeed, batteries using LTO have demonstrated long cycling life and high reliability [40–42]. Thus, it is very likely that LTO may soon also be exploited in large-scale commercial prototypes because the material is now available commercially.

Another appealing class of anode materials is that of the lithium metal alloys (e.g., lithium–silicon and lithium–tin alloys), which, in principle having a specific capacity which largely exceeds that of lithium–graphite, are among the most promising negative electrodes for advanced lithium-ion batteries. As shown by the pioneering work of Winter and Besenhard [43], lithium alloys cannot be used as such in lithium cells, the main issue being the large volume expansion–contraction that occurs during the charge–discharge processes, which in turn induces mechanical disintegration, and thus electrode failure, in the round of a few cycles. This issue has been addressed successfully by moving from bulk to nanostructures and, in particular, to nanocomposites [44,45]. Effectively, the practical validity of the nano concept has been demonstrated in a commercial battery, under the trade name of Nexelium [46]. Table 1 compares the pro and cons of the various anode materials presently under consideration for future development of lithium-ion batteries.

Particularly important also is the role of the cathode, which must be capable of providing lithium ions to assure the electrochemical process, as well as to accept them back in a reversible matter to assure the life of the battery. These characteristics were

TABLE 1 Summary of Characteristics and Properties of Anode Materials Designed for Lithium-Ion Batteries

Material	Theoretical capacity (mAh/g)	Advantages	Issues
$Li_4Ti_5O_{12}$	170	Negligible volume expansion, low cost, stable electrochemical operation, high thermal stability in the charged–discharged state; already in large-scale production	Low specific capacity, high voltage
Li_xSn_y	990	High capacity, low cost	Large volume changes, nanostructures required
Li_xSi_y	2000	Very high capacity, low cost	Large volume changes; nanostructures required; production announced for 2012 (Panasonic)
$Sn_xCo_yC_z$	500	High capacity, tested commercially	Cycle life to be demonstrated
C, MCMB	370	Field-tested operation	Operation outside electrolyte stability

TABLE 2   Summary of Characteristics and Properties of Cathode Materials Designed for Lithium-Ion Batteries

Material	Theoretical capacity (mAh/g)	Advantages	Issues
$LiNi_{0.5}Mn_{1.5}O_4$	145	High voltage, good specific capacity	Operation within the limits of the electrolyte stability domain
$LiFePO_4$	170	Basic low cost, intrinsic safety, environmental compatibility	Low electronic conductivity, low tap density, needs carbon coating
$LiCoO_2$	140	Field-tested operation	High cost

fulfilled by $LiCoO_2$, a material disclosed by Goodenough and co-workers in 1981 [47]; without this fundamental discovery, the success of the lithium-ion battery could not be achieved. However, although still the most used cathode for the commercial production of lithium-ion batteries, lithium cobalt oxide has some drawbacks, such as high cost and a certain degree of toxicity; hence, other cathode materials are currently under study.

The most qualified alternative to lithium cobalt oxide is lithium iron phosphate, $LiFePO_4$, another important success of Goodenough's group [48]. The wide interest in this material is motivated by its many appealing features, which include good capacity, a two-phase electrochemical process that evolves with a flat, 3.5 V vs. Li voltage, and, most significantly, a cost which in principle is much lower than that of $LiCoO_2$. On the other hand, $LiFePO_4$ suffers from a very high intrinsic resistance which requires special electrode preparations involving sophisticated coating processes to lead to enhancement of the electronic conductivity and, finally, allowing full capacity delivery. Also, commercial $LiFPO_4$ lithium-ion cells have a lower energy density.

Attention has also been directed toward lithium nickel manganese oxide, $LiNi_{0.5}Mn_{1.5}O_4$ [49]. This material, which adopts a spinel structure, is characterized by a two-phase electrochemical process reflecting in a flat voltage profile evolving around 4.5 V vs. Li. The theoretical specific capacity is on the same order as that of the conventional lithium cobalt oxide, $LiCoO_2$. However, the key difference is in the high operational voltage, which makes $LiNi_{0.5}Mn_{1.5}O_4$ a very competitive cathode for the progress of lithium-ion battery technology. Table 2 compares the pro and cons of the cathode materials presently under evaluation.

4   BEYOND THE HORIZON

The triumphal march of lithium-ion batteries did not stop with the success gained in the consumer electronic market. A new challenge is now opening for these unique power sources. The continuous decrease in the oil resources and the growing, alarming concern for the deterioration of the climatic conditions of our planet call with constantly increasing urgency for a larger use of green, alternative energy sources such as solar and wind. In addition, the replacement of polluting, internal combustion cars with more efficient, controlled-emission vehicles, such as hybrid vehicles, plug-in hybrid vehicles, and ultimately, full electric vehicles, is another urgent need in our society. Since sun and wind are discontinuous energy sources and electric engines need to be powered, the success of these ecological renewals depends on efficient storage systems. In this respect, among the various possible alternatives, electrochemical batteries, and lithium batteries in particular, are the best choices, since they are able to convert stored chemical energy into electrical energy with high conversion efficiency and without toxic emission.

However, the lithium-ion batteries presently available, although commercial realities, are not yet at a high enough technological level to meet the powering requirements of hybrid or electric vehicles. The major concerns are the lack of safety, the high cost, and, the still too low energy density. The challenge is to move a step forward passing from the present intercalation chemistry to novel concepts that may lead to advanced batteries. This is not an easy goal to achieve. On the other hand, its implications, not only ecological, but also economical and political, are so important that many countries worldwide are budgeting tremendous amounts of funding to stimulate research and development in the lithium battery technology. Catalyzed by this large money flow progress has been rapid, and many new battery systems have been proposed in recent years. Using the colorful but forceful words of Fletcher [50,51]: "Advanced-battery start-ups started to pop up as mushrooms after a spring rain." Regretfully, this intense activity also gave rise to a series of patent conflicts and endless legal battles. No doubt Alessandro Volta is turning in his grave: His dispute with Luigi Galvani that affected him so much is nothing compared to the fierce litigation among lithium battery manufacturers.

Fortunately, this violent atmosphere has not influenced too badly the lithium battery science community, which has continued to report progress in the field. As result, many innovative materials, concerning all three battery components—anode, cathode, and electrolyte—are now being investigated intensively with the goal of upgrading battery performance to automotive levels. However, although notable successes have been achieved, the turning point has not yet been attained. One may estimate that to assure a valid, single-charge driving range for a common subcompact passenger electric car, based on presently available lithium-ion technology, the weight of the battery would largely exceed any acceptable practical limit. Clearly, revolutionary batteries, having levels of energy density two or three times higher than those offered by conventional systems, are urgently needed. The achievement of this ambitious task is the challenge of the present (and future?) history of lithium batteries. Attention is focused on systems based on electrode combinations that in theory assure a quantum jump in energy density. These are primarily the lithium–sulfur [52,53] and lithium–air [54,55] batteries, pioneered by Abraham and others [52–55], having a theoretical energy density on the order of 2600 and 5000 Wh/kg, respectively [52–58].

The road to practical applications of these *superbatteries*, however, is still long, since many issues have to be overcome before they can be considered practically feasible. These aspects are treated competently and exhaustively in other chapters of this book, hence are not discussed here. We may then conclude by stating that the history of the lithium battery is presently facing a second, new age. As the first one led to a revolution in the consumer electronic market, the second is expected to favor an epoch-making change in energy renewal and vehicle transportation. If this will eventually be verified, lithium batteries will once again play a key role in the improvement of our society and the condition of our planet.

REFERENCES

1. B. Scrosati, *J. Solid State Electrochem.*, **15**, 1623 (2011).
2. K. Kordesch and W. Taucher-Mautner, in *Encyclopedia of Electrochemical Power Sources*, Elsevier, New York, 2009, Vol **4**, p. 62.
3. C. F. Holmes, in *Encyclopedia of Electrochemical Power Sources*, Elsevier, New York, 2009, Vol. **4**, p. 76.
4. C. F. Holmes, *J. Power Sources*, **97**, 739, (2001); B. B. Owens and B. Scrosati, *Cardiostimolazione*, **1**, 147 (1983).
5. K. Nisho, in *Encyclopedia of Electrochemical Power Sources*, Elsevier, New York, 2009, Vol **4**, p. 83.
6. R. Yazami and H. Touhara, in *Encyclopedia of Electrochemical Power Sources*, Elsevier, New York, 2009, Vol. **4**, p. 93.
7. D. Linden, *Handbook of Batteries*, 2nd ed., McGraw-Hill, New York, 1995, p. 14.69.
8. D. Linden, *Handbook of Batteries*, 2nd ed., McGraw-Hill, New York, 1995, p. 14.76.
9. W. S. Harris, Ph.D. dissertation, University of California, Berkeley, UCRL-8381, 1958.
10. M. S. Whittingham, *Prog. Solid State Chem.*, **12**, 41(1978).
11. P. J. Nigrey, D. MacInnes, Jr., D. P. Nairns, A. G. MacDiarmid, and A. J. Heeger, *J. Electrochem. Soc.*, **128**, 1651 (1981).
12. B. Scrosati, A. Padula, and G. C. Farrington, *Solid State Ionics*, **9–10**, 447 (1983).
13. M. S. Whittingham, *Chem. Rev.*, **104**, 4272 (2004).
14. P. Kurzeweil and K. Brandt, in *Encyclopedia of Electrochemical Power Sources*, Elsevier, New York, 2009, Vol. **5**, p. 1.
15. E. Peled, D. Golodnitski, G. Ardel, and V. Eshkenazy, *Electrochim. Acta*, **40**, 2197 (1995).
16. K. M. Abraham, J. L. Goldman, and M. D. Dempsey, *J. Electrochem. Soc.*, **128**, 2493 (1981) (see also K. M. Abraham et al., in *Proceedings of the Symposium on Power Sources for Biomedical Implantable Applications and Ambient Temperature Lithium Batteries*, B. B. Owens and N. Margalit, Eds., The Electrochemical Society, Pennington, NJ, 1980, 80-4, p. 384.
17. K. M. Abraham, *J. Power Sources*, **14**, 179 (1985).
18. B. B. Owens, in *Advances in Electrochemistry and Electrochemical Engineering*, P. Delahay and C. W. Tobias, Eds., Wiley-Interscience, New York, 1971, p. 1
19. B. B Owens, B. Scrosati, and P. Reale, in *Encyclopedia of Electrochemical Power Sources*, Elsevier, New York, 2009, Vol. **4**, p. 120.
20. B. B. Owens, P. Reale, and B. Scrosati, *Electrochem. Commun.*, **9**, 694 (2007).
21. C. C. Liang, in *Application of Solid Electrolytes*, T. Takahashi and A. Ozawa, Eds., JEC Press, Cleveland, OH, 1980, p. 60.
22. R. Kanno and M. Maruyama, *J. Electrochem. Soc.*, **148**, A742 (2001).
23. G. C. Farrington, B. Dunn, and J. O. Thomas, in *High Conductivity Solid Ionic Conductors*, T. Takahashi, Ed., World Scientific, Singapore, 1989, p. 32.7.

24. N. Kamaya, K. Homma, Y. Yamakawa1, M. Hirayama, R. Kanno, M. Yonemura, T. Kamiyama, Y. Kato, S. Hama, K. Kawamoto, and A. Mitsui, *Nat. Mater.*, (2011), doi 10.1038/NMAT3066.
25. M. B. Armand, *Adv. Mater.*, **2**, 278 (1990).
26. F. M. Grey, *Solid Polymer Electrolytes*, VCH, Weinheim, Germany, 1991.
27. G. Kobe, *Electric and Hybrid Vehicle Technology*, Automotive Industries, Sept. 1996.
28. K. M. Abraham and M. Alamgir, *J. Electrochem. Soc.*, **137**, 1657 (1990).
29. D. W. Murphy and J. N. Carides, *J. Electrochem. Soc.*, **126**, 349 (1979).
30. S. Basu, U.S. Patent 4,304,825 Dec. 8, 1981.
31. B. Scrosati, *J. Electrochem. Soc.*, **139**, 2776 (1992).
32. M. Lazzari and B. Scrosati, *J. Electrochem. Soc.*, **127**, 773 (1980).
33. M. Lazzari and B. Scrosati, U.S. Patent 4,464,447 Aug. 7, 1984.
34. J. J. Auborn and L. Barberio, *J. Electrochem. Soc.*, **134**, 638 (1987).
35. T. Nagaura and K. Tazawa, *Prog. Batteries Solar Cells*, **9**, 20 (1990).
36. S. Basu, U.S. Patent 4,423,125.
37. Yoshino, U.S. Patent 4,668,595.
38. S.-K. Jeong, M. Inaba, P. Abe, and Z. Ogumi, *Electrochem. Soc.*, **148**, A989 (2001).
39. T. Ohzuku, T. Ueda, and N. Yamamoto, *J. Electrochem. Soc.*, **142**, 1431 (1995).
40. S. Panero, D. Satolli, M. Salomon, and B. Scrosati, *Electrochem. Commun.* **2**, 810 (2000).
41. T. Ohzuku, K. Ariyoshi, S. Yamamoto, and Y. Makimura, *Chem. Lett.*, **12**, 1270 (2001).
42. P. Reale, S. Panero, B. Scrosati, J. Garche, M. Wohlfart-Meherens, and M. Wachtler, *J. Electrochem. Soc.*, **151**, 12 (2004).
43. M. Winter and J. O. Besenhard, *Electrochim. Acta*, **45**, 31 (1999).
44. A. S. Aricò, P. Bruce, B. Scrosati, J.-M. Tarascon, and W. van Schalkwijk, *Nat. Mater.*, **4**, 366 (2005).
45. P. Bruce, B. Scrosati, J.-M. Tarascon, and W. van Schalkwijk, *Angew. Chem. Int. Ed.*, **20**, 2939 (2008).
46. H. Inoue, International Meeting on Lithium Batteries, IMLB 2006, Biarritz, France, June 18–23, 2006, Abstr. 228.
47. K. Mizushima, P. C. Jones, and J. B. Goodenough, *Solid State Ionics*, **7**, 314 (1981).
48. A. K. Padhi, K. S. Nanjundaswamy, and J. B. Goodenough, *J. Electrochem. Soc.*, **144**, 1188 (1997).
49. J. Hassoun, P. Reale, and B. Scrosati, *J. Mater. Chem.* **17**, 3668 (2007).
50. S. Fletcher, *Bottled Lightning*, Hill and Wang, New York, 2011.
51. B. Scrosati, *Nature (London)*, **473**, 448 (2011).
52. R. D. Rauh, K. M. Abraham, G. F. Pearson, J. K. Surprenant, and S. B. Brummer, *J. Electrochem. Soc.* **126**, 523 (1979).
53. R. D. Rauh, G. F. Pearson, and S .B. Brumner, *Proc. 12th IECEC*, 1977, p. 283.
54. K. M. Abraham and Z. Jiang, *J. Electrochem. Soc.*, **143**, 1 (1996).
55. K. M. Abraham and Z. Jiang, U.S. Patent 5,510,209.
56. H.-J. Ahn, W.-W. Kim, and G. Cheruvally, in *Encyclopedia of Electrochemical Power Sources*, Elsevier, New York, 2009, Vol. **5**, p. 155.
57. S. J. Visco, E. Nimon, and E. L. C. De Longhe, in *Encyclopedia of Electrochemical Power Sources*, Elsevier, New York, 2009, Vol. **4**, p. 376.
58. X. Ji and L. F. Nazar, *J. Mater. Chem.* **20**, 9821 (2010).

CHAPTER 3

ADDITIVES IN ORGANIC ELECTROLYTES FOR LITHIUM BATTERIES

Susanne Wilken, Patrik Johansson, and Per Jacobsson

1 INTRODUCTION

In a lithium battery the electrolyte is ideally an inert transport medium for ions during their travel between the positive and negative electrodes under the charge and discharge processes. Depending on the cell design and purpose, different electrolyte concepts may be used: liquid, gel, solid polymer, or ionic liquid–based electrolytes.

Lithium Batteries: Advanced Technologies and Applications, First Edition.
Edited by Bruno Scrosati, K. M. Abraham, Walter van Schalkwijk, and Jusef Hassoun.
© 2013 John Wiley & Sons, Inc. Published 2013 by John Wiley & Sons, Inc.

To provide optimal functionality, the electrolyte should have at least the following characteristics: ability to allow for efficient ion transport (conductivity) without electron transfer, high electrochemical stability over a wide electrochemical potential window, and high thermal stability over a defined temperature range. Preferably all components should also be renewable, nontoxic, ecologically friendly, and low in cost.

Since 1990, [1] the commercially dominating lithium-ion conducting electrolytes have been either liquids or gels. The former consist basically of 1 M lithium hexafluorophosphate (LiPF$_6$) in a combination of linear and cyclic organic solvents such as ethylene carbonate (EC) and dimethyl carbonate (DMC), together with a number of additives with specific functionalities. Gel-type electrolytes in principle consist of the same liquid electrolyte immobilized in a swollen polymer matrix. Variations exist for specific applications. The main reasons for the dominance of the foregoing concepts are a high oxidation potential (>5.1 V vs. Li$^+$/Li), excellent solubility of the salt in organic solvents, and excellent conductivity (1 M LiPF$_6$ EC/DMC 1 : 1 ≈ 11 mS/cm at 20°C) [2]. Although completely new electrolyte concepts are slowly entering the scene, these liquid- or gel-type electrolytes based on organic liquids will play an important role over foreseeable time. In this chapter we focus on the strategies used to improve the performance of the base electrolyte using designed additives.

When discussing possible improvements in the electrolyte it is important to realize that the exact requirements on each cell type and chemistry are application dependent and that today's lithium-ion batteries have been developed and optimized for the portable electronics market. Thus when extending the market segment to, for example, stationary applications, not all electronic devices need to be able to work, for example, at $-20°C < T < 80°C$ or in an even larger temperature range, as in vehicle applications; also a 5-V range stability is needed for only a few applications. There is a sharp difference between trying to design ideal electrolytes that should work over a maximum electrochemical potential window and be compatible with most battery chemistries—and the attempt to design the optimal electrolyte for a specific application. The latter defines additional requirements for the electrolyte: specific performance levels in density and weight; safety, which is even more important in large-scale applications; temperature flexibility; mechanical stability; durability; and so on.

1.1 Shortcomings of Standard Liquid or Gel Electrolytes

Independent of application, all electrolytes based on LiPF$_6$ in organic solvents have severe shortcomings. Chemical and electrochemical decomposition pathways of such electrolytes exemplified for LiPF$_6$ EC/DMC are sketched in Figure 1; the right-hand side shows decomposition related to the salt, while the left-hand side represents problems with the organic solvents. The first and perhaps best-known shortcoming is the natural chemical dissociation equilibrium of the PF$_6^-$ anion producing lithium fluoride (LiF) and phosphorus pentafluoride (PF$_5$), which means that fewer lithium ions are available for charge transport and thus a decline in performance. This equilibrium can be shifted thermally toward the right side (Fig. 1) at as low temperatures as 80°C. More important, PF$_5$ is a strong Lewis acid that attacks the electrolyte solvents, causing decomposition under release of gases [7]. This poses a severe safety risk in terms

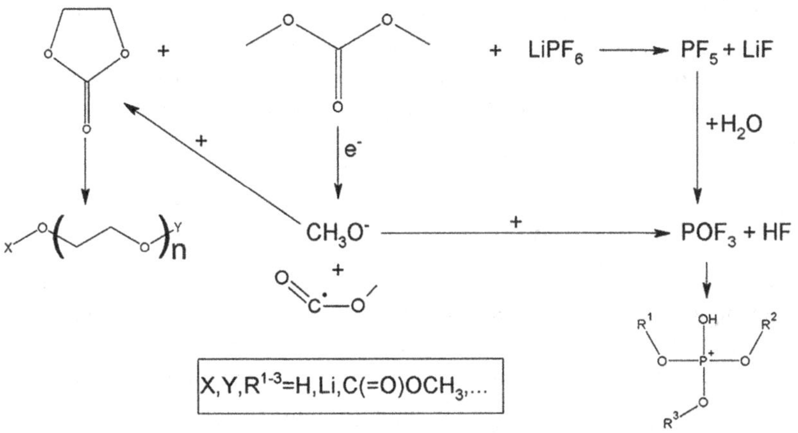

FIGURE 1 Chemical and electrochemical decomposition of LiPF$_6$-based electrolytes in organic solvents. The right-hand side shows the chemical reactions of the salt; the natural salt dissociation equilibrium and following reactions with trace amounts of water lead to HF formation. The bottom line of equations shows the electrochemically induced low potential DMC reduction and consequent reaction of MeOLi with EC that leads to chemical evolution of different oligomer series [3,4]. The reactions with POF$_3$ originating from the salt lead to formation of phosphates [5]. (From [6].)

of possible explosions of inflated cells and subsequent release of toxic decomposition products [8].

The risk scenario caused by released gases is enhanced further by the chemical instability of LiPF$_6$ toward minute traces of water and alcohols, traces that are always present in any electrolyte. This leads to formation of HF and POF$_3$ according to

$$\text{LiPF}_6(s) + \text{H}_2\text{O}(g) \rightarrow \text{LiF}(s) + \text{POF}_3(g) + 2\text{HF}(g)$$

Not only is the LiPF$_6$ salt unstable, but the organic solvents are also part of the decomposition of the standard electrolyte. Used because of their excellent ability to solvate lithium salts and relatively good electrochemical stability, major drawbacks are due to their very physical properties in a state of aggregation: melting at relevant temperatures [T_m(EC) $= 36°$C, T_m(DMC) $= 5°$C], boiling at elevated temperatures [T_B(DMC) $= 91°$C]; very low viscosity and thus a risk of leakage on the cell level; and finally, high flammability [T_{Fp}(DMC) $= 18°$C] [9].

In addition, organic solvents may decompose chemically at the catalytic electrode surfaces, as represented in the second line of the reactions shown in Figure 1. This effect is obvious for Li metal and any electrolyte chemistry. It has also received considerable attention when introducing graphite as an anode material in combination with propylene carbonate, which co-intercalates into graphite and causes exfoliation under the production of propene gas [10], but in principle all electrode and electrolyte chemistries are affected. Strongly oxidizing cathodes such as LiMn$_2$O$_4$ and LiCoO$_2$ catalytically destabilize the redox chemistry of the electrolyte, causing film and gas formation during cycling or storage at elevated temperatures, which increases safety concerns [11,12].

In all there are thus a number of battery electrolyte shortcomings that are of varying importance and criticality, depending on application as well as the overall battery chemistry. For these "old" electrolytes, especially when it comes to the development of new concepts or new chemistries, it is important, with a full understanding of the failure origins and how to mitigate the shortcomings, to avoid failures.

1.2   The Advent of Additives

Much effort has focused on finding the perfect electrolyte for rechargeable lithium batteries by changing the salt and/or solvent, but so far no other electrolyte system has proven to be as balanced when meeting the requirements for optimal cell performance. Instead, it is the installation of external safety devices and control equipment, the use of highly purified components, and not least the functionalization of the base electrolyte by the addition of small amounts of chemicals, additives, which allow for the state-of-the-art systems of today. Many books and review articles describe the pros and cons of different liquid electrolytes, primarily salts and solvents, in detail [13–15]; the more recent also include additives as a key to electrolyte research [9,16–19].

The concept of *functional electrolytes* was introduced in 1997 by Yoshio, Abe, and Yoshitake (Ube Chemical Co.) [20–22]. It defines electrolyte design as a combination of highly purified compounds with slight amounts of different additives that address the stated shortcomings, and furthermore, add functionality. In 2003, Ue et al. (Mitsubishi Chemicals Co.) introduced *role-assigned electrolytes* which further categorize the additives according to their working assignment [23,24]. The following general categories were suggested: (1) anode passivation film-forming agents, (2) cathode protection agents, (3) overcharge protection agents, (4) wetting agents, (5) flame-retardant agents, and (6) others.

A general aspect not to forget in electrolyte design is the trade-off between battery performance and the effect of the additive in relation to the cost issue. In practice, only very small amounts of additives, usually up to 5 wt%, are used, so as not to interfere with the unproblematic parts of battery functionality. Nevertheless, additives provide important solutions to specific shortcomings of the neat electrolyte and thus effectively improve cell performance: safety, life length, and so on. It is the final application that defines an "acceptable level" of performance decrease vs. improved specific properties. The use of additives, creating more complex electrolytes, is also affected by the changes in regulations on battery recycling and ecologically friendly components, which are getting stricter [25].

1.3   Additive Criteria and Development Process

The fundamental search for suitable additive candidates can be described as an iterative process with the aim of finding chemicals that should meet the following general criteria:

- Stable in the electrolyte
- Easily soluble in the electrolyte
- Noncoordinating (except if beneficial to Li^+ transport)

- Thermally stable over a defined range (unless the additive takes effect in a changed aggregate state)
- Electrochemically stable on both the anode and cathode sides (unless the additive takes part in the formation of a protective film)

The common strategy of this search is reflected in what Yoshio et al. call "successive steps for developing the electrolyte additives" [19]. Their first screening step in the search for electrode-related additives is based on an evaluation through quantum mechanical calculations to obtain the orbital energies of potential structures, which are in turn connected with the oxidation–reduction potentials and thus give a measure of the electrochemical stability window. A number of additives are ranked in this way based on their stability toward specific electrode chemistries. After synthesis, the analysis of the physical properties of the surface layer is the second screening step. That includes analysis of the conductivity, the oxidation–reduction potential, and the viscosity of the electrolyte–additive mixtures. Charge–discharge and battery performance tests using various electrolytes are done in the third step, and finally, postmortem analysis using several characterization techniques to determine the effectiveness of the additive.

Depending on the type of additive, various development strategies may be used. A little broader classification of additives is visualized in Figure 2. Obviously, the choice of chemicals must be vast, and even the very definition of what constitutes an additive becomes ambiguous when salts or solvents are used as additives.

In this chapter we restrict ourselves to presenting a few selected families of additives in the categories of thermal or chemical stability through $LiPF_6$ salt stabilizers, overcharge protectors, and flame retardants. Another important class (Fig. 2) is that of film-formation additives, which form and control the important ion-permeable thin layers on the anode [solid electrolyte interphase (SEI)] and on the cathode [solid permeable interphase (SPI)]. More specifically, the layers function as extended electrolytes enabling kinetically stable cell operation and therefore need to be taken into account for optimal electrolyte design. These film-forming additives have been reviewed thoroughly, and because of space limitations we refer the reader to those

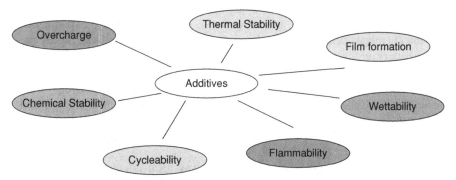

FIGURE 2   Additive categories with respect to target function, which can be electrode related (e.g., film formation, overcharge) or related to intrinsic electrolyte properties (e.g., chemical or thermal stability).

reviews [16–24]. Due to the added complexity when using many additives in the same base electrolyte, an issue of increasing importance, we also discuss synergy effects between additives.

2 LiPF$_6$ SALT STABILIZERS

To prolong battery lifetime and increase the efficiency while maintaining the basic chemistry of LiPF$_6$-based electrolytes, vast amounts of experimental work have been carried out on the degradation and decomposition mechanisms. As outlined above, even outside abuse conditions, the reactions in a battery are complex and span an entire range of chemical, electrochemical, and physical (thermal) processes. Among others, the chemical processes involve interactions between the ions and solvent molecules, or possible catalytic effects of electrodes; electrochemical processes involve the reduction–oxidation of solvent molecules; and relevant physical conditions are storage, temperature, and mechanical stress.

The typical approach to degradation studies is either postmortem analysis of such battery components as surface films on electrodes or separators after exposure to a variety of conditions, such as cycling, storage, and abuse; or in situ recording of decomposition products in model systems. The techniques used are manifold, so it is not surprising that the findings scatter broadly [6,7,12,26–47]. Experimentally deducted degradation mechanisms are the focal point of an ongoing discussion in the open literature, beginning with concrete details such as the exact onset temperature of thermal degradation of LiPF$_6$ and ranging to the complex question of the SEI as part of the mechanism [48]. Formation of the SEI layer through reduction of solvent molecules during the first charge is an example of the fact that not all decomposition is disadvantageous.

Independent of the above, the common findings on degradation of LiPF$_6$-based electrolytes can be summarized to, first, the presence of oligomeric species on electrode surfaces, and second, very low water–alcohol tolerance and resulting degradation reactions with solvent and electrode materials. The bridging parts of most suggested mechanisms emphasize the role of (1) the chemical dissociation equilibrium of LiPF$_6$, (2) storage conditions, (3) lithium methoxide (CH$_3$OLi) as a key compound, and (4) impurities always present. Here we give an example of different views arising, despite both being due to data from thorough experiments. First, the chemical–electrochemical decomposition was studied thoroughly by the lithium battery group in Amiens [5,6,26,27,32,36,43,44]. In a scenario for LiPF$_6$ EC/DMC degradation presented recently, DMC reduction at low potential is assigned an initiating role, as sketched in Figure 1 [44]. Second, the thermal decomposition of a similar electrolyte alone, as well as that in contact with MCMB graphite and various cathode surface particles, was studied by Lucht et al., and the autocatalytic reactions were assigned to the formation of PF$_5$ and related fluorophosphates, shown in Figure 3 for the pure electrolyte [41,49–52].

Although identifying the degradation pathway itself is outside the scope of this chapter, keeping the focus on additives, it is nevertheless important to realize that without a clear view of the mechanism at hand, suitable countermeasures are

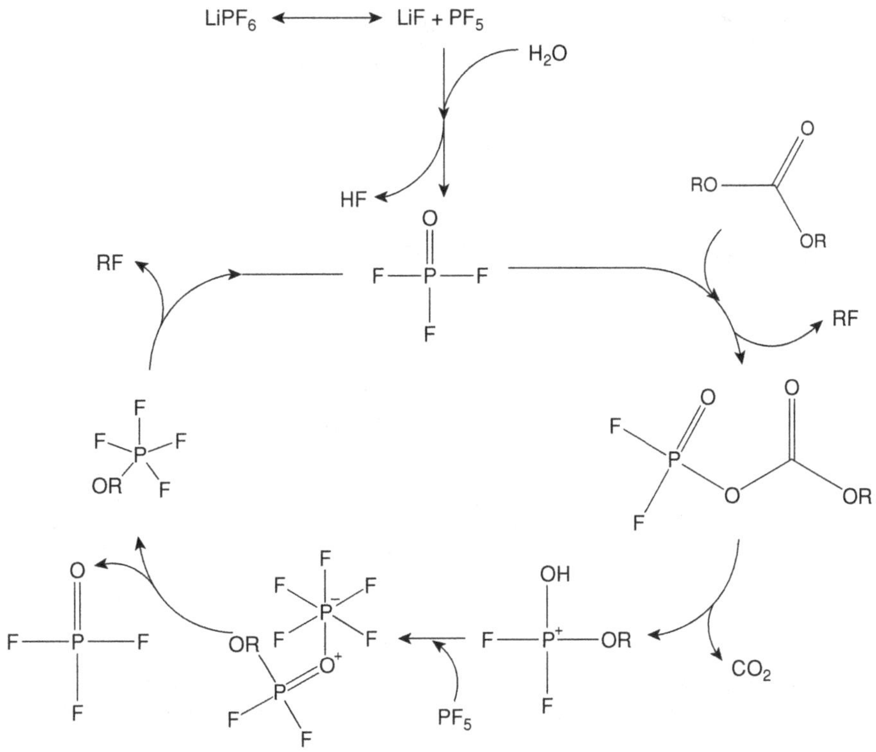

FIGURE 3 Autocatalytic reaction scheme of thermally induced decomposition of LiPF$_6$-based electrolytes: Trace impurities of water react with LiPF$_6$/PF$_5$ to generate POF$_3$, which reacts with the solvent molecules to POF$_2$OR under CO$_2$ and RF release. Chemical rearrangement generates more POF$_3$ and RF [53]. (From [54].)

difficult to suggest: for example, defining precise requirements for additives. In the following we therefore present a mixture of strategies, all of which aimed at increasing the thermal/chemical stability of conventional electrolytes: (1) hindering the PF$_5$ development and deactivating PF$_5$, (2) scavenging impurities such as water and HF, and (3) anion receptors. In addition, increasing or changing the SEI formation will be treated implicitly, as there are severe effects on the overall performance. Structures of example compounds are shown in Figure 4.

2.1 Hindering and Deactivating PF$_5$

One way to reduce the PF$_6^-$ decomposition is by hindering PF$_5$ evolution by controlling the dissociation equilibrium. This was investigated straightforwardly by adding excess LiF to the electrolyte [55]. Hiroi et al. showed that 0.05 wt% LiF added to a conventional electrolyte significantly reduced the gas generation upon cycling. Problematic with this approach is the reduced Li$^+$ diffusion into the anode caused by the insulating effect of LiF in the SEI layer [28].

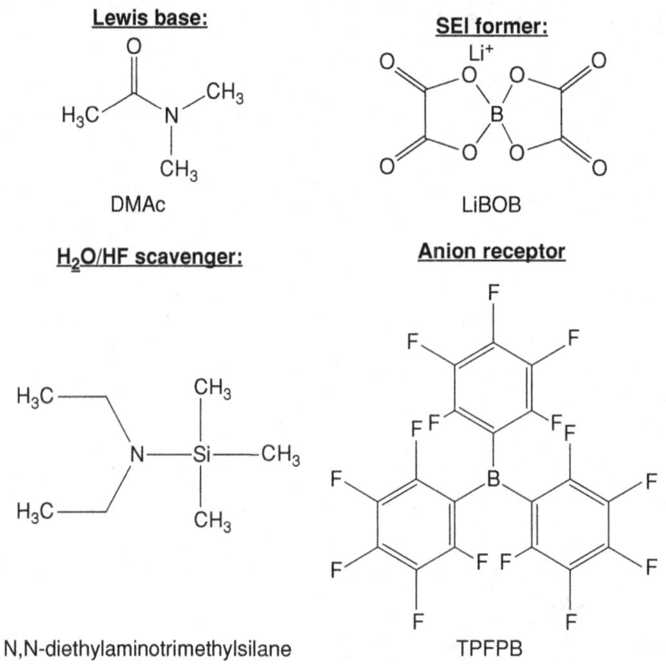

FIGURE 4 Additives improving the thermal and chemical stability of conventional electrolytes. Different strategies involve PF_5 deactivation utilizing Lewis bases, suppression of solvent decomposition by formation of a stable SEI layer, mitigation of acidic and hydrolysis reactions by HF or water scavenging, and increasing the ion-pair dissociation by addition of anion receptors.

Furthermore, a key role in the electrolyte decomposition reactions has been attributed to the Lewis acidity of PF_5; therefore, several Lewis bases were investigated as possible additives and their ability to form complexes with PF_5 was determined. Pyridine, hexamethoxycyclotriphosphazene (HMPN), hexamethylphosphoramide (HMPA), and dimethylacetamide (DMAc) have been reported as candidates [50,56]. Xiao et al. observed no decomposition products by nuclear magnetic resonance (NMR) spectroscopy and gas chromatography/mass spectrometry GC/MS upon storage of mixed standard electrolytes (1 to 10% additive) at 85°C for 1 to 52 weeks, depending on the base; the formation of a HMPA–PF_5 complex was also identified via NMR. In addition, Xu et al. recently reported that 1% DMAc significantly improves the cyclic performance of a $LiFePO_4$/graphite cell at 60°C. The improved thermal stability of the electrolyte and modifications of both SEI and SPI components were investigated by x-ray photoelectron spectroscopy, and fourier transform infrared spectroscopy and supported by density functional theory (DFT) calculations [57]. Similar to HMPN and HMPA, tris(2,2,2-trifluoroethyl)phosphite (TTFP), which is also known as a flame retardant (see Section 4), was shown to inhibit discoloration of 1.2 M $LiPF_6$ propylene carbonate (PC)/EC/EMC during storage at

60°C for 2 weeks, associated with the formation of a PF$_5$ complex [58]. Additionally, improved cycling performance of graphite/lithium nickel-based mixed oxide cells at 60°C, associated with the formation of a preferable SEI, was observed [59].

Following that train of thought, additives designed primarily to improve the SEI have been suggested as thermal stabilizers. The addition of lithium bis(oxalate)borate (LiBOB) was shown to increase the thermal stability of conventional electrolytes upon storage at 85°C for several months, and no decomposition products were observed by NMR [60,61]. However, vinylene carbonate (VC), known for its excellent anode film-forming properties [62], was shown to have no effect on the stability of a conventional electrolyte, as NMR and infrared studies showed similar decomposition products with and without VC after storage [63]. Amides other than HMPA, such as 1-methyl-2-pyrrolidinone or fluorinated carbamates, were also investigated, employing the nitrogen lone-pair electrons as the weak-base counterpart to the PF$_5$ activity [64,65].

2.2   Impurity Scavenging

Even with the precautions taken in modern lithium-battery production lines, trace amounts of impurities such as water and HF are always present. Compounds that are highly reactive toward H$_2$O and HF have therefore been explored as electrolyte stabilizers. As an example of a bit of serendipity discovery, aromatic isocyanates such as phenyl isocyanate, first investigated as SEI stabilizers against the PC solvent reduction, were also shown to scavenge H$_2$O and HF and function as a weak Lewis base [66,67].

Since HF is well known to destabilize electrode materials, in particular causing Mn^{2+} dissolution from oxides and subsequent capacity fading during cycling at elevated temperatures, cathode protection agents were investigated and are thus indirect electrolyte stabilizers [68,69]. Silicon compounds such as hexamethylsiloxane and tetraethyl orthosilicate were suggested to react with trace amounts of HF to form Si and Si–F complexes to form a stable SPI [70]. Aurbach reported that 1 wt% organosilicon compounds (R$_4$Si) also improve the cycling behavior at 60°C for MCMB graphite/LiCoO$_2$ cells [71].

Hexamethyldisilazane (HMDS) was used as a H$_2$O scavenger in a conventional electrolyte, and a drastic decrease of Mn dissolution at 80°C was observed in storage experiments [72]. The principle of breaking an N–Si bond to capture both HF and H2O was also employed in other N–Si-based additives, such as *N,N*-diethylaminotrimethylsilane, as reported by Takechi and Shiga [73]. As well as being a PF5 scavenger (above), DMAc was reported to be a successful thermal stabilizing additive in contact with various cathode materials [51,74]. Furthermore, amide- and carbodiimide-based compounds were proposed as acid scavengers. Saidi et al. demonstrated the suppression of HF formation by neutralization titration of stored conventional electrolytes with added water and butylamine; similarly, Takechi et al. showed it using *N,N'*-dicyclohexylcarbodiimide (DCC) [75,76]. An improvement in capacity retention of the LiMn$_2$O$_4$ cathode was also shown in both reports.

Completely different approaches to HF scavenging include the use of quartz separators or electrode coatings [77,78]. Sharabi et al. assign the improved cycling performance of $LiCoPO_4/Li$ in a conventional electrolyte to the high reactivity toward HF of the quartz separators used. On the other hand, surface coatings utilizing nanosized oxide particles such as ZrO_2, Al_2O_3, and SiO_2 reduce the local acidity of the electrolyte near the material surface and thereby improve the capacity retention of $LiCoO_2$ cathodes significantly, as reported by, for example, Chen and Dahn [79]. Another route is the temporary suppression of the hydrolysis reaction as reported by Kawamura et al. [80]. Yet another strategy to provide cathode protection from acidic electrolyte decomposition attack is to use additives that decompose prior to the electrolyte components and form a stabilizing SPI. For the latter strategy we refer to reviews by Zhang and Ue [17,18].

2.3 Anion Receptors

Anion receptors (ARs) are designed primarily with a different performance target rather than to increase the stability of electrolytes; they strongly coordinate the anions, reducing the attraction between Li^+ and the anions (PF_6^-, F^-) and, consequently, increase the Li^+ available for transport in the electrolyte. Further motivation for the deployment of ARs is, however, that the limiting power of lithium batteries often originates in decreased Li^+ diffusion in both the bulk electrode materials and SEI/SPI layers. As LiF is a major component of the SEI layers, ARs capable of dissolving LiF are of great importance in decreasing the thickness of the SEI layer [28]. With respect to the thermal stability of the electrolytes and their governing dissociation equilibrium of $LiPF_6$, it seems at first counterintuitive to add ARs that dissolve LiF— and thus generate more PF_5. The critical point is to use controlled amounts of weak ARs, which complex both F^- and PF_6^- and therefore can fulfill both purposes.

The chemistry of ARs is often based on electron-deficient boron-based additives, including borate, borane, and boronate compounds with various fluorinated aryl or alkyl groups. The addition of tris(pentafluorophenyl)borane (TPFPB) gave rise to improved cycling performance and power capability associated with enhanced Li^+ diffusion according to the reaction TPFPB + LiF → TPFPB − F^- + Li^+ [81–83]. At the same time, the interfacial impedance of the test cells was increased significantly by formation of PF_5, by the decomposition reaction TPFPB + PF_6^- → TPFPB − F^- + PF_5. Chen and Amine suggested using an excess of alternative ARs to dissolve LiF while substantially suppressing the decomposition of PF_6^- [84]. This first DFT computational study compared the fluoride affinities depending on possible substitution groups of boron-based backbone compounds, and in a follow-up, study, preselected ARs were synthesized and bis(1,1,1,3,3,3-hexafluoroisopropyl)pentafluorophenylboronate was found to improve the capacity retention [85]. However, the DFT calculations employed had a poor correlation to the electrochemical performance—probably due to kinetic rather than thermodynamic control.

Very recently, Cai et al. also reported the improved thermal stability of $LiFePO_4$ in combination with higher discharge retention and better cycling performance after the addition of 1% tris(trimethylsilyl)borate (TMSB). The TMSB–PF_6^- and TMSB–F^- complexes formed are suggested to improve both ion-pair dissociation and LiF solvation, without the negative interplay between each that is expected [86].

To summarize and give a perspective on ARs: While a promising route in general, a problem with all boron-based compounds is their expected delicate and expensive synthesis.

3   OVERCHARGE PROTECTORS

Any battery system, perhaps a lithium-based system particularly, is very sensitive to thermal and electrochemical abuse. The electrochemical abuse refers primarily to the charge–discharge mechanism, which involves constant reduction and oxidation of cell components already thermodynamically unstable. Kinetic stability is typically achieved through passivating surface layers on both electrodes (SEIs and SPIs; see Section 1.3), nevertheless, overcharge is a critical risk that arises during the operation of all types of lithium-ion batteries. During overcharging, extra lithium ions are accidentally subtracted from the cathode, and lithium metal is deposited on the anode surface instead of being intercalated. The chemical and electrochemical reactions that follow can lead to an irreversible degradation of the cell [87–89]. Gas release and a prompt increase in cell temperature (and pressure) can trigger self-sustaining thermal runaway, which may lead to fire and also to explosion.

Since most charging protocols use the cell voltage as the parameter to determine the end of charge, battery packs with cells with unequal capacities put in series are especially vulnerable to overcharge. Commercial solutions of state-of-the-art battery packs are regulated with expensive electronic safety devices such as integrated-circuit controls or positive-temperature-coefficient resistors, which lower the overall energy density. The demand for higher-capacity devices, an even higher safety level, and lower manufacturing and maintenance costs are all motivations for exploring built-in chemical solutions. If realized successfully, the overcharge protection mechanism can also be used during regular charging protocols of battery packs and not just as an emergency solution. Ultimately, in-series-connected cells with unequal levels of State-of-Charge (SOC) and capacity could be charged as a whole, under the condition that the single-cell maximum voltages are regulated. The extra current for a cell that reaches the top of its SOC during pack charging could be taken care of by reversible overcharge means, and costly manufacturing and repair processes that perform cell balancing could be reduced substantially. Thus, the overcharge protector additive chemistry and physics has a value far beyond just safety.

The working agenda of electronic devices is basically to monitor external voltage and bypass the current flow through an external circuit. Major advantages of this technology include its broad versatility for different types of energy storage devices, due to the fast and easy electronic setting of the appropriate voltage values independent of any battery chemistry. An additional advantage is that hybridization of a joint battery–supercapacitor system becomes rather straightforward. Moreover, the electric work and heat generated by bypassing the current through a wire can be used in the thermal management of a cell, which can be very useful when, for example, dealing with undercharge in a cold environment. Thus, it is a reversible and fail-safe technology. Major drawbacks of this technology include an increase in weight, volume, complexity, and maintenance.

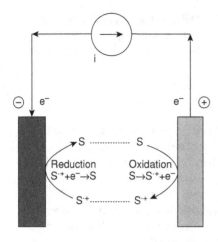

FIGURE 5 Redox shuttle mechanism during overcharge. The neutral shuttle molecule (S) is oxidized at the positive electrode to form a radical ($S \rightarrow S^{\cdot +} + e^-$) at the molecule's characteristic oxidation potential; subsequently, the radical diffuses through the electrolyte and is reduced on the negative electrode ($S^{\cdot +} + e^- \rightarrow S$). Thereafter, the neutral molecule diffuses back to the positive electrode and the cycle can be repeated. In this way no extra lithium metal is plated on the negative electrode during overcharge. (From [18].)

Overcharge protectors, the very chemicals that are included in an electrolyte to prevent the overcharge reactions of both electrodes, can basically be divided into two solution classes, depending on their working mechanism: redox shuttles and shutdown additives. Redox shuttles (S) are chemicals that are oxidized on the positive electrode to form a radical cation ($S^{\cdot +}$) that diffuses through the electrolyte ($S \rightarrow S^{\cdot +} + e^-$). On the negative electrode the radical is reduced again, effectively "shuttling" the charge away without plating extra lithium onto the electrodes ($S^{\cdot +} + e^- \rightarrow S$), as illustrated in Figure 5. Advantages of the technique are its low impact on weight and volume and the cost of the cell. Its major benefit compared to the use of shutdown additives is the reversibility of the process, leaving the cell active and fully functional following an overcharge. However, only a maximum current can be shuttled, which means that thermal runaway *can* be triggered *despite* redox shuttles being employed. Furthermore, the redox reactions of the molecules are potential dependent, and therefore different battery chemistries, or even different applications, may require different additives. Finally, these additives have to be balanced very carefully between the complexity of the electrolyte: possible positive effects, including synergistic ones, and the disadvantageous with respect to cell performance.

Shutdown additives, on the other hand, cause either (1) generation of a large amount of gas, which increases the internal cell pressure up to a point where the current interrupt device kicks in and the cell is deactivated, or (2) polymerization on the electrode surface, effectively hindering further supply of lithium ions and creating an open-circuit situation. Both modes of operation are illustrated in Figure 6. Only very small amounts of chemicals are used to achieve the desired effects. This is a considerable advantage, since it reflects the absence of a weight or volume increase and relatively low cost (depends on the chemical). Moreover, it is a fail-safe mechanism. On the other hand, since the gassing and polymerization processes are electrochemical reactions that take place at a characteristic potential, different additives need to be found for different battery chemistries and applications, which increases the cost. Without designed synergetic effects, the additive does not contribute in other ways to the cell performance. The major disadvantage is the total irreversibility.

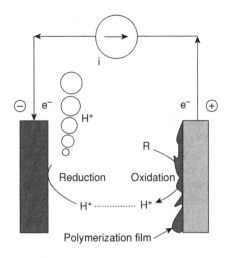

FIGURE 6 Shutdown additive working mechanism. The additive breaks down at its characteristic potential and forms an isolating polymer film on the positive electrode. The liberated protons diffuse to the negative electrode and form hydrogen gas, thus increasing the cell pressure, and the CID may deactivate the cell. (From [18].)

The comparison noted shows above that chemical additives can provide efficient solutions to shortcomings of lithium-ion batteries. Comparing the two classes, shuttle-type additives are limited in their maximum current as defined by their diffusion coefficients and concentrations, possible and economic, while shutdown additives are more reliable sources of definite shutdown and thus can be considered "safer." On the other hand, interest in reversible overcharge protection mechanisms is high due to (1) their use in battery pack charging processes to reduce production cost and maintenance, and (2) cell recycling and component recovery. Therefore, combination of both types of additives in the same electrolyte seems a useful approach as long as the shutdown additives are activated at a higher potential. This can be considered a synergetic effect and is therefore discussed at greater length in Section 5. The three strategies implemented at present, including electronic devices, are summarized in Table 1.

TABLE 1   Summary of Overcharge Protection Strategies: Electronic vs. Chemical Solutions

	Electronic device	Redox shuttles	Shutdown additive[a]
Working mechanism	Electronic regulation via integrated circuits and positive temperature coefficient resistors; reversible	Overcharge current transfer by extra redox molecules; reversible	Permanent deactivation through gassing and insulator coating on electrodes; irreversible
Weight, volume, cost	−	+	++
Thermal management	+	−	−
Effect on battery performance	0	−	−
Fail-safe	+	−	+

Source: Inspired by Chen et al. [90].
[a] Additives are generally superior regarding weight, volume and cost but show some negative impact on battery performance.

Specific requirements for each additive class are outlined in more detail below, and a few examples of chemicals are given.

3.1 Redox Shuttles

The idea of redox shuttles (Fig. 7) is not new; in 1988, Behl and Chin proposed that the oxidation reaction of iodine, $3I^- - 2e \rightarrow I_3^-$, takes place at 3.25 V vs. Li^+/Li and hence that it is suitable as a redox shuttle for the 3V class of lithium-ion batteries [91, 92]. A few other important examples of 3V additives that have been studied have been listed [9,90], but in this chapter we focus on 4V class additives. Independent of the targeted characteristic potential, Narayanan et al. proposed a theoretical analysis of the overcharge mechanism and a few concrete requirements for potential candidates [93]:

- The redox potential should be slightly higher (0.1 to 0.2 V) than the formal potential of the cathode at the end of charge.

- The redox reactions during overcharge should be kinetically reversible on the negative and positive electrodes (with typical electrochemical rate constants above 10^{-5} cm/s).

- To minimize mass transport limitations, the diffusion coefficients and solubilities of redox species should be as high as possible.

General to any redox shuttle, the limiting current density i_{lim} of one-electron transfer scales with initial concentration C_0 and diffusion coefficient D of radical and neutral species: $i_{lim} \approx F^*C_0/L^*(1/D_R + 1/D_0))$ [94,95]. With $D_0 \approx kT/\eta$, the diffusion coefficients of the neutral molecules are inversely proportional to the electrolyte viscosity η. Typical values are $D_0 = 10^{-5}$ to 10^{-6} cm^2/s in 1 M LiPF$_6$ EC/DMC (1 : 2), which is similar to those of solvated lithium ions. Due to solvents aimed at cation (really, lithium-ion) solvation, the radical cation diffusion coefficient is typically one order of magnitude reduced vs. the neutral shuttle. Therefore, an electrolyte with a shuttle molecule can tolerate rather large currents of $i_{lim} = 8$ mA/cm^2 given a concentration of $C_0 = 0.2$ mol/dm^3, a separator thickness of $L = 25$ μm, and

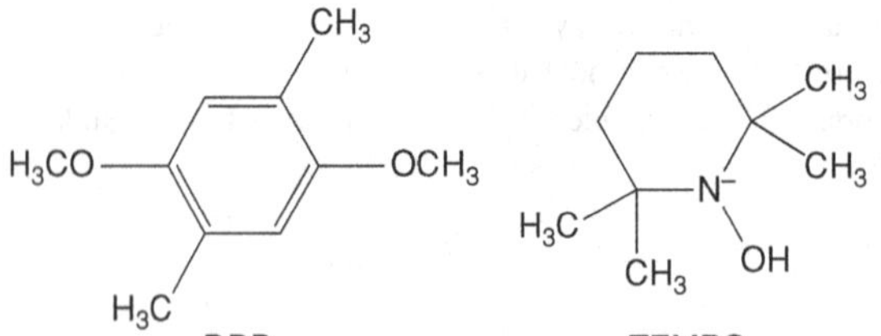

FIGURE 7 Examples of aromatic and nonaromatic redox shuttles and their redox potentials: DDB and its derivatives with different substitutions at various positions constitute the most widely studied family of redox shuttles. TEMPO and its derivatives use the nitrous oxide radical to shuttle the charge.

$D_R = 10^{-6}$ cm^2/s [18]. On the other hand, these target values are in general difficult to obtain, due to the low solubility and high molecular weight of the shuttles.

With the general criteria in mind, most redox shuttles reported in the literature have first been selected according to their redox potentials. Computational screening using quantum chemical methods has proven a useful tool in the search for candidates [96]. After synthesis, typical base characterizations of the additive include cyclic voltammetry, ac/dc capacity measurements for a large number of cycles, long-term stability tests during cycling as well as inactivity, as well as rate or heat flux calorimetry to monitor heat flow during the overcharge process. To determine the limiting current, different concentrations of shuttle species in the base electrolyte are studied with respect to viscosity, solubility, and diffusion using, for example, ultraviolet–visible spectroscopy [95]. Given a precise target definition and well-developed tools for testing, it is surprising how few examples of the urgently needed stable 4V class redox shuttles have been reported in the literature. The lack of understanding on the molecular level regarding the functionality of chemical groups and their impact on, for example, stability and oxidation mechanisms is one reason for the slow progress in the field despite the long-known concepts. For a recent review, see Chen et al. [90]. Below we provide some examples of redox shuttles, divided into aromatic and nonaromatic compounds to simplify structural comparisons.

Aromatic Compounds   Aromatic compounds have been shown to be promising candidates, in particular 2,5-di-*tert*-butyl-1,4-dimethoxybenzene (DDB) with its redox potential of 3.96 V vs. Li$^+$/Li suitable for LiFePO$_4$-based batteries [97–100]. Dahn et al. demonstrated the excellent stability of the redox shuttle for more than 200 cycles of 100% overcharge at C and C/2 rates in LiFePO$_4$/graphite cells for different electrolyte chemistries [98]. DDB has a low solubility in common electrolytes; the maximum was found at a concentration of 0.2 M in 0.5 M LiBOB PC/DEC (1:2), yielding $D = 1.6 \times 10^{-6}$ cm^2/s and $i_{lim} = 2.3$ mA/cm^2. Several groups have tried a systematic exchange of the side chains to the benzene ring to increase the solubility and oxidation potential in order to target LiCoO$_2$ chemistries. Feng et al. reported improved solubility and reasonable overcharge protection for 4-*tert*-butyl-1,2-dimethoxybenzene (TDB) in a standard electrolyte [101]. However, a direct comparison to DDB showed less stability under the same testing conditions [102]. Halogenation was supposed to increase the redox potential, and Taggougui et al. reported the fluorinated derivative 2,5-difluoro-1,4-dimethoxybenzene (F$_2$DMB) to have a redox potential of 4.4 V vs. Li$^+$/Li with a high diffusion and limiting current $D = 1.85 \times 10^{-5}$ cm^2/s and $i_{lim} = 3.7$ mA/cm^2. However, cycling experiments using a Li/Li$_4$Ti$_5$O$_{12}$ cell showed a drastic decrease in positive electrode capacity, associated with a layer formation reaction on the cathode, preventing intercalation [103]. As another example, 1-bromo-2,5-dimethoxybenzene, previously reported as a suitable redox shuttle [104], also fails its overcharge mechanisms after a few cycles [100]. Another approach to substitutions on a DDB basic frame was based on its relative stability during a first oxidation reaction, but irreversible decomposition during a second oxidation for working potentials above 4.2 V vs. Li$^+$/Li [106]. Based on this notion, 3,5-di-*tert*-butyl-1,2-dimethoxybenzene (DBDB) and further derivatives were recently suggested [107,108]. All of the above creates material for the ongoing debate about structure–property correlations of redox shuttles [105].

The currently highest redox potential synthesized and reported is tetraethyl-2,5-di-*tert*-butyl-1,4-phenylene diphosphate (TEDBPDP) with an oxidation potential of 4.8 V vs. Li/Li$^+$ [109]. It was shown to function for 10 cycles at a C/10 rate in a Li/LiMn$_2$O$_4$ cell and is thus the first successful redox shuttle for high-voltage cathode materials. Due to its phosphate groups, it can also act as a flame retardant; 5% TEDBPDP increased the onset temperature of exothermal reactions by up to 260°C.

Especially worth noting is 2-(pentafluorophenyl)-tetrafluoro-1,3,2-benzodioxaborole (PFPTFBB), shown to be stable up to 4.43 V vs. Li$^+$/Li and proposed to be bifunctional as an anion receptor, due to its strong Lewis acidity [110]. Five wt% PFPTFBB in a graphite/LiNi$_{0.8}$Co$_{0.15}$Al$_{0.05}$O$_2$ cell functioned for 170 cycles of 100% overcharge at C/5 at 55°C. One disadvantage is the expected expensive and difficult synthesis of PFPTFBB, which is currently under investigation by, for example, Weng et al. [111].

Nonaromatic Compounds   As the first, and perhaps the best known example, 2,2,6,6-tetramethylpiperidine-1-oxyl (TEMPO) has a redox potential of 3.52 V vs. Li$^+$/Li, suitable for LiFePO$_4$ chemistries with appropriate substitutions [112]. A mixed TEMPO LiBOB organic liquid electrolyte showed stable behavior for 120 cycles of 100% overcharge in Li$_4$Ti$_5$O$_{12}$/LiFePO$_4$ cells. The diffusion was found to be comparable to that of DDB with $D = 1.8 \times 10^{-6}$ cm^2/s. However, long-term cycling using Li/Li$_4$Ti$_5$O$_{12}$ cells showed capacity fading, and insufficient electrochemical stability of TEMPO was suggested [113]. In a rather fair comparison between DDB and TEMPO given by Moshurchak et al. [114], DDB was found to be more stable toward low potentials and TEMPO toward high potentials.

In a quite different route, lithium borate cluster salts Li$_2$B$_{12}$F$_{12-x}$H$_x$ ($x = 1$ to 12) with redox potentials between 4.2 and 4.7 V vs. Li$^+$/Li have been studied [115, 116]. In addition to their intrinsic high oxidation potentials, the built-in advantages over aromatic compounds are their excellent solubilities in organic solvents, thermal stabilities up to 400°C, and moisture tolerance. This, together with being a salt providing lithium-ion charge carriers, makes these borate cluster salts potentially suitable even to directly replace LiPF$_6$ in high-voltage battery applications [117]. However, as with most boron cages, delicate and expensive synthesis is the major drawback of these compounds and hinders commercialization for a wide variety of applications.

A more futuristic strategy to increase the limiting current density of shuttle additives is a two- or-more-electron transfer redox compound that makes use of several close oxidation states of the same neutral molecule. First, several oxidation states increase the number of charge carriers that can be shuttled in total. Second, and perhaps more important, if before thermal runaway occurs, the last oxidation state causes complete decomposition of the molecule under formation of an SPI layer (as has been reported of insufficiently stable species), these additives could be used as an internal fail-safe mechanism, thus increasing the safety level of redox shuttles enormously. However, to our knowledge no such materials have yet been reported in the literature.

3.2   Shutdown Additives

Compared to redox shuttles, a lot of work on shutdown-type additives (Fig. 8) has been presented, exclusively in patent applications and not in the open literature. Hence, industrial relevance is demonstrated and these additives have been used in practical battery manufacturing since 2000 [13,118]. Generally, the additives must meet the following requirements [119]:

- The oxidation potential should be lower than the oxidation potential of the electrolyte solution and preferably above the final standard full charging voltage, while still lower than the voltage of complete removal of lithium from the positive electrode.
- There should be no effect on the charge–discharge cycle life during normal operation.
- The oxidation reaction rate should be as high as possible.

Similar to the development of redox shuttles, quantum chemical calculations are also here useful screening tools for candidates, and subsequently, the experimental verification methods are also similar. In addition, imaging and spectroscopic techniques such as scanning electron microscopy and matrix-assisted laser desorption/ionization–time of flight mass are used to characterize the films formed on the cathode after successful shutdown operation.

Most candidates for shutdown additives are aromatic; biphenyl (BP), in particular, has been studied thoroughly [101,120–128]. After reaching its oxidation potential of 4.54 V vs. Li$^+$/Li, BP forms a thin cathode-covering film containing poly(p-phenylene) and generates hydrogen on the negative electrode. However, BP severely decreases cell performance during fully charged storage at elevated temperatures. To increase the oxidation potential and thus stability during long-term storage, partially hydrogenated compounds such as cyclohexylbenzene (CHB), hydrogenated terphenyl (H-TP), hydrogenated dibenzofuran (H-DBF), and tetralin have been proposed [129]. A fundamental study of the correlation between the degree of hydrogenation and the oxidation potential as well as the reaction mechanisms involved was presented by Shima et al. [130]. Tobishima et al. screened several aromatic compounds and gave a fair comparison for BP, CHP, and H-DBF, with the conclusion that BP is inferior with respect to oxidation potential and lithium cycling efficiency [119]. Feng et al. studied benzene derivatives and found that the film formed by xylene with an oxidation potential of 4.6 V vs. Li$^+$/Li covers the overcharged LiCoO$_2$ cathode completely, even at low concentrations (5%) and short overcharge times, indicating

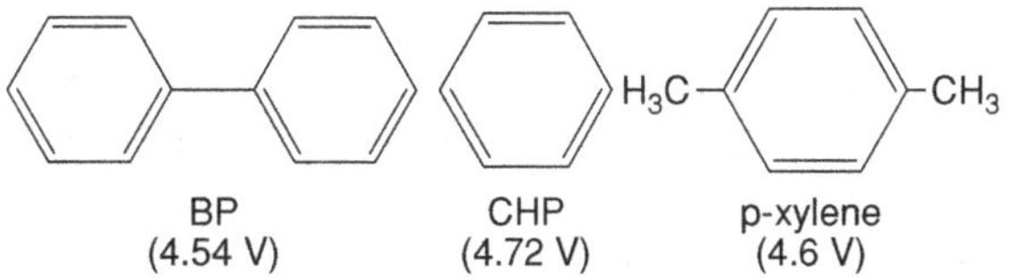

BP   CHP   p-xylene
(4.54 V)   (4.72 V)   (4.6 V)

FIGURE 8   Examples of shutdown additives: aromatic compounds such as biphenyl, cyclohexylbenzene, or xylene, all with oxidation potentials above 4.3 V.

fast reaction kinetics [131,132]. However, slight capacity fading during normal operation was reported, which is in line with the generally reported weakness of this class of additives: its poor long-term operation and storage due to slow and irreversible oxidation. For more examples of shutdown additives, we refer the reader to the review by Zhang [17].

Given the different approaches tried, the many failures outlined, and the continued new development of battery chemistries, there is still an urge to develop both new redox shuttle and shutdown types of additives. In this development task, a challenge beyond the optimal chemistry is how to approach the competitive/complementary role of electronic mitigation. When it comes to the need for battery safety, perhaps no chemistry or physical circuit alone can be considered sufficiently fail-safe.

4 FLAME RETARDANTS

Due to the severe safety concerns associated with organic electrolytes, related primarily to their high flammability with flashpoints below, sometimes far below, 100°C, large attempts have been made to locate additives that lower the flammability. The aim is to find compounds that are flame retardants (FRs), chemicals that suppress continued combustion after being exposed to an external source of heat, spark, or flame. The comparative parameter is the self-extinguishing time (SET): the time an electrolyte takes to extinguish the flame after an external flame source has been withdrawn. There is no standardized test or unit system specific to the SET, but based on Underwriters' Laboratory flammability standard 94, Test for Flammability of Plastic Materials for Parts in Devices and Appliances, procedures using ball-shaped cotton, fiberglass wicks, paper separators, or glassy filters have been developed [133]. According to Xu et al., an electrolyte is *nonflammable* if SET < 6 s/g, *flammable* if SET > 20 s/g, and *retarded* if 6 s/g < SET < 20 s/g [134].

Furthermore, standard physical chemistry analysis techniques such as differential scanning calorimetry (DSC), thermal gravimetric analysis, and accelerated rate calorimetry (ARC) are used, primarily to compare the thermal stability of electrolytes with and without additives. In particular, the lowering of the heat of reaction in contact with charged electrodes, an often observed favorable effect of flame-retardant additives, can be measured easily using DSC and ARC. In Figure 9 we show three classes of chemicals investigated as FR agents, given classical phosphates, mainly P(V) compounds (Section 4.1), cyclic phosphazenes (Section 4.2), and ionic liquids, just recently emerging as FRs (Section 4.3).

4.1 Classical Phosphates

Phosphorus(V) compounds are known as effective FRs in polymeric materials [135], so alkyl phosphates have therefore also been suggested as additives for lithium battery electrolytes. However, besides FR effectiveness first at rather high contents, 15 to 20 wt%, resulting in increased viscosity of the electrolyte and thus decreased ionic conductivity, this chemical family (4.1a, Fig. 9) in general suffers from poor anodic stability. One of the most studied FRs is trimethylphosphate (TMP) [134,136–140],

4.1. Classical Phosphates

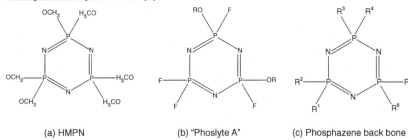

(a) P(V)   (b) P(V) fluorinated   (c) P(III) fluorinated phosphite

R=   ME: TMP   BMP   TFP   TTFP
     Et: TEP
     Bu: TBP
     PH: TPP

4.2. Cyclic Phosphazenes P(V)

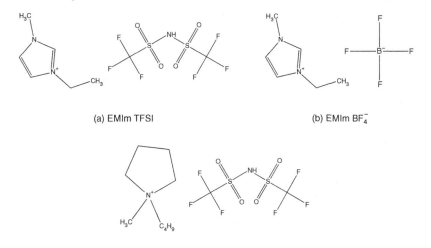

(a) HMPN   (b) "Phoslyte A"   (c) Phosphazene back bone

4.3. Ionic Liquids

(a) EMIm TFSI   (b) EMIm BF$_4^-$

(c) BMPyr TFSI

FIGURE 9   Examples of flame-retardant agents: Different chemical classes include phosphates or phosphites and phosphazenes employing the radical hydrogen scavenging property of phosphorus radicals to hinder flame propagation. Ionic liquids are utilized due to their low vapor pressure, decreasing the flashpoint of the electrolyte.

which at concentrations above 35 wt% has a SET < 20 s/g and no flash point [141]. TMP has functional limitations due to cointercalation with Li^+ into graphite but can still be used with an appropriate additional additive that forms an anode passivating film or with amorphous carbon anodes [142]. The proposed working mechanism as a FR follows the radical scavenging idea [137]:

- Exposed to an external heat source, TMP evaporates: TMP(l) → TMP(g).
- TMP breaks down in the flame to phosphorus radical species: TMP(g) → [P].
- The radical species scavenge H· radicals (the main active agent of combustion chain branching reactions): [P]· → [P]H.
- The chain decomposition reaction is hindered due to a deficiency of H· radicals: RH → R + H.

Other members of the alkyl phosphate family, such as triphenylphosphate (TPP) or tributylphosphate (TBP), are suggested to employ the same modus operandi and have been thoroughly studied [143,144]. TMP and triethylphosphate (TEP) were shown by direct flame tests to retain their flame-retarding property even in gel-polymer electrolytes [145,146]. Changing to fluorinated phosphate additives (4.1b, Fig. 9) such as tris(2,2,2-trifluoroethyl)phosphate (TFP) or bis(2,2,2-trifluoroethyl)methylphosphate (BMP) brought improved FR effectiveness at a lower wt% and better anodic stability than for 4.1a [147–151]. A fair comparison using the same base electrolyte and testing standards has been given [134,148–150], and the SET results are summarized in Figure 10. It is also important to note the general

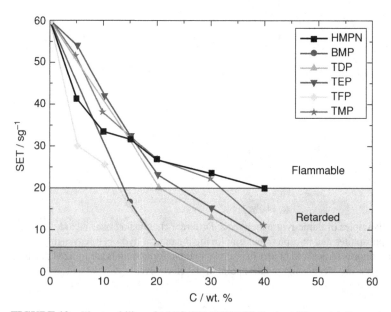

FIGURE 10   Flammability of 1 M $LiPF_6$ EC/EMC (1 : 1 wt%) containing various flame-retardant additives. (From [18].)

performance trade-offs. The first cycle anodic coulombic efficiencies of graphite/nickel-based cells with 20 wt% additives follows:

$$HMPN > TFP > BMP > TEP > TDP \gg TMP$$

As yet another family, P(III) compounds such as tris(2,2,2-trifluoroethyl)phosphate (TTFP) (4.1c, Fig. 9) have also been reported as effective, with an added advantage of forming good SEI layers and the possibility of acting as a Lewis base deactivating PF_5 (see Section 2.1) [58,59,140]. Used as a cosolvent (15 wt% in 1 M $LiPF_6$ PC/EC/EMC 3:3:4) the electrolyte became nonflammable while the ionic conductivity was retained at 80% of that of the base electrolyte. Additionally, TTFP increased the cycling efficiency of the electrolyte by suppressing PC decomposition, and cells (lithium nickel-based mixed oxide/graphite) could be cycled at 60°C for 200 cycles without visible capacity loss [59].

4.2 Cyclic Phosphazenes

Another class of phosphorus(V) compounds are the cyclic phosphazenes, with a higher phosphorus content due to the ring structure and stability with respect to graphitic anodes. The following working mechanism as a flame retardant is suggested [152]:

- Exposed to an external heat source the phosphazene decomposes and forms both a phosphate ester and phosphorus and halogenated radical species.
- The phosphate ester forms a passivation layer, a carbide film on the electrode, intercepting the oxygen supply.
- The radical species scavenge H· radicals: [P]· → [P]H.
- The chain decomposition reaction is hindered due to a deficiency of H· radicals: RH → R + H.

In particular, hexamethoxycyclotriphosphazene (HMPN) (4.2a, Fig. 9) has been studied thoroughly; when 1.5 wt% was added to a conventional electrolyte, the exothermal reaction heat was reduced considerably at elevated temperatures [153, 154]. Compared to TMP and BMP, the FR ability is not as effective, but on the other hand, there is no negative effect on the graphite anode. While capacity fading of the cells is almost negligible for 100 cycles, capacity utilization decreases by 10% after 100 cycles in a nickel-based mixed oxide/graphitic composite cell when the HMPN content reaches 30 wt% and retards flammability (Fig. 10) [134]. This is a typical example of the trade-off between the desired effect and performance in the area of FRs. Moreover, as discussed in Section 2.1, HMPN has also been considered as a salt stabilizer, due to the weak basicity of the iminophosphazene nitrogen centers. Indeed, decomposition products were not observed for 3 wt% HMPN for 2000 h of storage [56]. Unfortunately, the addition of only 10 wt% HMPN shows decomposition features in ^{1}H NMR after only 800 h of storage.

Partially fluorinated cyclic phosphazenes called Phoslytes (trademark of the Bridgestone Corporation) introduced by Otsuki et al. improve FR ability without reversibly affecting cell performance [155–157]. The fluorine side chains reduce the viscosity, while variability of the other side chains allows for tailoring of desired effects. General Phoslyte properties are (1) low viscosity (0.8 to 2.0 mPa·s); (2) high boiling point ($80°C < T_B < 400°C$), (3) low melting point ($T_m < -20°C$), and (4) reduced combustibility of the electrolyte. As an example, the addition of 5 to 10% Phoslyte A (4.2b, Fig. 9) to a standard electrolyte reduces the ionic conductivity only slightly (from 7.6 mS/cm to 7.2 mS/cm) while even exhibiting positive effects on low-temperature behavior, keeping the electrolyte liquid at $-20°C$ and improving the wettability [158,159]. Electrochemical stability is maintained for 0.3 to 5 V, and C/V tests show slight improvement in cyclability. In flammability tests an addition of 5 to 10% renders the base electrolyte inflammable, and ARC and DSC data show a shift in the thermal runaway of 20 to 30°C in contact with fully charged $LiCoO_2$ [160]. The flash point of the base electrolyte determined by American Society of Testing and Materials (ASTM) methods is shifted to higher temperatures, shown to have a considerable general impact on large-scale battery production, as safety regulations on storage quantities and handling permissions depend partially on flash point values. In general, Phoslyte A seems to improve cell performance while fulfilling its FR purpose.

A good comparison of the suitability of different FRs in practical applications and an advanced thermal abuse test in 18650 cells is given by Doughty et al. [161]. However, this study shows that neither TPP, the fluorinated version of TFP (both 4.1's above), or the Phoslytes are suitable. A hot-box test measuring the self-heating rate under exposure to air and an ignition source while cycling showed that none of the additives improve onset and runaway temperatures. Reduced flammability was observed for some of them: 5% Phoslyte C had the highest impact as measured after 95 cycles, but nonflammability was not reached.

The U.S. Department of Energy (DOE) project "Novel Phosphazene-Based Compounds to Enhance Electrolyte Safety and Stability for High Voltage Applications" (01/2009-) specifically evaluates FRs with respect to their use with advanced higher-voltage electrode couples, their safety performance under abuse conditions, and possible enhanced cell life by exchanging side-chain structures on the phosphazene backbone (4.2c, Fig. 9). The novel synthesized compounds have achieved inherent stability and nonflammability, very low vapor pressure, and good lithium salt dissolution capacity. Remaining disadvantages are the high viscosities and the need to attenuate N/Li^+ attraction—as free lithium ions are strongly bound, quasi-irreversibly at the nitrogen members of the phosphazene rings, which decreases the population of free charge carriers. Efforts to realize phosphazene-based ionic liquids that specifically address the N/Li^+ association are under way [162]. A thorough characterization of two compounds (SM4 and SM5) based on methoxyethoxide, ethoxide, and isopropoxide side chains, addressing conductivity, viscosity, flash point, and electrochemical properties, was presented by Sazhin et al. [163]. Both SM4 and SM5 were found to increase the flash point slightly. Conductivities were found at 0.7 S/m for 10% additive content. Whereas the electrochemical window was improved at 30% content compared to the base electrolyte, SEI formation was found to be nearly unaffected when using SM5 in 1.2 M $LiPF_6$ EC/MEC (2 : 8). A new technique was presented to determine the film-formation ability of the electrolytes, as well as a stability test to extend beyond the electrochemical window, worth noting for future comparisons with other possible FRs.

Fei and Allcock recently explored the possibility of polymeric side chains on the phosphazene backbone and found that 25% hexa(methoxyethoxyethoxy) cyclotriphosphazene (MEE trimer) and 25% poly[bis(methoxyethoxyethoxy) phosphazene] (MEEP) in $LiCF_3SO_3$/PC forms a uniform gel electrolyte with conductivities of 2.5×10^{-3} S/cm, while reducing the flammability by one-third compared to the base electrolyte [164].

4.3 Ionic Liquids as Additives

Ionic liquids (ILs), a class of room-temperature liquids composed entirely of ions, have recently attracted much attention as nonflammable electrolytes [165]. Suggested favorable properties in general include low heats of reaction with active materials, thus possibly resulting in enhanced safety as well as low concentration polarization if used as additives [166,167]. Due to the vast number of IL cation–anion combinations, it is difficult to formulate concrete materials that fulfill the desired properties due purely to their structure. We would like to stress that the exploration of ILs as possible additives is just emerging. Here we simply highlight a few studies on specific properties separately, as no thorough overall assessments like those for the other FRs have been made.

First, regarding the reactivity of ILs toward electrodes, Wang et al. found that the ILs BMMImTFSI, Pip_{14}TFSI, and 1114NTFSI indeed show a reduced self-heating rate compared to those of organic electrolytes [168]. Other ILs, EMIm-TFSI (4.3a, Fig. 9) and Pyr_{13}-FSI, showed a reactivity similar to that of an organic solvent, while EMImFSI showed even worse properties when in contact with charged electrode materials (Li_1Si, $Li_7Ti_4O_{12}$, or $Li_{0.45}CoO_2$).

Second, a basic ARC study on some EMIm-based ILs found no exothermal events for ILs with BF_4^-, TFSI, or DEP anions [169]. However, changing the anions to FSI, TCM, or $[B(CN)_4]^-$ resulted in exothermal reactions and self-heating behavior accompanied by pressure increases at temperatures above 300°C. Nevertheless, the latter ILs cannot be considered dangerous, as the onset temperatures are much higher than the working battery temperature range expected. Rather, this study verified the complexity of ILs as additives.

Overall, in one respect the ILs are FRs simply by reducing the content of flammable solvent, but the amount of ILs used as additives must be optimized carefully. This is due to the ILs' inherent high viscosities and thus low ionic conductivities. This, together with complex paths of synthesis, can be reasons that it seems difficult to find suitable ILs and why very little has been reported in the literature about successful implementations of ILs as FR additives. However, there are already a few examples. A study by Larush et al. included a broad choice of ILs (cations based on pyrrolidinium and imidazolium: HMIm, EMIm, BMPyr, and MEMPyr, and the anions BOB, TFSI, and FAP) and the impact of 10% IL on the thermal stability of a standard electrolyte [170]. The thermal stability of the electrolytes with and without immersed electrode materials (Li metal, $Li_{0.5}CoO_2$) was studied. The use of pyrrolidinium-type cations and FAP or TFSI anions was found to improve the stability of $Li_{0.5}CoO_2$ considerably, as the onset temperature of the thermal reactions was increased by 25°C and the heat evolution reduced.

As a second example, Guerfi et al. investigated the use of EMImTFSI, PMImTFSI, and HMImTFSI as additives in a standard electrolyte +2% VC and found the optimum IL content with respect to viscosity and ionic conductivity to be 30 to 40%. [171] In addition, charge–discharge tests showed improvement in the performance of a graphite anode; lithium intercalation occurred successfully as well as the reactions at the $LiFePO_4$ cathode. Very recently, Arbizzani et al. studied the thermal and flame-retarding properties of mixtures of a conventional electrolyte with BMPyrTFSI. All mixtures were found to be less volatile than the conventional electrolyte and at an IL content of 30% become inflammable [172]. However, longer flame exposure times caused all mixtures to ignite. An oxygen-poor combustion process of IL content initiated by organic solvent vapors is suggested. This is an important observation when determining the general safety behavior of electrolyte mixtures.

Also very recently, An et al. reported on 50% Pip_{13}TFSI in a standard electrolyte + 2% LiBOB and increased onset thermal decomposition temperature and nonflammability [173]. The electrochemical properties were found to be acceptable with 200 mAh/g reversible anode capacities at a 0.3 C rate. LiBOB was added to target the poor SEI formation ability of the ILs. Interestingly, as at first may seem against all logic, the reverse approach is also used: Classical FRs and carbonate solvents are suggested as additives to IL-based electrolytes. Here the working agenda is to improve the low ionic conductivity and bad film-formation ability of ILs. By adding both a carbonate solvent and an FR agent, low flammability and excellent solvent properties were obtained, for example, by Lalia et al. in their investigation of 20 wt% triethylphosphate and ethylene carbonate (TEP/EC 1 : 1) in 0.6 M LiTFSI/Pip_{13}TFSI graphite/$LiMn_2O_4$ cells [174].

5 SYNERGY EFFECTS BETWEEN ELECTROLYTE ADDITIVES

What by now should be very clear is the variety of additives needed to approach the "perfect" electrolyte, but unfortunately also, how additives often are counterproductive to electrolyte properties other than those specifically targeted. In addition, the complexity of using many different additives in batteries is an issue in itself, especially with respect to challenging recycling and other environmental aspects. Therefore, using both cost and environmental arguments on top of performance demands, there is an urge toward an optimized way of implementing functional additives. The development of additives addressing several targets seems a rational way forward. This can be realized either through a double functionality exhibited by an additive or by synergetic effects arising through the combined use of specific single functional additives.

5.1 Double-Functionality Additives

A few examples have been reported as the result of discoveries of additional benefits, such as the ability of Phoslytes to improve the low- and high-temperature behavior of electrolytes by changing the viscosity at low temperatures and increasing the flash point (see Section 4.2) [158]. Another case is that of additives built on direct chemical knowledge; for example some PF_5 stabilizers with a weak basicity are known to have FR properties (see Sections 2.1 and 4.2) [175]. Furthermore, a possible combination of overcharge protection and FR was suggested by Zhang et al., employing the radical scavenging properties of the phosphate groups of TEDBPDP (see Section 3.1) [109]. For the same purpose, but starting from the FR function, Feng et al. reported tri-(4-methoxythphenyl)phosphate (TMPP) in a standard electrolyte to be polymerized at 4.35 V vs. Li/Li^+ and thus also act as an overcharge protector [147]. Utilizing the electron deficiency of boron, a combined use of PFPTFBB as overcharge protector and anion receptor was suggested [110]. Korepp et al. reported that the overcharge protector 4-bromobenzyl isocyanate (Br-BIC), which polymerizes at 5.5 V vs. Li/Li^+, also slightly improves the charge–discharge performance of $MCMB/LiCoO_2$ cells [176]. The use of 25% allyl-tris(2,2,2-trifluoroethyl)carbonate (ATFEC) was reported to render an LiPF6/PC-based electrolyte inflammable while suppressing expected graphite exfoliation by PC [177]. Finally, the use of 0.1 wt% thiophene to target both high-voltage cyclability and thermal stability of standard electrolytes was suggested by Lee et al [178].

From the above it seems that finding two effects always renders positive synergies. To the contrary, the opposite may be more normal—but possibly less reported. For example, Shigematsu et al. reported on destructive interference with respect to the reactivity of FR/electrolyte mixes toward electrode materials [179]. FRs are expected to improve the thermal stability, however, $LiBF_4$/butyrolactone-based electrolytes containing TMP and TTFP additives with good FR effects caused a violent reaction in contact with charged graphite/Li_xCoO_2.

Thinking more creatively regarding how to combine beneficial properties systematically, one possibility with the full electrolyte in focus is to create lithium salts out of whatever additive compound is being used. One example is to use the lithium salt version of redox shuttle compounds. During "regular" battery operation such redox shuttles are dissociated by the organic solvents, and the additional Li^+ ions created increase the ion conductivity. During overcharge, however, the redox shuttle anionic compounds disintegrate and work as shuttles. Chen et al. reported lithium borate cluster salts such as $Li_2B_{12}F_9H_3$ as successful examples [116,180]. The same mechanism could in principle also be possible for FRs; in the normal battery working range temperature they would function as charge carriers, but upon any abnormal thermal event or other abuse conditions, disintegrate to provide the necessary radicals, gases, and so on. The latter approach is implemented in the DOE project "Bi-functional Electrolytes for Lithium-Ion Batteries" with the goal of designing and synthesizing novel lithium salts called flame-retardant ions (FRIons) containing functionalized boron and phosphorus [181]. A first successful synthesis was reported with a hybrid oxalate-phoshinate salt $Li(THF)[(C_2O_4)B(O_2PPh_2)_2]$ [182]. In a further step, the most promising FRIons will be functionalized as redox shuttles, called flame-retardant overcharge protectors (FROPs) and overcharge protecting ions (OPIONs). Realizations of the latter approaches have not yet been reported in the literature. For the general approach of designing additives in the shape of Li salts, there is, of course, the well-known example of LiBOB, first designed to be the electrolyte salt of choice. When used as an additive, LiBOB increases the thermal stability of the electrolyte or battery, forms a stable SEI, acts as an overcharge protector, and, of course, increases the lithium-ion conductivity. However, as discussed earlier, the definition of additive becomes ambiguous when salts or solvents are used [60,61,183].

Film formation at the electrodes is at the very heart of battery functionality. With synergies in mind, Abe et al. reported on the use of former overcharge agents as film formers on cathode materials in very low concentrations (0.1 wt%) to increase the cycling performance [124]. Benzene derivatives (e.g., biphenyl and o-terphenyl) could be oxidized electrochemically to form a very thin film on the cathode surface during battery cycling, with the resulting film to be electrically conductive: an electro-conducting membrane. However, at these low additive concentrations the overcharge protection effect is lost, and the search for synergy results in a changed target function. Another very broad range of possibilities arises from designing film-forming agents that have other modes of operation in addition to their primary target. The films created at the electrode surfaces, acting ion-permeably during normal battery operation, or at high voltages, under high temperatures could be constructed to melt permanently, and thus close the pores to stop any chemical reactions. This could be highly relevant as a final, nonreversible safety measure, possibly replacing or at least reducing the need for expensive electronic control devices.

5.2 Synergies of Single-Functionality Additives

The use of additives, together acting in synergy, while being different types of chemicals with either the same or different targets, is an area with very few examples reported. A major problem with this approach to the search for new additives is that there is no rationale as to how to screen for this property. Usually, screening work is done either by hardcore synthesis (low throughput) or computational efforts (high throughput), but in both cases a fundamental understanding of the reasons behind such synergies is of the essence, and this is lacking at the moment.

However, a few examples have been reported recently. Abe et al. found that combining propargyl methanesulfonate (PMS), commonly used as a stabilizer, and VC, used as a film former, results in both thin and dense SEIs and SPIs [184]. Based on an analysis of the electrochemical properties of the additives and the electrodes, the authors suggest that not only is the expected SEI from VC formed, but also a copolymerized SPI by the synergetic decomposition of the two additives. A higher total cyclability and reduced gas evolution was recorded than that found using each additive separately. The keys for producing the synergetic functions are suggested to consist of both a structural difference in the unsaturated moiety and a large difference in the reduction potential of the two additives.

A second strategy aimed at lowering the total additive content involves the synergy effects between different additives of the same type and target function. The best-known example of this strategy is the combined use of VC and VEC as film formers, but also the combined use of BP and CHP as overcharge protectors has also received attention [142,185]. Lee et al. found that using a CHB and BP combination resulted in a much larger oxidation current and a thicker polymeric film than if the contributions and responses for each were superimposed. The origin of the synergetic effects between CHB and BP are only speculated, based on their different electrochemical characteristics, and no unambiguous conclusions can yet be drawn. Santee et al. reported that the combined use of thermal stabilizing additives with different mechanisms—VC, LiBOB, and DMAc—results in a better performance of stored standard electrolytes at 70°C than that obtained when employing the single additives [186]. VC and LiBOB increase the thermal stability of the electrolyte by forming a stable SEI, LiBOB also decreases the decomposition of the electrolyte upon storage, and DMAc complexes PF_5 and thus stabilizes the electrolyte. Surface film analysis showed in addition that the ternary additive electrolyte forms a thinner SEI with a lower LiF content than that obtained when using single-additive electrolytes, suggesting a synergy effect.

6   CONCLUSIONS

Chemicals, added to the electrolyte in small amounts to provide solutions to known shortcomings of the lithium-ion battery technology, were the focus of this chapter. The various target functions of additives have been explained, and a few selected families of additives in the categories of thermal and chemical stabilities, overcharge protectors, and flame retardants were presented in more detail. With designing the perfect electrolyte as the final goal in mind, synergy effects between additives in the same base electrolyte is an issue of foremost importance. A few reported examples were discussed, and the difficulty in anticipating and screening for such effects was emphasized. With the vast choice of chemicals theoretically applicable in electrolytes, there is plenty of opportunity for more laboratory exercises and computational studies to be done, and for fair comparisons of the results, generalized testing standards would be highly beneficial to the field. Yet as we pointed out, ionic liquids, multiple-electron transfer redox shuttles, and progress in boron-based synthesis procedures are emerging fields of high potential.

As we touched upon, one general disadvantage with additives is that they are strongly linked to a specific battery design. Therefore, efficient additive screening and subsequent electrolyte design gets more complex and difficult with the many new battery materials appearing and application areas emerging. Nevertheless, we believe that any knowledge gained on additive functionality in conventional electrolytes brings battery technology forward. What a happy way to end!

REFERENCES

1. Y. Nishi, H. Azuma, and A. Omaru U.S. Patent 4,959,281(A) 1990.
2. J.-M. Tarascon and D. Guyomard, *Solid State Ionics*, **69**, 293 (1994).
3. T. Sasaki, T. Abe, Y. Iriyama, M. Inaba, and Z. Ogumi, *J. Power Sources*, **150**, 208 (2005).
4. D. Aurbach, A. Zaban, Y. Ein-Eli, I. Weissman, O. Chusid, B. Markovsky, M. Levi, E. Levi, A. Schechter, and E. Granot, *J. Power Sources*, **68**, 91 (1997).
5. S. Laruelle, S. Pilard, P. Guenot, S. Grugeon, and J.-M. Tarascon, *J. Electrochem. Soci.*, **151**, A1202 (2004).
6. G. Gachot, S. Grugeon, M. Armand, S. Pilard, P. Guenot, J.-M. Tarascon, and S. Laruelle, *J. Power Sources*, **178**, 409 (2008).

7. S. E. Sloop, J. K. Pugh, S. Wang, J. B. Kerr, and K. Kinoshita, *Electrochem. Solid-State Lett.*, **4**, A42 (2001).
8. A. Hammami, N. Raymond, and M. Armand, *Nature (London)*, **424**, 635 (2003).
9. K. Xu, *Chem. Rev.*, **104**, 4303 (2004).
10. P. Verma, P. Maire, and P. Novák, *Electrochim. Acta*, **55**, 6332 (2010).
11. K. Edström, T. Gustafsson, and J. O. Thomas, *Electrochim. Acta*, **50**, 397 (2004).
12. H. Maleki, G. Deng, A. Anani, and J. Howard, *J. Electrochem. Soci.*, **146**, 3224 (1999).
13. G. E. Blomgren, *J. Power Sources*, **119–121**, 326 (2003).
14. J. Barthel, and H. J. Gores, *Liquid Nonaqueous Electrolyte*, Wiley-VCH, New York, 1999, Chap. 7.
15. M. Nazri, D. Aurbach, and A. Schechter, Kluwer Academic, Norwell, MA, 2003.
16. D. Aurbach, J.-I. Yamaki, M. Saolmon, H.-P. Lin, E. J. Plichta, and M. Hendrickson, Kluwer Academic/Plenum, New York, 2002.
17. S. S. Zhang, *J. Power Sources*, **162**, 1379 (2006).
18. M. Ue, *Role-Assigned Electrolytes: Additives*, Springer-Verlag: New York, 2009.
19. M. Yoshio, H. Nakamura, and N. Dimov, *Development of Lithium-Ion Batteries: from the Viewpoint of Importance of Electrolytes*, Wiley-VCH; Hoboken, NJ, 2009.
20. H. Yoshitake, *Proceedings of the Battery and Power Supply in Techno-Frontier Symposium,* Japan Management Associates, F5-3, 1999.
21. M. Yoshio, H. Yoshitake, and K. Abe, *Electrochemical Society Extended Abstracts*, 2003-2, 2003.
22. K. Abe and H. Yoshitake, *Electrochemistry* (Tokyo), **72** (2004).
23. M. Ue, *Proceedings of the Battery and Power Supply in Techno-Frontier Symposium*, F5-2, 2003.
24. M. Ue, *Extended Abstracts of the Battery and Fuel Cell Materials Symposium*, Graz, Austria, Apr. 18–22, 2004, 53.
25. T. E. P. a. t. Council, Directive 2006/66/EC on batteries and accumulators and waste batteries and accumulators, *Official Journal of the European Union*, 2006.
26. R. Dedryveére, L. Gireaud, S. Grugeon, S. Laruelle, J.-M. Tarascon, and D. Gonbeau, *J. Phys. Chem. B*, **109**, 15868 (2005).
27. R. Dedryvére, S. Laruelle, S. Grugeon, L. Gireaud, J.-M. Tarascon, and D. Gonbeau, *J. Electrochem. Soc.*, **152**, A689 (2005).
28. A. M. Andersson and K. Edstrom, *J. Electrochem. Soc.*, **148**, A1100 (2001).
29. K. Kamakura, S. Shiraishi, and Z. Takehara, *J. Electrochem. Soc.*, **143**, 2187 (1996).
30. K. Kamakura, H. Tamura, S. Shiraishi, and Z. Takehara, *J. Electrochem. Soc.*, **142**, 340 (1995).
31. L. El Ouatani, R. Dedryvere, C. Siret, P. Biensan, and D. Gonbeau, *J. Electrochem. Soc.*, **156**, A468 (2009).
32. L. Gireaud, S. Grugeon, S. Laruelle, S. Pilard, and J.-M. Tarascon, *J. Electrochem. Soc.*, **152**, A850 (2005).
33. D. Aurbach and Y. Ein-Eli, *J. Electrochem. Soc.*, **142**, 1746 (1995).
34. Y. Ein-Eli, S. F. Devitt, D. Aurbach, B. Markovsky, and A. Schecheter, *J. Electrochem. Soc.*, **144**, L180 (1997).
35. D. Aurbach, Y. Ein-Eli, O. Chusid, Y. Carmeli, M. Babai, and H. Yamin, *J. Electrochem. Soc.*, **141**, 603 (1994).
36. S. Grugeon, S. Laruelle, R. Herrera-Hurbina, L. Dupont, P. Poizot, and J.-M. Tarascon, *J. Electrochem. Soc.*, **148**, A285 (2001).
37. A. Kominato, E. Yasukawa, N. Sato, T. Ijuuin, H. Asahina, and S. Mori, *J. Power Sources*, **68**, 471 (1997).
38. K. Kumai, H. Miyaschiro, Y. Kobayashi, K. Takei, and R. Ishikawa, *J. Power Sources*, **81–82**, 715 (1999).
39. R. Mogi, M. Inaba, Y. Iriyama, T. Abe, and Z. Ogumi, *J. Power Sources*, **119–121**, 597 (2003).
40. S. E. Sloop, J. B. Kerr, and K. Kinoshita, *J. Power Sources*, **119–121**, 330 (2003).
41. C. Campion, W. Li, and B. Lucht, *J. Electrochem. Soc.*, **152**, A2327 (2005).
42. P. Poizot, S. Laruelle, S. Grugeon, L. Dupont, and J.-M. Tarascon, *Nature (London)*, **407**, 496 (2000).
43. S. Grugeon, S. Laruelle, L. Dupont, F. Chevallier, P. L. Taberna, P. Simon, L. Gireaud, S. Lascaud, E. Vidal, B. Yrieix, and J.-M. Tarascon, *Chem. Mater.*, **17**, 5041 (2005).

44. G. G. Gachot, P. Ribiére, D. Mathiron, S. Grugeon, M. Armand, J.-B. Leriche, S. Pilard, and S. P. Laruelle, *Anal. Chem.*, **83**, 478 (2010).
45. K. S. Gavritchev, G. A. Sharpataya, A. A. Smagin, E. N. Malyi, and V. A. Matyukha, *J. Thermal Anal. Calorimetry*, **73**, 71 (2003).
46. D. D. MacNeil and J. R. Dahn, *J. Electrochem. Soci.*, **150**, A21 (2003).
47. J. S. Gnanaraj, E. Zinigrad, L. Asraf, H. E. Gottlieb, M. Sprecher, M. Schmidt, W. Geissler, and D. Aurbach, *J. Electrochem. Soci.*, **150**, A1533 (2003).
48. H. Yang, G. V. Zhuang, and P. N. Ross, Jr., *J. Power Sources*, **161**, 573 (2006).
49. C. L. Campion, W. Li, W. B. Euler, B. L. Lucht, B. Ravdel, J. F. DiCarlo, R. Gitzendanner, and K. M. Abraham, *Electrochem. Solid-State Lett.*, **7**, A194 (2004).
50. A. Xiao, W. Li, and B. L. Lucht, *J. Power Sources*, **162**, 1282 (2006).
51. W. Li and B. L. Lucht, *J. Power Sources*, **168**, 258 (2007).
52. W. Li and B. L. Lucht, *J. Electrochem. Soci.*, **153**, A1617 (2006).
53. G. M. Kosolapoff, *Organophosphorus Compounds*, Wiley, New York, 1950; p. 213.
54. B. L. Lucht, T. Markmaitree, and L. Yang, Thermal stability of lithium ion battery electrolytes, in *Encyclopedia of Inorganic Chemistry*, Wiley, Hoboken, NJ, 2011.
55. O. Hiroi, K. Hamano, Y. Yoshida, S. Yoshioka, H. Shiota, J. Aragane, S. Aihara, D. Takemura, T. Nishimura, M. Kise, H. Urushibata, and H. Adachi, U.S. Patent 6,305,540 B1, 2001.
56. W. Li, C. Campion, B. L. Lucht, B. Ravdel, J. DiCarlo, and K. M. Abraham, *J. Electrochem. Soci.*, **152**, A1361 (2005).
57. M. Xu, L. Hao, Y. Liu, W. Li, L. Xing, and B. Li, *J. Phys. Chem. C*, **115**, 6085 (2011).
58. S. S. Zhang, K. Xu, and T. R. Jow, *Electrochem. Solid-State Lett.*, **5**, A206 (2002).
59. S. S. Zhang, K. Xu, and T. R. Jow, *J. Power Sources*, **113**, 166 (2003).
60. K. Xu, U. Lee, S. Zhang, M. Wood, and T. R. Jow, *Electrochem. Solid-State Lett.*, **6**, A144 (2003).
61. A. Xiao, L. Yang, and B. L. Lucht, *Electrochem. Solid-State Lett.*, **10**, A241 (2007).
62. B. Simon and J.-P. Boeuve U.S. Patent 5,626,981, 1997.
63. H.-H. Lee, Y.-Y. Wang, C.-C. Wan, M.-H. Yang, H.-C. Wu, and D.-T. Shieh, *J. Appl. Electrochem.*, **35**, 615 (2005).
64. X. Wang, H. Naito, Y. Sone, G. Segami, and S. Kuwajima, *J. Electrochem. Soci.*, **152**, A1996 (2005).
65. W. Appel and S. Pasenok, U.S. Patent 6,159,640, 2000.
66. T. Jow, S. Zhang, K. Xu, and M. Ding U.S. Patent 6,905,762, 2005.
67. Z. S. Shui, *J. Power Sources*, **163**, 567 (2006).
68. D. Aurbach, B. Markovsky, G. Salitra, E. Markevich, Y. Talyossef, M. Koltypin, L. Nazar, B. Ellis, and D. Kovacheva, *J. Power Sources*, **165**, 491 (2007).
69. A. Du Pasquier, A. Blyr, P. Courjal, D. Larcher, G. Amatucci, B. Gérand, and J.-M. Tarascon, *J. Electrochem. Soci.*, **146**, 428 (1999).
70. B. Markovsky, A. Nimberger, Y. Talyosef, A. Rodkin, A. M. Belostotskii, G. Salitra, D. Aurbach, and H.-J. Kim, *J. Power Sources*, **136**, 296 (2004).
71. D. Aurbach, *J. Power Sources*, **146**, 71 (2005).
72. H. Yamane, T. Inoue, M. Fujita, and M. Sano, *J. Power Sources*, **99**, 60 (2001).
73. K. Takechi and T. Shiga U.S. Patent 6,235,431, 2001.
74. W. Li and B. L. Lucht, *Electrochem. Solid-State Lett.*, **10**, A115 (2007).
75. M. Saidi, F. Gao, J. Barker, and C. Scordilis-Kelley U.S. Patent 5,846,673, 1998.
76. K. Takechi, A. Koiwai, and T. Shiga U.S. Patent 6,077,628, 2000.
77. R. Sharabi, E. Markevich, V. Borgel, G. Salitra, D. Aurbach, G. Semrau, M. A. Schmidt, N. Schall, and C. Stinner, *Electrochem. Commun.*, **13**, 800 (2011).
78. Z. Chen, Y. Qin, K. Amine, and Y. K. Sun, *J. Mater. Chem.*, **20**, 7606 (2010).
79. Z. Chen and J. R. Dahn, *Electrochim. Acta*, **49**, 1079 (2004).
80. T. Kawamura, T. Sonoda, S. Okada, and J. Yamaki, *Electrochemistry* (Tokyo), **71**, (2003).
81. Z. Chen and K. Amine, *J. Electrochem. Soci.*, **153**, A1221 (2006).
82. X. Sun, H. S. Lee, X.-Q. Yang, and J. McBreen, *Electrochem. Solid-State Lett.*, **6**, A43 (2003).
83. M. Herstedt, M. Stjerndahl, T. Gustafsson, and K. Edström, *Electrochem. Commun.*, **5**, 467 (2003).
84. Z. Chen and K. Amine, *J. Electrochem. Soci.*, **156**, A672 (2009).
85. Y. Qin, Z. Chen, H. S. Lee, X. Q. Yang, and K. Amine, *J. Phys. Chem. C*, **114**, 15202 (2010).

86. Z. Cai, Y. Liu, J. Zhao, L. Li, Y. Zhang, and J. Zhang, *J. Power Sources*, **202**, 341 (2011).
87. T. Ohsaki, T. Kishi, T. Kuboki, N. Takami, N. Shimura, Y. Sato, M. Sekino, and A. Satoh, *J. Power Sources*, **146**, 97 (2005).
88. R. A. Leising, M. J. Palazzo, E. S. Takeuchi, and K. J. Takeuchi, *J. Power Sources*, **97-98**, 681 (2001).
89. R. A. Leising, M. J. Palazzo, E. S. Takeuchi, and K. J. Takeuchi, *J. Electrochem. Soci.*, **148**, A838 (2001).
90. Z. Chen, Y. Qin, and K. Amine, *Electrochim. Acta*, **54**, 5605 (2009).
91. W. K. Behl and D.-T. Chin, *J. Electrochem. Soci.*, **135**, 21 (1988).
92. W. K. Behl and D.-T. Chin, *J. Electrochem. Soci.*, **135**, 16 (1988).
93. S. R. Narayanan, S. Surampudi, A. I. Attia, and C. P. Bankston, *J. Electrochem. Soci.*, **138**, 2224 (1991).
94. T. J. Richardson and P. N. Ross, Jr., *Proc. Electrochem. Soc.*, **99-25** (2000).
95. T. J. Richardson and P. N. Ross, Jr., *J. Power Sources*, **84**, 1 (1999).
96. R. L. Wang and J. R. Dahn, *J. Electrochem. Soci.*, **153**, A1922 (2006).
97. C. Buhrmester, J. Chen, L. Moshurchak, J. Jiang, R. L. Wang, and J. R. Dahn, *J. Electrochem. Soci.*, **152**, A2390 (2005).
98. J. R. Dahn, J. Jiang, L. M. Moshurchak, M. D. Fleischauer, C. Buhrmester, and L. J. Krause, *J. Electrochem. Soci.*, **152**, A1283 (2005).
99. L. M. Moshurchak, C. Buhrmester, and J. R. Dahn, *J. Electrochem. Soci.*, **152**, A1279 (2005).
100. J. Chen, C. Buhrmester, and J. R. Dahn, *Electrochem. Solid-State Lett.*, **8**, A59 (2005).
101. J. K. Feng, X. P. Ai, Y. L. Cao, and H. X. Yang, *Electrochem. Commun.*, **9**, 25 (2007).
102. L. M. Moshuchak, M. Bulinski, W. M. Lamanna, R. L. Wang, and J. R. Dahn, *Electrochem. Commun.*, **9**, 1497 (2007).
103. M. Taggougui, B. Carré, P. Willmann, and D. Lemordant, *J. Power Sources*, **174**, 1069 (2007).
104. M. Adachi, K. Tanaka, and K. Sekai, *J. Electrochem. Soci.*, **146**, 1256 (1999).
105. T. Li, L. Xing, W. Li, B. Peng, M. Xu, F. Gu, and S. Hu, *J. Phys. Chem. A*, **115**, 4988 (2011).
106. Z. Chen and K. Amine, *Electrochim. Acta*, **53**, 453 (2007).
107. Z. Zhang, L. Zhang, J. A. Schlueter, P. C. Redfern, L. Curtiss, and K. Amine, *J. Power Sources*, **195**, 4957 (2010).
108. W. Weng, Z. Zhang, P. C. Redfern, L. A. Curtiss, and K. Amine, *J. Power Sources*, **196**, 1530 (2011).
109. L. Zhang, Z. Zhang, H. Wu, and K. Amine, *Energy Environ. Sci.*, **4**, 2858 (2011).
110. Z. Chen and K. Amine, *Electrochem. Commun.*, **9**, 703 (2007).
111. W. Weng, Z. Zhang, J. A. Schlueter, P. C. Redfern, L. A. Curtiss, and K. Amine, *J. Power Sources*, **196**, 2171 (2011).
112. C. Buhrmester, L. M. Moshurchak, R. L. Wang, and J. R. Dahn, *J. Electrochem. Soci.*, **153**, A1800 (2006).
113. M. Taggougui, B. Carré, P. Willmann, and D. Lemordant, *J. Power Sources*, **174**, 643 (2007).
114. L. M. Moshurchak, C. Buhrmester, R. L. Wang, and J. R. Dahn, *Electrochim. Acta*, **52**, 3779 (2007).
115. G. Dantsin, K. Jambunathan, S. V. Ivanov, W. J. Casteel, K. Amine, J. Liu, A. N. Jansen, and Z. Chen, *ECS Meeting Abstracts*, **502**, 223 (2006).
116. Z. Chen, J. Liu, A. N. Jansen, G. GirishKumar, B. Casteel, and K. Amine, *Electrochem. Solid-State Lett.*, **13**, A39 (2010).
117. G. GirishKumar, W. H. Bailey, III, B. K. Peterson, and J. W. J. Casteel, *J. Electrochem. Soci.*, **158**, A146 (2011).
118. A. Yoshino, *Proceedings of the 4th Hawaii Battery Conference*, ARAD Enterprises, Hilo, HI, Jan. 8, 2002.
119. S. Tobishima, Y. Ogino, and Y. Watanabe, *J. Appl. Electrochem.*, **33**, 143 (2003).
120. L. Xiao, X. Ai, Y. Cao, and H. Yang, *Electrochim. Acta*, **49**, 4189 (2004).
121. S. H. Choy, H. G. Noh, H. Y. Lee, H. Y. Sun, and H. S. Kim, U.S. Patent 6,921,612, 2005.
122. H. Mao and D. S. Mainwright, U.S. Patent 6,074,776, 2000.
123. H. Mao and U. v. Sacken, U.S. Patent 6,033,797 2000.
124. K. Abe, Y. Ushigoe, H. Yoshitake, and M. Yoshio, *J. Power Sources*, **153**, 328 (2006).
125. H. Mao and D. S. Wainwright, Canadian Patent 2,205,683, 1999.

126. H. Mao, Canadian Patent 2,163,187, 1995.
127. J. N. Reimers and B. M. Way, U.S. Patent 6,074,777, 2000.
128. M. Yoshio, H. Yoshitake, and K. Abe, *204th ECS Meeting Abstracts*, Orlando, FL, Oct. 12–16, 2003, Abstr. 280.
129. M. Q. Xu, L. D. Xing, W. S. Li, X. X. Zuo, D. Shu, and G. L. Li, *J. Power Sources*, **184**, 427 (2008).
130. K. Shima, K. Shizuka, M. Ue, H. Ota, T. Hatozaki, and J.-I. Yamaki, *J. Power Sources*, **161**, 1264 (2006).
131. X. M. Feng, X. P. Ai, and H. X. Yang, *J. Appl. Electrochem.*, **34**, 1199 (2004).
132. Q. Zhang, C. Qiu, Y. Fu, and X. Ma, *Chinese J. Chem.*, **27**, 1459 (2009).
133. Underwriters' Laboratory Inc. UL 94, Standard for Safety of Flammability of Plastic Materials for Parts in Devices and Appliances Testing. Accessed Feb. 11, 2011.
134. K. Xu, M. S. Ding, S. Zhang, J. L. Allen, and T. R. Jow, *J. Electrochem. Soci.*, **149**, A622 (2002).
135. S. L. Levan and J. E. Winandy, *Wood Fiber Sci.*, **22**, 113 (1999).
136. N. Yoshimoto, Y. Niida, M. Egashira, and M. Morita, *J. Power Sources*, **163**, 238 (2006).
137. X. Wang, E. Yasukawa, and S. Kasuya, *J. Electrochem. Soci.*, **148**, A1058 (2001).
138. X. Wang, E. Yasukawa, and S. Kasuya, *J. Electrochem. Soci.*, **148**, A1066 (2001).
139. H. Ota, A. Kominato, W.-J. Chun, E. Yasukawa, and S. Kasuya, *J. Power Sources*, **119-121**, 393 (2003).
140. X. L. Yao, S. Xie, C. H. Chen, Q. S. Wang, J. H. Sun, Y. L. Li, and S. X. Lu, *J. Power Sources*, **144**, 170 (2005).
141. M. Ue, Japanese Patent 3274102B, 1992.
142. X. Wang, C. Yamada, H. Naito, G. Segami, and K. Kibe, *J. Electrochem. Soci.*, **153**, A135 (2006).
143. Y. E. Hyung, D. R. Vissers, and K. Amine, *J. Power Sources*, **119-121**, 383 (2003).
144. M. C. Smart, F. C. Krause, C. Hwang, W. C. West, J. Soler, G. K. S. Prakash, and B. V. Ratnakumar, *ECS Meeting Abstracts*, **1102**, 1264 (2011).
145. B. S. Lalia, T. Fujita, N. Yoshimoto, M. Egashira, and M. Morita, *J. Power Sources*, **186**, 211 (2009).
146. N. Yoshimoto, D. Gotoh, M. Egashira, and M. Morita, *J. Power Sources*, **185**, 1425 (2008).
147. J. K. Feng, Y. L. Cao, X. P. Ai, and H. X. Yang, *Electrochim. Acta*, **53**, 8265 (2008).
148. K. Xu, S. Zhang, J. L. Allen, and T. R. Jow, *J. Electrochem. Soci.*, **149**, A1079 (2002).
149. K. Xu, M. S. Ding, S. Zhang, J. L. Allen, and T. R. Jow, *J. Electrochem. Soci.*, **150**, A161 (2003).
150. K. Xu, S. Zhang, J. L. Allen, and T. R. Jow, *J. Electrochem. Soci.*, **150**, A170 (2003).
151. M. S. Ding, K. Xu, and T. R. Jow, *J. Electrochem. Soci.*, **149**, A1489 (2002).
152. M. Otsuki and T. Ogino, *Flame-Retardant Additives for Lithium-Ion Batteries*. Springer-Verlag, New York, 2009.
153. C. W. Lee, R. Venkatachalapathy, and J. Prakash, *Electrochem. Solid-State Lett.*, **3**, 63 (2000).
154. J. Prakash, C. W. Lee, and K. Amine U.S. Patent 6,455,200, 2002.
155. M. Kajiwara, T. Ogino, T. Miyazaki, and T. Kawagoe, Japanese Patent 3055358, 2000.
156. M. Otsuki, S. Endo, and T. Ogino, Japanese Patent 083628, 2002.
157. M. Otsuki, T. Ogino, and K. Amine, *ECS Trans.*, **1**, 13 (2006).
158. M. Otsuki and T. Ogino, *Flame-Retardant Additives for Lithium-Ion Batteries*, Springer-Verlag, New York, 2009, p. 279.
159. M. Otsuki and T. Ogino, *Flame-Retardant Additives for Lithium-Ion Batteries*, Springer-Verlag, New York, 2009; p. 286.
160. M. Otsuki and T. Ogino, *Flame-Retardant Additives for Lithium-Ion Batteries*, Springer-Verlag, New York, 2009; p. 284.
161. D. H. Doughty, E. P. Roth, C. C. Crafts, G. Nagasubramanian, G. Henriksen, and K. Amine, *J. Power Sources*, **146**, 116 (2005).
162. K. L. Gering, Novel Phosphazene-Based Compounds to Enhance Electrolyte Safety and Stability for High Voltage Applications (INL), http://www1.eere.energy.gov/vehiclesandfuels/resources/fcvt_reports.html, 2010.
163. S. V. Sazhin, M. K. Harrup, and K. L. Gering, *J. Power Sources*, **196**, 3433 (2011).
164. S.-T. Fei and H. R. Allcock, *J. Power Sources*, **195**, 2082 (2010).
165. M. Armand, F. Endres, D. R. MacFarlane, H. Ohno, and B. Scrosati, *Nat. Mater.*, **8**, 621 (2009).
166. M. Ue, *Electrochemical Aspects of Ionic Liquids*. Wiley, Hoboken, NJ, 2005, Chap. 17.

167. A. Weber and G. E. Blomgren, *Advances in Lithium-Ion Batteries*, Kluwer Academic/Plenum, New York, 2002, Chap. 6.

168. Y. Wang, K. Zaghib, A. Guerfi, F. F. C. Bazito, R. M. Torresi, and J. R. Dahn, *Electrochim. Acta*, **52**, 6346 (2007).

169. R. Vijayaraghavan, M. Surianarayanan, V. Armel, D. R. MacFarlane, and V. P. Sridhar, *Chem. Commun.*, 6297 (2009).

170. L. Larush, V. Borgel, E. Markevich, O. Haik, E. Zinigrad, D. Aurbach, G. Semrau, and M. Schmidt, *J. Power Sources*, **189**, 217 (2009).

171. A. Guerfi, M. Dontigny, P. Charest, M. Petitclerc, M. Lagacé, A. Vijh, and K. Zaghib, *J. Power Sources*, **195**, 845 (2010).

172. C. Arbizzani, G. Gabrielli, and M. Mastragostino, *J. Power Sources*, **196**, 4801 (2011).

173. Y. An, P. Zuo, X. Cheng, L. Liao, and G. Yin, *Electrochim. Acta*, **56**, 4841 (2011).

174. B. S. Lalia, N. Yoshimoto, M. Egashira, and M. Morita, *J. Power Sources*, **195**, 7426 (2010).

175. S. Izquierdo-Gonzales, W. Li, and B. L. Lucht, *J. Power Sources*, **135**, 291 (2004).

176. C. Korepp, W. Kern, E. A. Lanzer, P. R. Raimann, J. O. Besenhard, M. Yang, K. C. Möller, D. T. Shieh, and M. Winter, *J. Power Sources*, **174**, 637 (2007).

177. S. Chen, Z. Wang, H. Zhao, H. Qiao, H. Luan, and L. Chen, *J. Power Sources*, **187**, 229 (2009).

178. K.-S. Lee, Y.-K. Sun, J. Noh, K. S. Song, and D.-W. Kim, *Electrochem. Commun.*, **11**, 1900 (2009).

179. Y. Shigematsu, M. Ue, and J.-I. Yamaki, *J. Electrochem. Soci.*, **156**, A176 (2009).

180. Z. Chen, A. N. Jansen, and K. Amine, *Energy Environ. Sci.*, **4**, 4567 (2011).

181. D. A.Scherson, J. Protasiewicz, I. Treufeld, and A. Shaffer, Bifunctional Electrolytes for Lithium-Ion Batteries, http://batt.lbl.gov/battfiles/BattReview2011/es068_scherson_2011_o.pdf. Accessed Dec. 2, 2011.

182. A. R. Shaffer, N. Deligonul, D. A. Scherson, and J. D. Protasiewicz, *Inorg. Chem.*, **49**, 10756 (2010).

183. K. Amine, J. Liu, I. Belharouak, S. H. Kang, I. Bloom, D. Vissers, and G. Henriksen, *J. Power Sources*, **146**, 111 (2005).

184. K. Abe, K. Miyoshi, T. Hattori, Y. Ushigoe, and H. Yoshitake, *J. Power Sources*, **184**, 449 (2008).

185. H. Lee, J. H. Lee, S. Ahn, H.-J. Kim, and J.-J. Cho, *Electrochem. Solid-State Lett.*, **9**, A307 (2006).

186. S. Santee, A. Xiao, L. Yang, J. Gnanaraj, and B. L. Lucht, *J. Power Sources*, **194**, 1053 (2009).

CHAPTER 4

ELECTROLYTES FOR LITHIUM-ION BATTERIES WITH HIGH-VOLTAGE CATHODES

Mengqing Xu, Swapnil Dalavi, and Brett L. Lucht

1 INTRODUCTION

Lithium-ion batteries have the best gravimetric and volumetric energy density of all rechargeable batteries produced commercially. The high energy density is due to a large potential difference between the electrodes (3 to 5 V per cell) and high-capacity electrode materials enabled by the use of nonaqueous electrolytes. Due to the high energy density, lithium-ion batteries are being pursued intensively for transportation applications, including hybrid electric vehicles (HEVs), plug-in hybrid electric vehicles (PHEVs), and electric vehicles (EVs). However, cost, safety, cycle and calendar life, energy, and power density are some of the major obstacles in adopting lithium-ion technology successfully for vehicle applications. Among the leading contributors to many of these obstacles are the interactions and reactions of the electrolytes with the electrode materials [1].

Lithium Batteries: Advanced Technologies and Applications, First Edition.
Edited by Bruno Scrosati, K. M. Abraham, Walter van Schalkwijk, and Jusef Hassoun.

It has been established that nonaqueous electrolytes are not thermodynamically stable on the surface of lithium metal or lithiated graphite, resulting in the reductive decomposition of electrolyte. Electrolyte decomposition products deposit on the electrode surface during the initial formation cycling and prevent further electrolyte reduction while allowing Li^+ conduction [2]. The interfacial deposition layers have been termed *solid electrolyte interphase* (SEI) [3]. An understanding of this complex component of lithium-ion batteries has been of significant interest since the inception of lithium-ion battery technology [4–7]. The effect of different factors, including graphite structure, electrolyte composition, and Li^+ solvation sphere, on the chemistry and formation mechanism of the SEI have been explored extensively [8].

However, the presence, structure and formation mechanisms of interfacial surface films on the cathode, the *cathode electrolyte interphase* (CEI), have received far less attention. Traditional cathode materials for lithium-ion batteries, including $LiCoO_2$, are typically charged to between 3.7 and 4.1 V vs. Li. Under these conditions standard electrolytes are considered oxidatively stable [9]. Residual surface Li_2CO_3 exists on most transition metal oxides, and the reaction between residual Li_2CO_3 with the acidic $LiPF_6$/carbonate electrolyte results in the deposition of organic electrolyte decomposition products [10–12]. Related investigations of lithium-ion cells after accelerated aging experiments revealed that the cathode surface undergoes significant changes [12–14]. Although many of the components of the CEI are known, the structure, thickness, and function of the CEI are unclear [15–17]. In addition, the mechanisms of CEI formation are poorly understood but are probably due to a complex combination of thermal and electrochemical reactions of the electrolyte with the cathode surface which are dependent on time, temperature, potential, and cathode structure [9].

There is significant current interest in increasing the energy density of lithium-ion batteries. One method to increase the energy density is to utilize cathodes that operate at higher voltages (>4.5 V vs. Li) and exhibit higher capacities. However, a major difficulty in using high-voltage cathodes is the instability of the standard electrolyte, $LiPF_6$, in organic carbonate solvents, in contact with the cathode surface at operating potentials over 4.5 V. One method to inhibit the detrimental reaction of the electrolyte with the cathode surface is the generation of inert surface coatings such as Al_2O_3, ZnO, and Bi_2O_3 to prevent the oxidation of the electrolyte [18–21]. Surface-coated cathodes have cyclability superior to that of uncoated material; however, the surface-coating method has a negative effect on the discharge capacity of the material and may be difficult to scale for commercial applications. Thus, the development of a thorough understanding of the reactions of an electrolyte with the surface of high-voltage cathode materials and the development of novel electrolytes capable of long-term reversible cycling to high voltage is of great current interest.

2 OXIDATION REACTIONS OF THE ELECTROLYTE WITH TRADITIONAL METAL OXIDE CATHODE MATERIALS

The presence of a CEI was uncovered early in the development of lithium-ion batteries. Thomas et al. first suggested that a film exists on the cathode [22]. An equivalent circuit which includes the formation of a surface layer on the electrode due to the oxidation of the electrolyte was proposed to simulate the ac impedance responses of an intercalation-type cathode in liquid electrolytes. Aurbach et al. studied the interfacial behavior of various cathode materials in $LiAsF_6$ ethylene carbonate (EC)/dimethyl carbonate (DMC) using electrochemical impedance spectroscopy (EIS) and found that the impedance spectra obtained reflected the step of lithium-ion migration through a surface layer in a manner very similar to that of the reversible lithium-ion intercalation–deintercalation process on carbonaceous materials for all of the cathode materials investigated: $LiNiO_2$, $LiCoO_2$, and $Li_xMn_2O_4$ [23].

The CEI formation process on the cathode surface is different from SEI formation on the surface of the graphite anode. The high oxidation potential of the cathode surface leads to oxidation reactions of the electrolyte components near the interface. In addition, the different polarities of the electrode result in a high concentration of Li^+ near the anode and a high concentration of PF_6^- near the cathode. However, the concentration of EC is high near both interphases [24]. The Li^+ is undergoing a desolvation process at the anode surface before intercalation of Li^+ into the graphite and a solvation process at the cathode surface as Li^+ deintercalate from the cathode surface [25].

The majority of the investigations of the oxidative decomposition of solvents and salt anions were carried out on nonactive electrodes [26–31]. Similar to the reductive decomposition on the anode surface, a mechanism has been proposed involving a single-electron process producing a cation radical from carbonate solvents (Scheme 1). Subsequent decomposition of the intermediates leads to gaseous as well as solid decomposition products, which form a surface layer on the electrode, as supported by electron spin resonance [32]. The surface of a charged $LiCoO_2$ cathode was identified as the source of radical cation generation since parallel experiments with electrolytes in the absence of a $LiCoO_2$ cathode produces no radical species. Aurbach et al. also characterized the surface species on $LiCoO_2$, $LiNiO_2$, and $Li_2Mn_2O_4$ cathodes with fourier transform infrared spectroscopy (FTIR) spectroscopy [23]. After cycling with an $LiAsF_6$ EC/DMC electrolyte from 3.0 to 4.4 V, all of the cathode surfaces were found to be covered with new chemical species, while the signals corresponding to residual Li_2CO_3 diminished. Since the residual Li_2CO_3 does not change when in contact with pure solvents, the salts probably play a crucial role in the formation of the new surface species. The FTIR spectra contain a variety of absorptions corresponding to C–H, C=O, and C–O bonds (Fig. 1).

The carbonyl functionality around 1800 to 1700 cm^{-1} is consistent with the presence of poly(ethylene carbonate) (PEC) from the ring-opening polymerization of EC (Scheme 2). Additional surface analysis of the cathode via x-ray photoelectron spectroscopy (XPS) and energy-dispersive x-ray analysis (EDAX), confirms that the salt anions and related impurities are also involved in surface layer generation [15]. EDAX detected a pronounced increased in the fluorine content of the surface

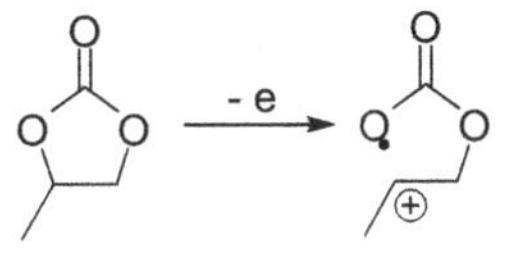

SCHEME 1   Single-electron electrochemical oxidation path for propylene carbonate. (From [29].)

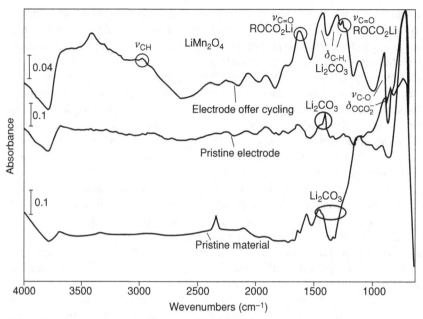

FIGURE 1   FTIR spectra measured in the diffusion reflectance mode from the pristine $Li_xMn_2O_4$ spinel electrode and after its cycling, as indicated. (From [23], with permission of the Electrochemcial Society.)

after the cathode was cycled in $LiPF_6$ or $LiAsF_6$-based solution, and appreciable concentrations of arsenic or phosphorus were also present on the surface. The cathode surface chemistry as detected by XPS correlated well with FTIR results, which confirmed the formation of organic and inorganic carbonates as well as lithium alkoxides.

Aurbach et al. employed EIS to monitor the thickness change of surface film formed on a cathode with prolonged cycling. They found that the surface layer on the cathode continued to grow during prolonged cycling, while the chemical composition was not altered significantly [15]. The thermal stability of interphase formed on cathodes had been investigated by Andersson et al. [33,34]. Accelerated aging of graphite/$LiNi_{0.8}Co_{0.2}O_2$ cells at temperatures ranging from 25 to 70°C

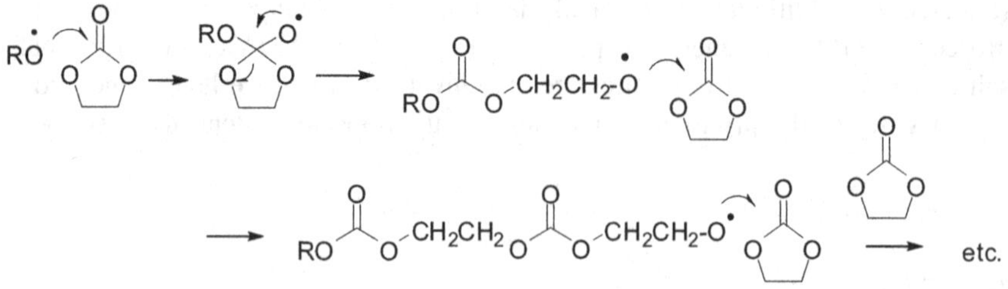

SCHEME 2   (From [15].)

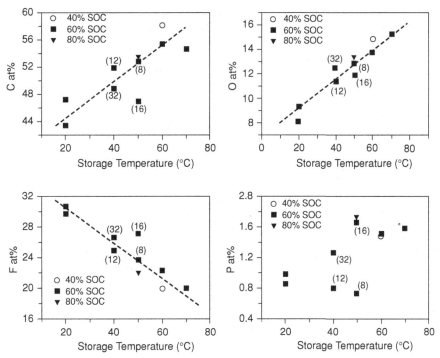

FIGURE 2   Relative amount (at %) of surface C, O, F, and P on cathode samples as a function of cell test temperature. Numbers in parentheses specify the storage duration in weeks. (From [33], with permission of the Electrochemical Society.)

followed by ex situ surface analysis by XPS and scanning electron microscopy (SEM) was reported. Interestingly, the same surface compounds were observed on cathodes regardless of test temperature, test duration, and state of charge. However, the concentration of the surface compounds increased with increasing temperature (Fig. 2). The mixture of organic species included polycarbonates and LiF, Li_xPF_y-type, and $Li_xPO_yF_z$-type compounds is very similar to the structure of the surface species generated via storage of electrolyte in the presence of cathode materials [12].

3   THERMAL REACTIONS OF THE ELECTROLYTE WITH THE SURFACE OF METAL OXIDE CATHODES

An investigation of the thermal stability of the surface layers formed on cathode materials ($LiNi_{0.8}Co_{0.2}O_2$, $LiCoO_2$, and $LiMn_2O_4$) through the combined use of SEM, thermal gravimetric analysis (TGA), and XPS was reported [12]. The cathode particles were found to enhance the thermal stability of the electrolyte at moderately elevated temperatures, 60 to 100 °C. However, the thermal reactions of the electrolyte partially remove the surface layer of Li_2CO_3 and deposit a complex mixture of decomposition products, including poly(ethylene oxide) (PEO), polycarbonate, $ROCO_2Li$,

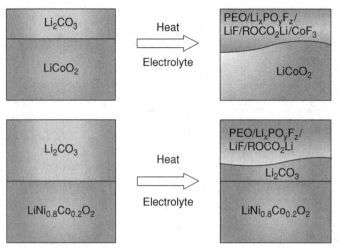

FIGURE 3 Depiction of reactions occurring on the surface of metal oxide cathode particles. (From [12], with permission of the Electrochemical Society.)

LiF, and $Li_xPO_yF_z$, on the cathode particles. The thermally generated surface film is similar to the surface film found in the aged cells, suggesting that thermal reactions between electrolyte and electrode materials are partially responsible for power fade and capacity loss in aged cells. The presence of Li_2CO_3 results in a stabilizing equilibrium via generation of $ROCO_2Li$, a Lewis base. The base stabilizes the electrolyte thermally by sequestering free PF_5 and preventing thermal decomposition of the bulk electrolyte. Higher temperature shifts the equilibrium, resulting in more Li_2CO_3 dissolution and more $Li_xPO_yF_z$ deposition (Fig. 3). The stabilizing equilibrium is disrupted in an open system or upon the incorporation of protic impurities, leading to decomposition of the electrolyte, increases in surface film thickness, and corrosion of the bulk metal oxide. The thickness of surface layers was estimated to be 4 to 8 nm from Ar^+ ion-sputtering experiments [12].

Related investigations by Dupre et al. established an interesting method to study the interphase formed on $LiNi_{0.5}Mn_{0.5}O_2$ by using ^7Li magic angle spinning NMR spectroscopy [17]. Evolution of the nuclear magnetic resonance (NMR) signal shows that the reaction of the active material with the electrolyte is extremely fast during initial exposure and tends to slow with longer exposure time. In addition, combined ^7Li NMR and XPS experiments on samples containing residual surface Li_2CO_3 soaked in electrolyte indicate that the Li_2CO_3 is dissolved rapidly followed by the generation of electrolyte decomposition products, including LiF, Li_xPF_y, and $Li_xPO_yF_z$. Combined SEM and transmission electron microscopy (TEM) analysis estimated an average thickness of 2 to 20 nm for the film after the reaction.

Since the thermal reaction of the electrolyte with the surface of the cathode materials is linked to the generation of acidic species (HF and PF_5) from $LiPF_6$ decomposition, inhibition of the surface reactions was investigated by the incorporation of Lewis basic additives [35–37]. Addition of low concentrations of dimethylacetamide (DMAc) slows the reaction of the electrolyte with the surface of metal oxide cathode materials significantly, inhibiting the deposition of electrolyte decomposition products and slowing the corrosion of the metal oxide surface.

4   FORMULATION OF ELECTROLYTES FOR HIGH-VOLTAGE MATERIALS

4.1   Chemistry of Cathodes at High Voltage

With the goal of utilizing high-energy Li-ion batteries for electric vehicles, there is growing interest in cathodes that operate at high voltages (>4.5 V) and/or exhibit high-capacities. Various high-voltage cathode materials have been investigated, including Li_2MnO_3–$LiNi_{0.5}Mn_{0.5}O_2$, $LiMn_{1.5}Ni_{0.5}O_4$, $LiCoPO_4$, and $LiNiPO_4$, with operating voltages between 4.7 and 5.2 V [1,18–21,38–45]. A major concern in utilizing high-voltage cathodes is the instability of the organic electrolytes at high voltage (i.e., >4.5 V). Cycling Li-ion cells to high voltage typically results in low coulombic efficiency and poor cycle life. The primary contributor to poor cycling efficiency is the electrochemical oxidation reaction of the electrolytes at high positive potentials, producing CO_2 and H_2O and PEC, as depicted in Figure 4 [46].

Recent investigations of EC oxidation on $LiNi_{0.5}Mn_{1.5}O_4$ cathodes stored at high potential using XPS and FTIR revealed that polycarbonate species are the major product on the surface [47]. The oxidation of EC at the surface of $LiNi_{0.5}Mn_{1.5}O_4$ potentials above 4.7 V vs. Li results in the generation of PEC (Scheme 3), as supported by XPS and FTIR analysis [47]. Further investigations were conducted of the oxidation reactions of electrolyte on $LiNi_{0.5}Mn_{1.5}O_4$ and platinum electrodes. A

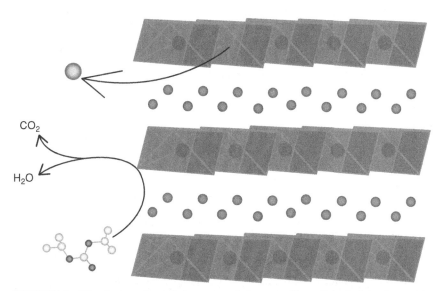

FIGURE 4   Dissolution of metal elements and decomposition of electrolyte at a high-voltage cathode surface.

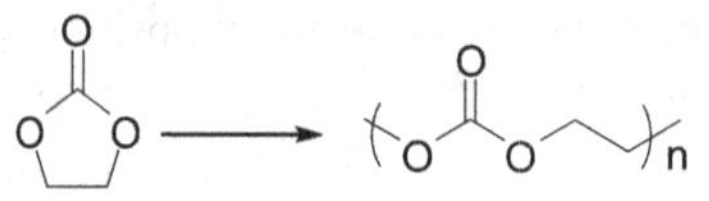

SCHEME 3   (From [47])

comparison between charge transferred on $LiNi_{0.5}Mn_{1.5}O_4$ and Pt. electrodes at 4.75 and 5.30 V is provided in Figure 5.

After subtraction of the charge associated with lithium intercalation on an $LiNi_{0.5}Mn_{1.5}O_4$ cathode, the residual charge transferred per unit area is similar to that of a Pt electrode, suggesting that at high potential the electrolyte oxidation reactions are similar on $LiNi_{0.5}Mn_{1.5}O_4$ and Pt electrodes. While the surface analysis of the Pt electrode was inconclusive, the structure of the products of electrolyte oxidation on $LiNi_{0.5}Mn_{1.5}O_4$ and Pt appear to be similar.

Recently, Xing et al. investigated the oxidative decomposition mechanism of PC with and without PF_6^- and ClO_4^- anions with density functional theory at the B3LYP/6-311++G(d) level [48]. The presence of PF_6^- and ClO_4^- anions significantly reduces the oxidative stability of PC, stabilizes the PC-anion oxidation decomposition products, and changes the order of the oxidation decomposition pathways. Interestingly, HF and PF_5 were detected at the initial step of $PC–PF_6^-$ oxidation, whereas $HClO_4$ formed during initial oxidation of $PC–ClO_4^-$. Understanding the mechanisms of solvent oxidation and the role of the salt in solvent oxidation at high potential should lead to the development of novel electrolyte systems that are compatible with high-voltage cathode materials.

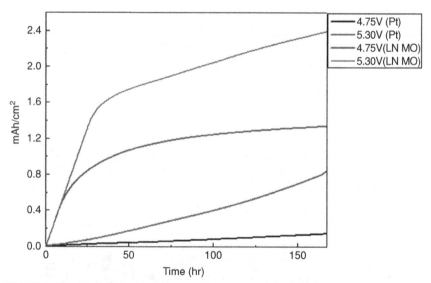

FIGURE 5   C/20 charging to 4.75 and 5.30 V and held at the same voltage on an $LNi_{0.5}Mn_{1.5}O_4$ electrode and a scan (5 mV/S) and hold (4.75 and 5.30 V) experiment on a Pt electrode in $LiPF_6$ in 3 : 7 EC/EMC.

4.2 Novel Organic Solvents with Greater Oxidative Stability: Sulfones, Nitriles, and Fluorinated Solvents

Since the oxidative stability of current $LiPF_6$/carbonate electrolytes is considered to be limited by the oxidative stability of the organic solvents, there has been significant interest in the development of novel organic solvents with high anodic stability. Three classes of solvents have been investigated for improved stability when cycled to a higher potential. However, incorporation of novel solvents with higher oxidative stability is challenging since the stability of the anode SEI is highly dependent on the structure and composition of the electrolyte. Altering the solvents in the electrolyte is likely to alter the nature of the anode SEI and may limit the application of these novel solvents.

Sulfones were one of the initial solvents investigated for anodic stability superior to that of carbonates [49,50]. Xu and Angell reported on the properties of ethylmethyl sulfone (EMS) in the presence of high-voltage cathode materials [49]. EMS solutions of high-stability salts such as $LiClO_4$ and LiTFSI are stable to 5.8 V vs. Li/Li$^+$ before onset of oxidation. Other sulfones were investigated, including trimethylene sulfone (TriMS), 1-methyltrimethylene sulfone (MTS), ethyl-*sec*-butyl sulfone (EsBS), ethylisobutyl sulfone (EiBS), ethylisopropyl sulfone (EiPS), and 3,3,3-trifluoropropylmethyl sulfone (FPMS) [49]. Sulfone solution of lithium salts were found to be stable to 5.5 V vs. Li on the surface of $Li_xMn_2O_4$. Interestingly, when the sulfones were mixed in different ratios with the linear carbonates DMC or ethylmethyl carbonate (EMC) to reduce the viscosity of electrolyte, the mixed-solvent electrolyte possessed very similar stability to that of the pure sulfone-based electrolyte, despite the lower oxidative stability of DMC and EMC (Fig. 6). This suggests that the less oxidatively stable carbonates are stabilized by the presence of

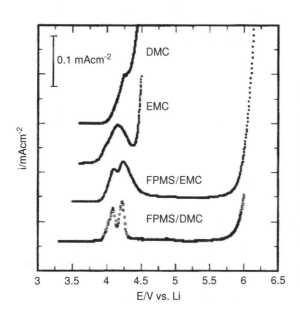

FIGURE 6   Oxidative stability of FPMS in an electrolyte based on its mixture with linear carbonates. Shown for comparison is the oxidative decomposition behavior of linear carbonates. Conditions: 1.0 M $LiPF_6$ solutions, scan rate 0.1 mV/s, Li as counter and reference electrodes, and cathode composite on the Al substrate as the working electrode. (From [60], with permission by the Electrochemical Society.)

sulfones. The formation of a passivating film on the cathode was proposed as a source for the improved stability of the carbonates. The more polar component appears to have a larger contribution to the formation of the cathode electrolyte interface. This is related to the reactivity of EC observed on the anode and the role of EC in anode SEI formation. The electrochemical windows of the sulfone-based and sulfone–carbonate mixed solvents were determined on an Al electrode. Thus, further investigations of sulfone-based electrolyte in the presence of high-voltage metal oxide cathodes and graphite anodes need to be conducted.

A more recent investigation of sulfone solvents blended with carbonates was reported by Abouimrane et al. supporting the superior anodic stability to neat carbonates by cyclic voltammography. The cycling performance of a $Li_4Ti_5O_{12}/$ $LiNi_{0.5}Mn_{1.5}O_4$ cell with a tetramethyl sulfone (TMS)/EMC electrolyte was investigated from 2.0 to 3.5 V. The cell delivered an initial capacity of 80 mAh/g and cycled fairly well for 1000 cycles under a 2 C rate, supporting the stability of the electrolyte to these electrode materials within this potential range [51].

Organic nitriles are another family of solvents that were investigated as an alternative to carbonates[52–56]. The most widely investigated solvent containing the nitrile functional group $–C\equiv N$ is acetonitrile ($CH_3C\equiv N$), which has been used in several electrochemical applications, including Li/SO_2 batteries and ultracapacitors, but its anodic stability in the presence of some lithium salts is limited to 3.8 V. Abu-Lebdeh and Davidson demonstrated that adiponitrile, $CN[CH_2]_4CN$ (ADN), used as a solvent or cosolvent, has high anodic stability. An electrolyte solution of $Li(CF_3SO_2)_2N$, LiTFSI, in EC/ADN has an electrochemical stability window of 6 V vs. Li/Li^+. However, the authors reported only on the cycling performance of $MCMB/LiCoO_2$ cells with EC/ADN (1 : 1, v/v), in 1 M LiTFSI, and 0.1 M LiBOB as electrolyte. Sebaconitrile, $N\equiv C(CH_2)_8C\equiv N$, has also been investigated as a solvent for high-voltage cathodes. A 1 M $LiBF_4$/EC/DMC/sebaconitrile (25 : 25 : 50, v/v) electrolyte exhibits excellent electrochemical stability above 6 V vs. Li/Li^+. The dinitrile-based electrolyte can support Li^+ deintercalation above 5.0 V in Li_2NiPO_4F, but cycling data were not presented. Although nitrile-based electrolytes exhibit a very high electrochemical stability window of 6.0 V as measured on inert electrolytes use with high-voltage cathode materials (>4.8 V) and graphitic anodes is difficult.

More recently, fluorinated solvents have been investigated as cosolvents or additives in electrolytes and are reported to decrease the electrolyte flammability and increase the electrolyte stability at high voltage (Fig. 7) [57]. The fluorinated carbonates methyl-2,2,2-trifluoroethylcarbonate and ethyl-2,2,2-trifluoroethyl carbonate have the best anodic stability ($\sim$5.8 to 5.9 V vs. Li/Li^+). However, the cyclability of 1 M $LiPF_6$ in EC/ethyl-2,2,2-trifluoroethyl carbonate in a Li metal/$LiNi_{0.5}Mn_{1.5}O_4$ cell is comparable to that in an EC–DMC electrolyte system.

4.3 Novel Additives for Cathode Surface Passivation

An alternative approach to developing bulk solvents with greater oxidative stability as a replacement for carbonate solvents is the development of novel cathode surface film-forming additives. Ideally, incorporation of an additive would result in the generation of a passivating film on the cathode similar in nature to the anode SEI. The additive

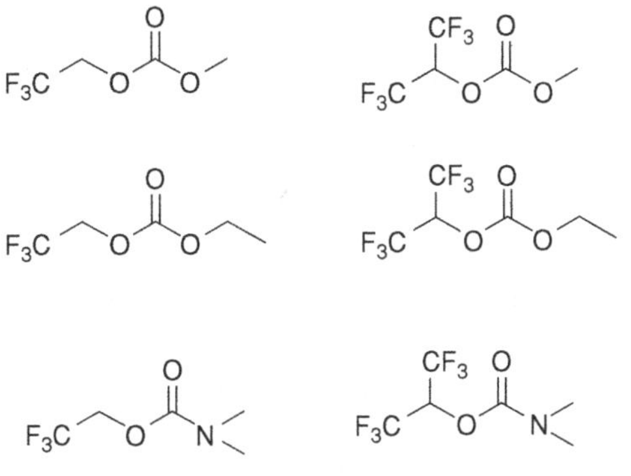

FIGURE 7   Structure of fluorinated solvents.

approach would result in smaller changes to the electrolyte and not affect anode SEI formation. Related additives have been widely investigated to improve the stability of the graphite anode SEI [58–61], and thus cathode film-forming additives are a reasonable extension. At this time few cathode film-forming additives have been investigated [62–69].

Computation methods have been employed to screen molecules for potential additives for the design of a cathode surface film. Molecular orbital theory is used to calculate the energies of the highest occupied molecular orbital, which should correlate with the potential of oxidative decomposition, and the lowest unoccupied molecular orbital, which should correlate with the potential of reductive decomposition. Vinylene carbonate (VC) and vinylethylene carbonate (VEC), which have been used widely in lithium-ion batteries, were investigated by this computational approach [58–61]. Han et al. employed density functional theory to calculate the ionization potential and the oxidation potential (E_{ox}) of 108 organic molecules that are potential electrolyte additives for overcharge protection in lithium-ion batteries [62]. The E_{ox} values calculated were found to be in close agreement with the experimental values, where the maximum deviation is only 0.15 V. Similar methods should be useful for predicting solvent decomposition potentials.

It was reported recently that unsaturated cyclic ethers or lactones can be preferentially oxidized on the cathode to form a stable cathode passivation layer, which inhibits further oxidative decomposition of electrolyte on the cathode surface [9] (Fig. 8). A common electrolyte (1.0 M LiPF$_6$ in 1:1:1 EC/DEC/DMC) has an anodic stability around 5.2 V without additives. Upon incorporation of 2% 2,5-DHF, the first cycle has a lower voltage threshold at 4.8 V, but a much higher anodic stability is observed during subsequent scans. This suggests that 2,5-DHF is sacrificially oxidized on the surface of glassy carbon electrode, generating a passivation layer that prevents further oxidation of electrolyte.

The cycling performance of Li$_{0.17}$Mn$_{0.58}$Ni$_{0.25}$O$_2$ half cells charged to 4.9 V with and without additives was also evaluated in the presence of additives. The addition of 0.5% 2,5-DHF or 1% GBL renders better cycling performance than that of the standard electrolyte . After 50 cycles, the capacity retention is 85%, 92%, and 89% for the standard electrolyte, and 0.5% 2,5-DHF and 1% GBL, respectively (Fig. 9).

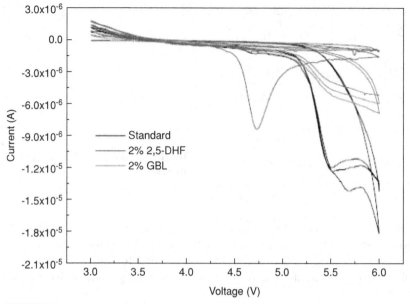

FIGURE 8 Anodic stability of 1 M $LiPF_6$ in 1 : 1 : 1 EC/DEC/DMC with and without additives. (From [9], with permission of the Electrochemical Society.)

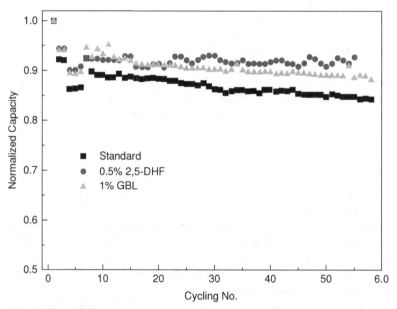

FIGURE 9 Capacity retention of $Li/Li_{1.17}Mn_{0.58}Ni_{0.25}O_2$ cells cycled from 2.0 to 4.9 V containing 1 M $LiPF_6$ in 1 : 1 : 1 EC/DEC/DMC with and without additives. (From [9], with permission of the Electrochemical Society.)

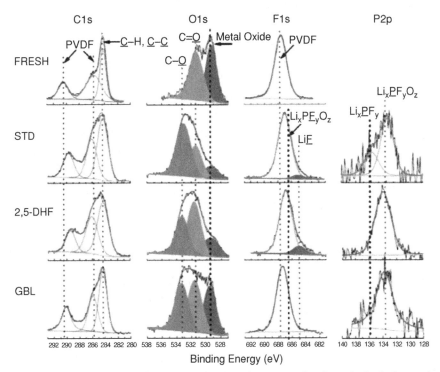

FIGURE 10 XPS spectra of fresh and cycled $Li_{1.17}Mn_{0.58}Ni_{0.25}O_2$ cathodes before and after cycling with 1 M $LiPF_6$ in 1 : 1 : 1 EC/DEC/DMC with and without additives. (From [9], with the permission of the Electrochemical Society.)

Ex situ surface analysis via XPS and FTIR was conducted on the cycled cathode to understand the chemical reactions on the surface. The surface layer is modified by the incorporation of 2,5-DHF and GBL, resulting in the generation of a thinner surface layer, as evidenced by lower concentrations of metal oxide and higher concentrations of the C–O and C=O species characteristic of electrolyte decomposition (Fig. 10).

Other cathode film-forming additives have been investigated, including thiophene, which was reported to improve the cycle life of lithium-ion cells at high voltages [63,64]. The improvement was attributed to the depression of the charge transfer resistance during cycling up to high voltage, due to the addition of thiophene. Another interesting additive, N-(triphenylphosphoranylidene)aniline (TPPA), was investigated for high voltage in lithium-ion batteries [65]. Cells with 0.2% TPPA achieved 94.6% of the initial capacity after 200 cycles, whereas cells containing traditional electrolyte have only 84.2% capacity retention. Boroxine compounds, $B_3O_3(OR)_3$, with different types of substituents $[R = CH_3, CH_2CH_3, (CH_2)_2CH_3,$ and $CH(CH_3)_2]$, have also been reported as additives that improve the performance of high-voltage lithium-ion batteries [66]. Cyclic voltammetry and electrochemical impedance spectroscopy revealed that the anodic stability of the electrolyte was improved up to 5.0 V vs. Li/Li^+ by the addition of triisopropoxyboroxine $[R = CH(CH_3)_2]$. While thiophene, TPPA,

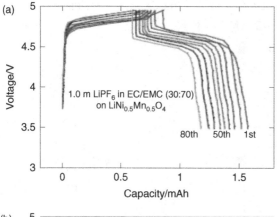

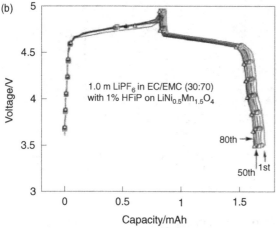

FIGURE 11 Voltage profiles of a 5 V class cathode $LiNi_{0.5}Mn_{1.5}O_4$ in a baseline electrolyte: (a) $LiPF_6/EC/EMC$ (30 : 70), and (b) at baseline with 1% HFiP as an additive. Only selected cycles are shown, for graphical clarity. (From [69], with the permission of the Electrochemical Society.)

and boroxine compounds have potential interest as additives to improve the cycling performance of high-voltage cathode materials, the experiments reported initially were conducted only with traditional cathode materials, $LiMn_2O_4$ or $LiCoO_2$, cycled to 4.4 V vs. Li/Li^+. Results have not been reported for high-voltage cathode materials, $LiNi_{0.5}Mn_{1.5}O_4$ or $Li_2MnO_3–LiMO_2$. However, the evidence that these additives react with the cathode surface and may generate passivation films is promising.

Beyond the additives discussed above, other inorganic additives have been reported to react sacrificially on the cathode surface [67,68]. Those additives include lithium bisoxalatoborate (LiBOB), lithium difluoroxalatoborate [$LiBF_2(C_2O_4)$], tetramethoxytitanium (TMTi), and tetraethoxysilane (TEOS). The capacity retention of $Li/Li_{1.17}Mn_{0.58}Ni_{0.25}O_2$ cells containing 1% LiBOB, $LiBF_2(C_2O_4)$, and TMTi have better cycling performance than cells containing the standard electrolyte. The cells with additives have significantly lower capacity loss during formation cycles. The improved performance was attributed to a decrease in oxygen evolution during formation cycles. A related improvement in the cycling performance of $Li/LiNi_{0.5}Mn_{1.5}O_4$ was reported for the addition of LiBOB [64].

Another additive that has attracted significant interest is tris(hexafluoro-isopropyl)phosphate (HFiP), which is based on phosphate ester structure with highly fluorinated alkyl arms [69]. This additive in a carbonate-based electrolyte at 1% effectively stabilizes the interfacial impedance and capacity retention of a high-voltage spinel $LiNi_{0.5}Mn_{1.5}O_4$ cathode Fig. 11. After 200 cycles between 3.5 to 4.95 V, the capacity retention of a $Li/LiNi_{0.5}Mn_{1.5}O_4$ cell is around 58% for a baseline electrolyte and almost 90% for a 1% HFiP-added electrolyte. Although the capacity fading still exists with the presence of HFiP, there is significant improvement over the baseline electrolyte. The improvement was attributed to stabilization of the baseline electrolyte and lattice stabilization and surface modification of the $LiNi_{0.5}Mn_{1.5}O_4$ interface.

5 SUMMARY

Lithium-ion batteries have been commercialized for over a decade; however interest in the development of lithium-ion batteries for electric vehicles has increased interest in lithium-ion batteries with greater energy density. The use of higher-voltage cathode materials is one method of increasing the energy density of lithium-ion batteries. However, the higher oxidation potential results in detrimental reactions of the cathode with common electrolytes.

The reactions of metal oxide cathode materials with standard $LiPF_6$/alkyl carbonate electrolytes has been reviewed. The detrimental reactions of the cathode surface with the electrolyte is a complex mixture of thermal and electrochemical reactions. Investigation of novel electrolytes for high-voltage cathodes has recently been initiated. The generation of novel electrolytes with greater oxidative stability is promising but is complicated by changes in the anode SEI-forming reactions required in lithium-ion batteries. The development of novel cathode film-forming additives has also been reported and will probably lead to the development of workable electrolyte systems for high-voltage cathodes.

REFERENCES

1. A. Manthiram, *J. Phys. Chem. Lett.*, **2**, 176 (2011).
2. E. Peled and H. Straze, *J. Electrochem. Soc.*, **124**, 1030 (1977).
3. E. Peled, *J. Electrochem. Soc.*, **126**, 2047 (1979).
4. D. Aurbach, I. Weissman, A. Schechter, and H. Cohen, *Langmuir*, **12**, 3991 (1996).
5. A. Schechter, D. Aurbach, and H. Cohen, *Langmuir*, **15**, 3334 (1999).
6. K. Xu, U. Lee, S. Zhang, and T. R. Jow, *J. Phys. Chem. B*, **110**, 7708 (2006).
7. Y. Wang, S. Nakamura, M. Ue, and P. Balbuena, *J. Am. Chem. Soc.*, **123**, 11708 (2001).
8. K. Xu, *Chem. Rev.*, **104**, 4303 (2004).
9. L. Yang and B. L. Lucht, *Electrochem. Solid-State Lett.*, **12**, A229 (2009).
10. M. Koltypin, D. Aurbach, L. Nazar, and B. Ellis, *J. Power Sources*, **174**, 1241 (2007).
11. M. Xu, L. Hao, Y. Liu, W. Li, W. Xing, and B. Li, *J. Phys. Chem. C*, **115**, 6085 (2011).
12. W. Li and B. L. Lucht, *J. Electrochem. Soc.*, **153**, A1617 (2006).
13. M. Xu, W. Li, and B. L. Lucht, *J. Power Sources*, **193**, 804 (2009).
14. A. M. Andersson and K. Edstrom, *J. Electrochem. Soc.*, **148**, A1100 (2001).
15. D. Aurbach, K. Gamolsky, B. Markovsky, G. Salitra, Y. Gofer, U. Heider, R. Oesten, and M. Schmidt, *J. Electrochem. Soc.*, **147**, 1322 (2000).

16. N. Liu, H. Li, Z. Wang, X. Huang, and L. Chen, Electrochem. *Solid-State Lett.*, **9**, A328 (2006).
17. N. Dupre, J. -F. Martin, J. Oliveri, P. Soudan, D. Guyomard, A. Yamada, and R. Kanno, *J. Electrochem. Soc.*, **156**, C180 (2009).
18. J. Liu and A. Manthiram, *J. Electrochem. Soc.*, **156**, A833 (2009).
19. J. Liu and A. Manthiram, *J. Phys. Chem. C*, **113**, 15073 (2009).
20. J. Liu and A. Manthiram, *J. Electrochem. Soc.*, **156**, A66 (2009).
21. J. Liu and A. Manthiram, *Chem. Mater.*, **21**, 1695 (2009).
22. M. G. S. R. Thomas, P. G. Bruce, and J. B. Goodenough, *J. Electrochem. Soc.*, **132**, 1521 (1985).
23. D. Aurbach, M. D. Levi, E. Levi, H. Teller, B. Markvosky, and G. Salitra, *J. Electrochem. Soc.*, **145**, 3024 (1999).
24. S.-P. Kim, A. C. T. van Duin, and V. B. Shenoy, *J. Power Sources*, **196**, 8590 (2011).
25. K. Xu and A. v. Cresce, *J. Mater. Chem.*, **21**, 9849 (2011).
26. K. Xu, S. Ding, and T. R. Jow, *J. Electrochem. Soc.*, **148**, A267 (2001).
27. M. Ue, M. Takeda, M. Takehara, and S. Mori, *J. Electrochem. Soc.*, **144**, 2684 (1997).
28. B. Rasch, E. Cattaneo, P. Novak, and W. Vielstich, *Electrochim. Acta*, **36**, 1397 (1991).
29. K. Kanamura, S. Toriyama, S. Shiraishi, and Z. Takehara, *J. Electrochem. Soc.*, **142**, 1383 (1995).
30. F. Ossola, G. Pistoia, R. Seeber, and P. Ugo, *Electrochim. Acta*, **33**, 47 (1988).
31. V. R. Koch, L. A. Dominey, C. Nanjundish, and M. J. Ondrechen, *J. Electrochem. Soc.*, **143**, 798 (1996).
32. S. Matsuta, Y. Kato, T. Ota, H. Kurokawa, S. Yoshimura, and S. Fujitani, *J. Electrochem. Soc.*, **148**, A7 (2001).
33. A. M. Andersson, D. P. Abraham, R. Hassch, S. Maclaren, J. Liu, and K. Amine, *J. Electrochem. Soc.*, **149**, A1358 (2002).
34. T. Eriksson, A. M. Andersson, A. G. Bishop, C. Gejke, T. Gustafsson, and J. O. Thomas, *J. Electrochem. Soc.*, **149**, A69 (2002).
35. W. Li, C. L. Campion, B. L. Lucht, B. Ravdel, and K. M. Abraham, *J. Electrochem. Soc.*, **152**, A1361-A1365 (2005).
36. C. L. Campion, W. Li, and B. L. Lucht, *J. Electrochem. Soc.*, **152**, A2327–A2334 (2005).
37. W. Li and B. L. Lucht, *J. Power Sources*, **168**, 258–264 (2007).
38. S.-H. Kang, V. G. Pol, I. Belharouak, and M. M. Thackeray, *J. Electrochem. Soc.*, **157**, A267 (2010).
39. S.-H. Kang and M. M. Thackeray, *Electrochem. Commun.*, **11**, 748 (2009).
40. J. R. Croy, S.-H. Kang, M. Balasubramanian, and M. M. Thackeray, *Electrochem. Commun.*, **13**, 1063 (2011).
41. M. Kunduraci, J. F. Al-Sharab, and G. G. Amatucci, *Chem. Mater.*, **18**, 3582 (2006).
42. T. Muraliganth and A. Manthiram, *J. Phys. Chem. C*, **114**, 15530 (2010).
43. D.-H. Seo, H. Gwon, S.-W. Kim, G. Kim, and K. Kang, *Chem. Mater.*, **22**, 518 (2010).
44. C. V. Ramana, A. Sit-Salah, S. Utsunomiya, U. Becker, A. Mauger, F. Gendron, and C. M. Julien, *Chem. Mater.*, **18**, 3788 (2006).
45. C. A. J. Fisher, V. M. H. Prieto, and M. S. Islam, *Chem. Mater.*, **20**, 5907 (2008).
46. L. Xing, W. Li, C. Wang, F. Gu, M. Xu, C. Tan, and J. Li, *J. Phys. Chem. B*, **113**, 16596 (2009).
47. L. Yang, B. Ravdel, and B. L. Lucht, *Electrochem. Solid-State Lett.*, **13**, A95 (2010).
48. L. D. Xing, O. Borodin, G. D. Smith, and W. S. Li, *J. Phys. Chem. A* (2011), dx.doi.org/10.1021/jp2006153n.
49. K. Xu and C. A. Angell, *J. Electrochem. Soc.*, **145**, L70 (1998).
50. K. Xu and C. A. Angell, *J. Electrochem. Soc.*, **149**, A920 (2002).
51. A. Abouimrane, I. Belharouak, and K. Amine, *Electrochem. Commnu.*, **11**, 1073 (2009).
52. K. Xu, S. Ding, and T. R. Jow, *J. Electrochem. Soc.*, **146**, 4172 (1999).
53. H. J. Santner, K. C. Moller, J. Ivanco, M. G. Ramsey, F. P. Netzer, S. Yamaguchi, J. O. Besenhard, and M. Winter, *J. Power Sources*, **119–121**, 368 (2003).
54. Q. Wang, P. Pechy, S. M. Zakeeruddin, I. Exnar, and M. Gratzel, *J. Power Sources*, **146**, 813 (2005).
55. Y. Abu-Lebdeh and I. Davidson, *J. Electrochem. Soc.*, **156**, A60 (2009).
56. M. Nagahara, N. Hasegawa, and S. Okada, *J. Electrochem. Soc.*, **157**, A748 (2010).
57. M. L. P. Le, F. Alloin, P. Strobel, A. Moussa, and B. Langlois, 219th ECS Meeting, 2011, Abstr. 382.
58. B. Simon and J. P. Boeuve, *U.S. Patent*, 5,626,981, 1997.

59. D. Aurbach, K. Gamolsky, B. Markovsky, Y. Gofer, M. Schmidt, and U. Heider, *Electrochim. Acta*, **47**, 1423 (2002).

60. G. Chen, G. V. Zhuang, T. J. Richardson, G. Liu, and P. N. J. Ross, *Electrochem. Solid-State Lee*, **8**, A344 (2005).

61. Y. S. Hu, W. H. Kong, Z. X. Wang, H..Li, X. Huang, and L. Q. Chen, *Electrochem. Solid-State Lett.*, **7**, A442 (2004).

62. Y.-K. Han, J. Jung, S. Yu, and H. Lee, *J. Power Sources*, **187**, 581 (2009).

63. K.-S. Lee, Y.-K. Sun, J. Noh, K. S. Song, and D.-W. Kim, *Electrochem. Commun.*, **11**, 1900 (2009).

64. A. Abouimrane, S. A. Odom, H. Wu, W. Weng, Z. Zhang, J. S. Moore, and K. Amine, *MRS Fall* Meeting, Boston, MA, 2010, Abstr. KK4.9.

65. J.-N. Lee, G.-B. Han, M.-H., Ryou, D. J. Lee, J. Song, J. W. Choi, and J.-K, Park, *Electrochim. Acta*, **56**, 5195 (2011).

66. T. Horino, H. Tamada, A. Kishimoto, J. Kaneko, Y. Iriyama, Y. Tanaka, and T. Fujinami, *J. Electrochem. Soc.*, **157**, A677 (2010).

67. L. Yang, T. Markmaitree, and B. L. Lucht, *J. Power Sources*, **196**, 2251 (2011).

68. S. Dalavi, M. Xu, B. Knight, and B. L. Lucht, *Electrochem. Solid-State Lett.*, **15**, A28 (2012).

69. A. v. Cresce, and K. Xu, *J. Electrochem. Soc.*, **158**, A337 (2011).

CHAPTER 5

CORE–SHELL STRUCTURE CATHODE MATERIALS FOR RECHARGEABLE LITHIUM BATTERIES

Seung-Taek Myung, Amine Khalil, and Yang-Kook Sun

1 INTRODUCTION

The prospect of drastic climate change and the ceaseless fluctuation of fossil fuel prices are primary motivators to reduce the use of fossil fuels and to discover new energy conversion and storage systems that are able to limit carbon dioxide generation. Among existing systems, lithium-ion batteries (LIBs) are recognized as the most appropriate energy storage system because of their high energy density and thus space savings in applications. However, current LIBs do not meet the performance and safety requirements for use in green plants and hybrid electric vehicles. Improvement in the cell chemistry is necessary to develop a safe and reliable high-performance electrode.

One of the simplest means of improvement is modification of the outer surface of active materials. Nanoscale coating of the active material has proven to be substantially effective in improving cycle life and safety issues [1–10]. However, uniformity of the coating layers remains an issue; some parts of the active materials are covered by nanolayers, but others are directly exposed to the electrolyte, and this gives rise

Lithium Batteries: Advanced Technologies and Applications, First Edition.
Edited by Bruno Scrosati, K. M. Abraham, Walter van Schalkwijk, and Jusef Hassoun.
© 2013 John Wiley & Sons, Inc. Published 2013 by John Wiley & Sons, Inc.

to severe capacity fade. Specifically, active materials are not readily encapsulated by nanoscale coating layers because nanoparticles have a tendency to crystallize on the surface of active materials. Thus, the formation of a core–shell structure is a promising way to realize uniform encapsulation of the active materials by foreign materials.

One point that should be considered is the similarity of crystallographic structures of both core and shell materials. A heterostructure between the core and the shell may cause phase segregation or separation after the formation of core–shell particles. For rechargeable lithium batteries, criteria for electrode materials that should be satisfied are capacity and safety. For this reason, it is desirable to adopt high-capacity materials as the core and thermally stable materials as the shell. Examples of such high-capacity core materials are $Li[Ni_{0.8}Co_{0.1}Mn_{0.1}]O_2$, $Li[Ni_{0.8}Co_{0.2}]O_2$, and $Li[Ni_{0.8}Co_{0.15}Al_{0.05}]O_2$, whose reversible capacities approach 200 mAh/g in the range 2.5 to 4.3 V vs. Li/Li^+, although they exhibit poor thermal stability. Shell materials should have a high exothermic decomposition temperature to ensure their thermal stability.

The use of $Li[Ni_{0.5}Mn_{0.5}]O_2$ is suitable for this purpose because the tetravalent Mn provides significant structural and thermal stability at a highly delithiated state [11,12], although the reversible capacity is lower than that of Ni-rich compounds. Therefore, the synergistic effect of high capacity from the core and good thermal stability from the shell is attractive with respect to the achievement of both sustainable high capacity and reliable safety. These core and shell combinations are desirable since the core and shell materials possess the same crystallographic structure: for example, layer structure ($R\overline{3}m$), spinel structure ($Fd\overline{3}m$), olivine structure ($Pnma$), and so on.

Fabrication of spherical core–shell structure particles with hollow interiors has attracted considerable attention in recent years because of the particles' potential use as low-density capsules for photonic crystals, catalysts, diagnostics, and pharmacology [13–16]. A number of methods were developed for the synthesis of core–shell structured powders: interface assembly strategies [17,18], the layer-by-layer self-assembly process [19], the hydrothermal precipitation method [20], and the template method [21]. Spherical nanometer- and submicrometer-sized core–structure particles were produced effectively using the preparation methods noted above [13–21]. For lithium battery applications, active materials are required to achieve a higher energy density because the inner space of a battery is limited, and this necessitates active materials with a high tap density with spherical morphology. To satisfy those prerequisites, the coprecipitation method is believed to be one of the most effective approaches to synthesizing spherical core–shell structure particles with high interior and exterior densities.

2   LAYER-STRUCTURED CORE–SHELL

Figure 1 shows a procedure for the synthesis of microscale spherical core–shell particles. First, Ni-rich hydroxide is formed as the core, with a particle diameter of several micrometer. The core is then completely encapsulated by a $[Ni_{0.5}Mn_{0.5}](OH)_2$

Ni-rich hydroxide

Coprecipitation onto
core by Ni-Mn hydroxide

Formation of
core-shell hydroxide

Incorporation of Li salt
at high temperature

Formation of lithiated core-shell
cathode material
Core: high capacity
Shell: good thermal stability

> 10 μm

FIGURE 1   Formation process of microscale core–shell structured $Li[(Ni_{0.8}Co_{0.1}Mn_{0.1})_{1-x}(Ni_{0.5}Mn_{0.5})_x]O_2$. (From [22], with permission; copyright © 2005 American Chemical Society.)

shell with a thickness of less than 2 μm. Finally, the calcination of the core–shell hydroxide and lithium salt gives rise to the formation of spherical core–shell $Li[(Ni_{1-x-y}Co_xMn_y)_{0.8}(Ni_{0.5}Mn_{0.5})_{0.2}]O_2$ ($x = 0$ to 0.2, $y = 0 - 0.1$) powders, of which Ni-rich $Li[Ni_{1-x-y}Co_xMn_y]O_2$ ($x = 0$ to 0.2, $y = 0$ to 0.1) is the core and $Li[Ni_{0.5}Mn_{0.5}]O_2$ is the shell.

Figure 2 shows the formation process of the core–shell hydroxide. After preparation of the spherical core hydroxide (Fig. 2a), sedimentation of the shell hydroxide onto the core hydroxide is possible by extending the piling time (Fig. 2b to f), yielding variations in shell thickness. The difference in the chemical composition between the core and the shell also results in a clear separation of the core and the shell in a particle. Even though the shell was thickened to 1.2 μm (Fig. 2f), there was no remarkable change in core size for core–shell hydroxides compared to the core shown in Figure 2a. Pores were not observed in the core and shell parts, indicating that the synthesized core–shell hydroxide particles are quite dense.

Powder x-ray diffraction (XRD) data proved the coexistence of the core and shell phases in the core–shell particles. The core powders, Ni-rich hydroxide, exhibited a typical XRD pattern of layer-type $M(OH)_2$ with the $R\bar{3}c$ space group (Fig. 2g). The appearance of small shoulder peaks in the diffraction patterns indicated the presence of the shell with the core in the core–shell particles. The diffraction intensity became stronger as the shell thickness increased (Fig. 2h to k), which was consistent with the scanning electron microscope (SEM) results.

Calcination of the core and core–shell hydroxides with lithium salt resulted in different morphology, particularly the primary particle shape (Fig. 3a to f). As shown, there were no pores in the hydroxides, and the primary particle size was found to be tens of nanometers. Interestingly, growth of the primary particle size is evident

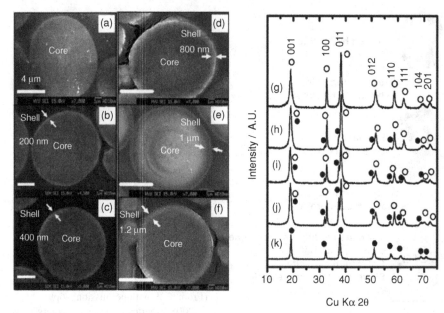

FIGURE 2 Cross-sectional SEM images of (a) $[Ni_{0.8}Co_{0.1}Mn_{0.1}](OH)_2$ core and $[Ni_{0.8}Co_{0.1}Mn_{0.1}](OH)_2$ core–$[Ni_{0.5}Mn_{0.5}](OH)_2$ shell with increasing shell thickness: (b) 200 nm, (c) 400 nm, (d) 800 nm, (e) 1 μm, and (f) 1.2 μm. Powder XRD patterns of (g) $[Ni_{0.8}Co_{0.1}Mn_{0.1}](OH)_2$ core and $[Ni_{0.8}Co_{0.1}Mn_{0.1}](OH)_2$ core–$[Ni_{0.5}Mn_{0.5}](OH)_2$ shell with increasing shell thickness: (h) 800 nm, (i) 1 μm, (j) 1.2 μm, and (k) $[Ni_{0.5}Mn_{0.5}](OH)_2$ shell. Open circles indicate $[Ni_{0.8}Co_{0.1}Mn_{0.1}](OH)_2$ and closed circles denote $[Ni_{0.5}Mn_{0.5}](OH)_2$. (From [23], with permission; copyright © 2006 American Chemical Society.)

in Figure 3a to f, and the core–shell products were clearly divided by an interface. The contrast between the core and shell parts is further indication of the formation of core–shell particles. The original spherical particle shape and the thickness of the shell were maintained as designed after high-temperature calcination. Elemental diffusion is likely to occur during high-temperature calcination, and this may cause the core–shell morphology in the particles to disappear. As shown in Figure 3a to f, however, control of the calcination conditions prevented interdiffusion of the elements in the core–shell particle, and this, in turn, made it possible to maintain the original core–shell particle morphology.

The XRD patterns showed that the calcined products had a well-ordered layer structure with an *R-3m* space group. The relative diffraction intensity varied with the thickness of the shell; a thicker shell yielded a lower diffraction intensity, implying that the presence of $Li[Ni_{0.5}Mn_{0.5}]O_2$ in the shell leads to a relatively lower diffraction intensity. Because of the structural similarity between $Li[Ni_{0.8}Co_{0.1}Mn_{0.1}]O_2$ and $Li[Ni_{0.5}Mn_{0.5}]O_2$, calcination resulting from $[Ni_{0.8}Co_{0.1}Mn_{0.1}](OH)_2$ and LiOH exhibited overlapping diffraction patterns (Fig. 3g to k). The foregoing results clearly indicate coexistence of the core and shell in one particle without phase segregation and separation. In addition, control of the amount of shell hydroxide piled on the core

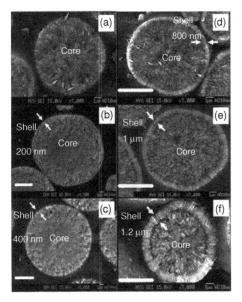 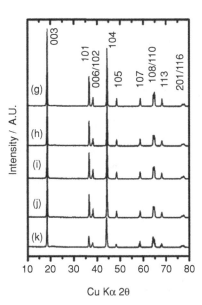

FIGURE 3   Cross-sectional SEM images of (a) Li[Ni$_{0.8}$Co$_{0.1}$Mn$_{0.1}$]O$_2$ core and Li[Ni$_{0.8}$Co$_{0.1}$Mn$_{0.1}$]O$_2$ core–Li[Ni$_{0.5}$Mn$_{0.5}$]O$_2$ shell with increasing shell thickness: (b) 200 nm, (c) 400 nm, (d) 800 nm, (e) 1 μm, and (f) 1.2 μm. Powder XRD patterns of (g) Li[Ni$_{0.8}$Co$_{0.1}$Mn$_{0.1}$]O$_2$ core and Li[Ni$_{0.8}$Co$_{0.1}$Mn$_{0.1}$]O$_2$ core–Li[Ni$_{0.5}$Mn$_{0.5}$]O$_2$ shell with increasing shell thickness: (h) 800 nm, (i) 1 μm, (j) 1.2 μm, and (k) Li[Ni$_{0.5}$Mn$_{0.5}$]O$_2$ shell. (From [23], with permissions; copyright © 2006 American Chemical Society.)

surface allows different shell thicknesses from nano- to micrometers. As expected, the coprecipitation technique is the only possible way to synthesize spherical and dense core–shell powders.

The core materials, Li[Ni$_{0.8}$Co$_{0.1}$Mn$_{0.1}$]O$_2$ and Li[Ni$_{0.8}$Co$_{0.2}$]O$_2$, exhibited discharge capacities of 200 mAh/g (Fig. 4a and b; Li[Ni$_{1/3}$Co$_{1/3}$Mn$_{1/3}$]O$_2$ also shows approximately 200 mAh/g up to 4.5 V). Compared with the core material, the coreshell material delivers a slightly lower discharge capacity because the reversible capacity of Li[Ni$_{0.5}$Mn$_{0.5}$]O$_2$ is approximately 150 mAh/g at a voltage cutoff limit of 4.3 V. Thus, as the Li[Ni$_{0.5}$Mn$_{0.5}$]O$_2$ shell thickens, the charge and discharge capacity decreases. The presence of microscale Li[Ni$_{0.5}$Mn$_{0.5}$]O$_2$ appears to increase the operation charge voltage. The effect of the operation voltage on the charge of the core–shell material is greater than on that of the core material. This finding can be attributed to the intrinsic properties of Li[Ni$_{0.5}$Mn$_{0.5}$]O$_2$, which has a higher charge voltage than that of the core material. The core–shell cells have superior cyclability during cycling (Fig. 4c to f). In this case, the Li[Ni$_{0.5}$Mn$_{0.5}$]O$_2$ outer shell [26], which has stable cyclability, is associated with an improvement in cycling behavior. If the core materials (including Li[Ni$_{1/3}$Co$_{1/3}$Mn$_{1/3}$]O$_2$) appear on the surface of the core–shell particles or exist as impurities in the final product, such extraordinary cycling behavior would not be achieved, due to the structural instability of the core materials. Since the Li[Ni$_{0.5}$Mn$_{0.5}$]O$_2$ shell is surrounded completely by the core, it

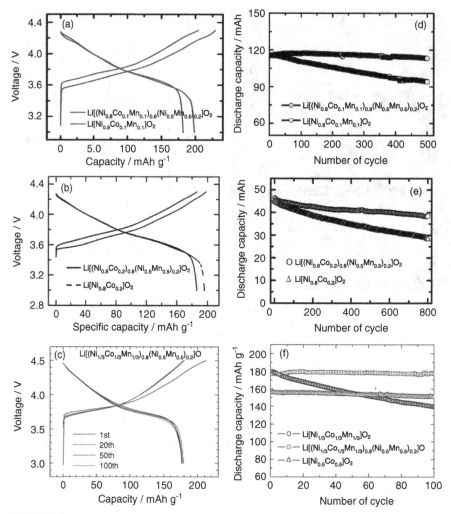

FIGURE 4 Charge and discharge curves of (a) and (d) Li[(Ni$_{0.8}$Co$_{0.1}$Mn$_{0.1}$)$_{0.8}$ (Ni$_{0.5}$Mn$_{0.5}$)$_{0.2}$]O$_2$ (from [22], with permission; copyright © 2005 American Chemical Society), (b) and (e) Li[(Ni$_{0.8}$Co$_{0.2}$)$_{0.8}$(Ni$_{0.5}$Mn$_{0.5}$)$_{0.2}$]O$_2$ (from [24], with permission; copyright © 2006 American Chemical Society), and (c) and (f) Li[(Ni$_{1/3}$Co$_{1/3}$Mn$_{1/3}$)$_{0.8}$ (Ni$_{0.5}$Mn$_{0.5}$)$_{0.2}$]O$_2$ (from [25], with permission; copyright © 2010 Elsevier).

can prevent the potential HF attack of the core materials by the electrolyte during long-term cycling, thereby suppressing Co dissolution and retaining capacity during cycling.

LiNiO$_2$ and its derivatives suffer from poor thermal instability [28,29], as shown in Figure 5a and b. For the deeply charged Li$_{1-\delta}$[Ni$_{0.8}$Co$_{0.1}$Mn$_{0.1}$]O$_2$ and Li$_{1-\delta}$[Ni$_{0.8}$Co$_{0.2}$]O$_2$ electrodes, the onset temperature of an exothermic peak was at about 190°C, above which an abrupt exothermic reaction was observed at about 220 and 210°C for each electrode, respectively. Meanwhile, the charged core–shell

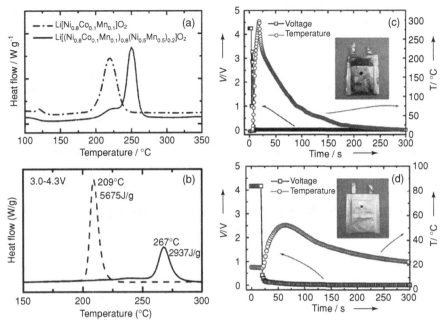

FIGURE 5   Differential scanning calorimetry traces of delithiated (a) Li[(Ni$_{0.8}$Co$_{0.1}$Mn)$_{0.8}$ (Ni$_{0.5}$Mn$_{0.5}$)$_{0.2}$]O$_2$ (solid line) and Li[Ni$_{0.8}$Co$_{0.1}$Mn$_{0.1}$]O$_2$ (dashed line) (from [22], with permission; copyright © 2005 American Chemical Society); (b) delithiated Li[(Ni$_{0.8}$Co$_{0.2}$)$_{0.8}$ (Ni$_{0.5}$Mn$_{0.5}$)$_{0.2}$]O$_2$ (solid line) and Li[Ni$_{0.8}$Co$_{0.2}$]O$_2$ (from [24], with permission; copyright © 2006 American Chemical Society) at charged state to 4.3 V. Voltage vs. cell surface temperature plots in the lithium-ion cells of (c) Li[Ni$_{0.8}$Co$_{0.1}$Mn$_{0.1}$]O$_2$/C and (d) core–shell Li[(Ni$_{0.8}$Co$_{0.1}$Mn$_{0.1}$)$_{0.8}$(Ni$_{0.5}$Mn$_{0.5}$)$_{0.2}$]O$_2$/C cells during nail penetration tests at 4.3 V. Insets contain lithium-ion cell images of (c) Li[Ni$_{0.8}$Co$_{0.1}$Mn$_{0.1}$]O$_2$/C and (d) core–shell Li[(Ni$_{0.8}$Co$_{0.1}$Mn$_{0.1}$)$_{0.8}$(Ni$_{0.5}$Mn$_{0.5}$)$_{0.2}$]O$_2$/C cells after nail penetration at 4.3 V (from [27], with permission; copyright © 2006 Electrochemical Society).

electrodes showed an exothermic reaction at 250 and 270°C, respectively, and had reduced heat generation compared with the core and the Li$_{1-\delta}$[Ni$_{0.8}$Co$_{0.1}$Mn$_{0.1}$]O$_2$ and Li$_{1-\delta}$[Ni$_{0.8}$Co$_{0.2}$]O$_2$ electrodes. Thus, a thermally stable shell material greatly improved the thermal stability of the core–shell material.

Nail penetration of the charged lithium-ion cell is one of the best safety tests. Once the battery is internally shorted by the nail, the charged electric energy is suddenly converted into heat. The electroconducting metallic nail contacts the highly oxidized active material directly. For the Li[Ni$_{0.8}$Co$_{0.1}$Mn$_{0.1}$]O$_2$ cathode (Fig. 5c), the temperature suddenly rose to 300°C and the cell voltage dropped rapidly within a few seconds, causing a short circuit in the cell as a result of the metallic nail penetration. Then the Li[Ni$_{0.8}$Co$_{0.1}$Mn$_{0.1}$]O$_2$/C cell ignited abruptly within a few seconds. This abrupt exothermic reaction indicated the poor thermal properties of Li[Ni$_{0.8}$Co$_{0.1}$Mn$_{0.1}$]O$_2$. Therefore, the cell containing Li[Ni$_{0.8}$Co$_{0.1}$Mn$_{0.1}$]O$_2$ suffered from explosive ignition in the nail penetration test. In contrast, the cell containing a

$Li[(Ni_{0.8}Co_{0.1}Mn_{0.1})_{0.8}(Ni_{0.5}Mn_{0.5})_{0.2}]O_2$ cathode did not ignite in the nail penetration test (Fig. 5d) and its maximum temperature was only about 50°C when the highly oxidized active material made direct contact with the metallic nail. Divalent Ni in $Li[Ni_{0.5}Mn_{0.5}]O_2$ is oxidized electrochemically to trivalent Ni at 4.3 to 4.4 V. Thus, exothermic decomposition of the material is delayed up to temperatures as high as 280°C [28,29]. Thus, the superior thermal properties of the shell are thought not to reflect a thermal event for the spherical core–shell materials. The synergetic effects (i.e., the high capacity from the Ni-based layer–structured core materials and the reliable thermal stability from the $Li[Ni_{0.5}Mn_{0.5}]O_2$ shell) of microscale core–shell particles suggest a new possibility for the development of safe advanced lithium-ion batteries with high energy density and long cycle life.

3 LAYER-STRUCTURED CORE–SHELL PARTICLES WITH A CONCENTRATION GRADIENT

After reviewing carefully the analysis results discussed above, a structural mismatch was found between the core and the shell; voids of tens of nanometers between the core and the shell were found in the core–shell powders (Figs. 3f and 6a). Nickel-rich compounds (core material) were believed to undergo a volume change of approximately 9 to 10%, whereas the shell volume change was only 2 to 3% during deintercalation. The structural mismatch can cause gradual separation between the core and the shell material, and the core can be isolated from the shell in the core–shell particle, leading to a blocking of Li^+ on the electron path. The problem can readily be solved by encapsulation of the core with a concentration gradient shell of microscale size (Fig. 6b).

To synthesize the concentration gradient core–shell materials, Ni-rich hydroxide, $[Ni_{0.8}Co_{0.1}Mn_{0.1}](OH)_2$ or $[Ni_{0.8}Co_{0.2}](OH)_2$, was first coprecipitated in the reactor (Fig. 6c). The precipitates had a spherical morphology like that of the scheme shown in Fig. 6b, and coprecipitation continued by changing the Ni, Co, and Mn concentrations as shown in Fig. 6d.

Thus, Ni-rich hydroxides with a concentration gradient shell were formed (Fig. 7a). The chemical composition in the bulk section was kept constant. After going through the interface region, the relative intensity of Ni, Co, and Mn changed drastically, due to the presence of a concentration gradient of transition metal elements. This result was associated with the presence of the concentration gradient shell because the chemical composition at the interface was close to that of the core material. This condition was maintained even after calcination. Thus, the chemical shell composition varied from $Li[Ni_{0.8}Co_{0.2}]O_2$ at the core to $Li[Ni_{0.55}Co_{0.15}Mn_{0.30}]O_2$ at the outer surface of the shell, giving an average composition of $Li[Ni_{0.72}Co_{0.18}Mn_{0.10}]O_2$. Unlike the conventional core–shell particle (Figs. 3f and 6a), no structural mismatch occurred in the core–shell interface (inset in Fig. 7b). The oxidation state of Ni near the surface was slightly higher than 2+, and the binding energy observed for Mn was 4+. Note that the Co element was always stabilized to Co^{3+} in this layer structure. Further sputtering of the surfaces of our material did not change the binding energies of Ni and Mn (Fig. 7c and 7d). Even though the concentration-gradient particles were

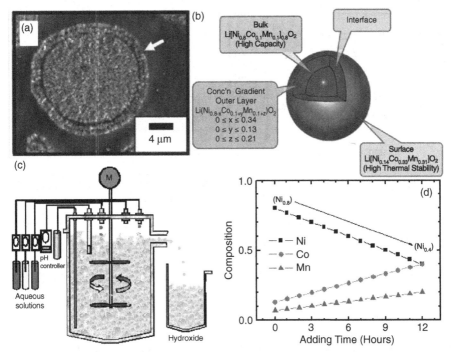

FIGURE 6 (a) Cross-sectional SEM image of conventional core–shell architecture particle, which shows a flaw between the core and the shell; (b) schematic diagram of a particle with Ni-rich core surrounded by concentration-gradient outer layer; (c) schematic drawing of coprecipitation reactor; (d) concentration variation of transition metal–containing aqueous solution during formation of concentration-gradient shell hydroxide. (Adapted from [30].)

etched by up to 2.1 μm in depth, which corresponds to the interface region, there were no apparent changes in the binding energies of Ni and Mn. For the composition in the bulk of the particles, $Li[Ni_{0.8}Co_{0.1}Mn_{0.1}]O_2$ or $Li[Ni_{0.8}Co_{0.2}]O_2$, the oxidation state of each Ni, Co, and Mn ion would be preferentially trivalent. However, some of the manganese near the interface may have been tetravalent, as observed in the x-ray photoelectron spectroscopy spectra (Fig. 7d). Therefore, in the final lithiated oxide material, both the concentration and the oxidation state of Ni, Co, and Mn changed from the bulk to the surface of the particle.

Three kinds of concentration gradient core–shell materials were compared: $Li[Ni_{0.83}Co_{0.07}Mn_{0.10}]O_2$, $Li[Ni_{0.72}Co_{0.18}Mn_{0.10}]O_2$, and $Li[Ni_{0.64}Co_{0.18}Mn_{0.18}]O_2$ (Fig. 8a to c). The Li intercalation stability of the concentration gradient $Li[Ni_{0.83}Co_{0.07}Mn_{0.10}]O_2$ was remarkably improved by complete encapsulation of Ni-rich $Li[Ni_{0.90}Co_{0.05}Mn_{0.05}]O_2$ with a stable outer shell layer having a lower Ni content ($Li[Ni_{0.68}Co_{0.12}Mn_{0.20}]O_2$), and maintained a discharge capacity of 196 mAh/g (only 7.9% capacity loss) after 50 cycles. Similarly, our concentration gradient materials showed an excellent 96% capacity retention during the same cycling period (Fig. 8d and f), similar to that of the cell based on the surface composition only. This result clearly indicates that our cathode material can provide high capacity with

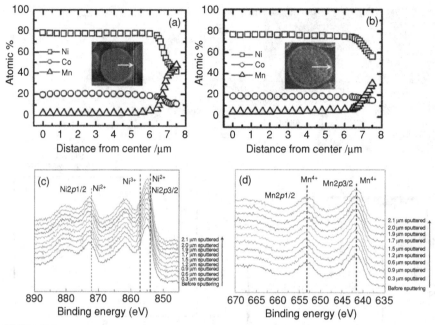

FIGURE 7 (a) EPMA compositional change from a cross section of the core–shell hydroxide with a concentration-gradient shell particle; (b) EPMA compositional change from a cross section of lithiated core–shell oxide with a concentration-gradient shell particle. (From [31], with permission; copyright © 2010 Wiley-VCH.) X-ray photoelectron spectroscopic data for the concentration-gradient $Li[Ni_{0.64}Co_{0.18}Mn_{0.18}]O_2$: (c) Ni 2p and (d) Mn. (Adapted from [32].)

a long cycle and calendar life even at high temperature and high cutoff voltages. By contrast, the core $Li[Ni_{0.90}Co_{0.05}Mn_{0.05}]O_2$ showed a rapid decrease in capacity, leading to a capacity retention of only 64.6% (145.4 mAh/g) over the same cycling period, resulting mainly from the structural instability of the core materials.

Again, thermal properties are critical when these electrode materials are employed in commercial applications such as electric vehicles and energy storage systems. As seen in the differential scanning colorimetry profiles, all Ni-rich electrode materials displayed lower onset exothermic temperatures, below 190°C, with a considerable amount of heat generated (Fig. 9). Notably, the heat generated was significantly reduced by forming a concentration gradient of $Li[Ni_{0.83}Co_{0.07}Mn_{0.10}]O_2$ (Fig. 9a).

A similar tendency was also observed for the charged $Li_{1-\delta}[Ni_{0.72}Co_{0.18}Mn_{0.10}]O_2$ and $Li_{1-\delta}[Ni_{0.64}Co_{0.18}Mn_{0.18}]O_2$ (Fig. 9b and c). It is clear that the exothermic reaction was heavily dependent on the chemical composition of the surface. The surface compositions were as follows: $Li[Ni_{0.68}Co_{0.12}Mn_{0.20}]O_2$ for $Li[Ni_{0.83}Co_{0.07}Mn_{0.1}]O_2$, $Li[Ni_{0.55}Co_{0.15}Mn_{0.30}]O_2$ for $Li[Ni_{0.72}Co_{0.18}Mn_{0.10}]O_2$, and $Li[Ni_{0.46}Co_{0.23}Mn_{0.31}]O_2$ for $Li[Ni_{0.64}Co_{0.18}Mn_{0.18}]O_2$. One common point for both the charged $Li_{1-\delta}[Ni_{0.72}Co_{0.18}Mn_{0.10}]O_2$ and $Li_{1-\delta}[Ni_{0.64}Co_{0.18}Mn_{0.18}]O_2$ is that the

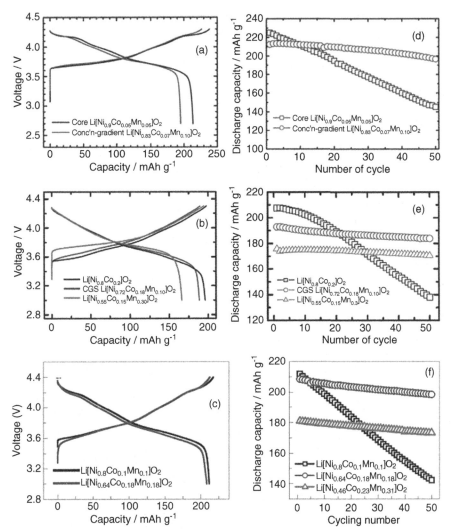

FIGURE 8 (a) Initial charge and discharge voltage profiles and cyclability for core Li[Ni$_{0.90}$Co$_{0.05}$Mn$_{0.05}$]O$_2$ and concentration-gradient Li[Ni$_{0.83}$Co$_{0.07}$Mn$_{0.10}$]O$_2$ at (a) 25°C and (d) 55°C (adapted from [33]); (b) initial charge–discharge curves for the Li/Li[Ni$_{0.8}$Co$_{0.2}$]O$_2$, CGS Li[Ni$_{0.72}$Co$_{0.18}$Mn$_{0.10}$]O$_2$, and Li[Ni$_{0.55}$Co$_{0.15}$Mn$_{0.30}$]O$_2$ cells and (e) their cyclability at 55°C (from [31], with permission; copyright © 2010 Wiley-VCH); (c) initial charge–discharge curves for Li/Li[Ni$_{0.8}$Co$_{0.1}$Mn$_{0.1}$]O$_2$, CGS Li[Ni$_{0.64}$Co$_{0.18}$Mn$_{0.18}$]O$_2$, and Li[Ni$_{0.46}$Co$_{0.23}$Mn$_{0.31}$]O$_2$ cells and (f) their cyclability at 55°C (Adapted from [32].)

Mn content in the surface part occupies 30% of the transition metal layer, in which the average oxidation state of Mn was tetravalent. Meanwhile, the surface oxidation state of Mn for Li[Ni$_{0.68}$Co$_{0.12}$Mn$_{0.20}$]O$_2$, which is the surface composition of the Li[Ni$_{0.83}$Co$_{0.07}$Mn$_{0.10}$]O$_2$, would be close to trivalent. As reviewed earlier in the chapter, tetravalent Mn on the surface would provide significant structural stability

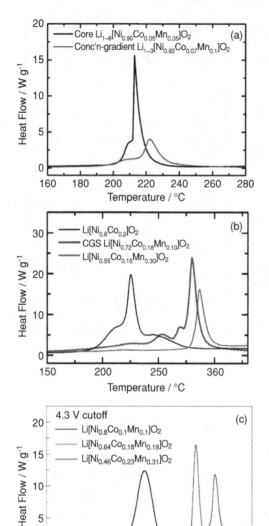

FIGURE 9 (a) DSC results of the core $Li_{1-\delta}[Ni_{0.90}Co_{0.05}Mn_{0.05}]O_2$ and CGS $Li_{1-\delta}[Ni_{0.83}Co_{0.07}Mn_{0.10}]O_2$ (adapted from [33]), (b) $Li_{1-\delta}[Ni_{0.8}Co_{0.2}]O_2$, CGS $Li_{1-\delta}[Ni_{0.72}Co_{0.18}Mn_{0.10}]O_2$, and $Li_{1-\delta}[Ni_{0.55}Co_{0.15}Mn_{0.30}]O_2$ (from [31], with permission; copyright © 2010 Wiley-VCH), and (c) $Li_{1-\delta}[Ni_{0.8}Co_{0.1}Mn_{0.1}]O_2$, CGS $Li_{1-\delta}[Ni_{0.64}Co_{0.18}Mn_{0.18}]O_2$, and $Li_{1-\delta}[Ni_{0.46}Co_{0.28}Mn_{0.31}]O_2$ (adapted from [32]).

such that it would affect the improved thermal stability. Hence, the surface chemical state of Mn in the concentration gradient particle was pivotal in both the retention of high capacity during cycling and the improvement in the thermal properties, which directly influence battery safety.

This concentration gradient core–shell concept is a plausible approach to increasing the Ni content in the spherical particles, which spontaneously provides more capacity originating from the Ni redox couple. Additionally, the intrinsic drawback of the core–shell particle, which is a structural mismatch between the core and shell, was solved by introduction of the present scheme. Even though the average oxidation state of Mn was close to trivalent in the bulk part, the valence gradually changed to tetravalent on the surface, thereby conferring structural and thermal stability and giving rise to high-capacity retention upon cycling.

4   SPHERICAL CORE–SHELL Li[($Li_{0.05}Mn_{0.95})_{0.8}(Ni_{0.25}Mn_{0.75})_{0.2}]_2O_4$ SPINEL

The widespread practical use of the lithium manganese spinel electrode has so far been hindered by severe capacity fading on cycling, especially when operating at temperatures above ambient (e.g., above 40°C) [34]. This limitation is due to manganese dissolution, which in turn is associated with the disproportionate reaction of Mn^{3+}. Mn^{2+} is known to dissolve into the electrolyte solution while Mn^{4+} remains in the bulk of the electrode. Both these events are deleterious for the performance of the electrode and consequently, of the battery that uses it [35]. Similar to the layered materials described earlier, surface modification using nanolayers provides a shield that protects against manganese dissolution, combined with a scavenging action for the fluoride anions eventually present in the electrolyte solution. However, this beneficial effect may be contrasted with an increase in the battery impedance resulting from the insulating nature of the coating media.

A new strategy, which is innovative core–shell morphology, is described in Figure 10a. In this process a lithium-rich manganese spinel, $Li_{1.1}Mn_{1.9}O_4$, was the core, and a high-voltage lithium nickel manganese spinel, $LiNi_{0.5}Mn_{1.5}O_4$, was the shell. The average manganese oxidation state in the shell compound was Mn^{4+}, with no presence of Mn^{3+}; thus, conditions favorable to the disproportionate process were not present in the shell.

In addition, the $LiNi_{0.5}Mn_{1.5}O_4$ shell compound was not electrochemically active in the 4-V range of operation of the $Li_{1.1}Mn_{1.9}O_4$ core compound. Consequently, any contact with the electrolyte was prevented and the $Li_{1.1}Mn_{1.9}O_4$ could safely operate with no risk of manganese dissolution. The synthesis began with a spherical $MnCO_3$ core that was completely encapsulated by a $(Ni_{0.25}Mn_{0.75})CO_3$ shell, and finally transformed, via thermal lithiation of the core carbonate, into the microscale, spherical $Li_{1.1}Mn_{1.9}O_4$ core–$LiNi_{0.5}Mn_{1.5}O_4$ shell morphology. The scanning electron micrograph (SEM) images shown in Figure 10b and c illustrate the morphology of the precursor and of the final material, respectively. The images show that the shell has an uninterrupted thickness on the order of 2 μm that completely and uniformly covers the core, thus being the unique and primary feature of this material. The XRD pattern confirmed that the material adopted the typical single–phase spinel structure shown in Figure 10d.

The initial specific capacity of the pristine $Li_{1.1}Mn_{1.9}O_4$ electrode was higher than that of the core–shell electrode. This was not surprising but rather expected, since the shell component was not electrochemically active in the voltage range of cell operation (i.e., around 4 V). For the C/Li[($Li_{0.05}Mn_{0.95})_{0.8}(Ni_{0.25}Mn_{0.75})_{0.2}]_2O_4$ full cell, the operation voltage on discharge was maintained during cycling (Fig. 11a). Moreover, the capacity retention of the core–shell particles during extensive cycling

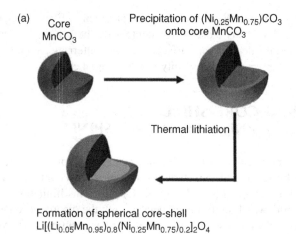

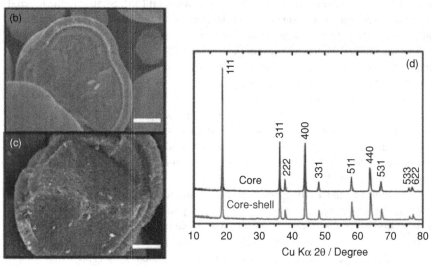

FIGURE 10 (a) Scheme of the formation sequence of the core–shell spinel structure; SEM images of fractured (b) core–shell precursor (scale bar: 5 μm), and (c) lithiated spinel powder (scale bar: 5 μm); (d) powder XRD patterns of core $Li_{1.1}Mn_{1.9}O_4$ and core–shell $Li[(Li_{0.05}Mn_{0.95})_{0.8}(Ni_{0.25}Mn_{0.75})_{0.2}]_2O_4$. (Adapted from [36].)

was approximately 94% of its initial capacity, although the $C/Li_{1.1}Mn_{1.9}O_4$ cell had a disappointing cycling performance at high temperature.

Figure 11b shows the charge–discharge voltage profiles of a lithium cell galvanostatically cycled at 60°C over 3.0 to 5.0 V, a voltage range where both components are electrochemically active. The two representative mixed charge–discharge curves were clearly visualized: one evolving with a 5-V plateau associated with the shell component and a subsequent curve evolving with a 4-V plateau associated with the core component. The total capacity amounted to 116.8 mAh/g, 26 mAh/g

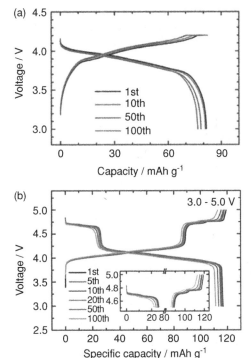

FIGURE 11 (a) C/Li[(Li$_{0.05}$Mn$_{0.95}$)$_{0.8}$ (Ni$_{0.25}$Mn$_{0.75}$)$_{0.2}$]$_2$O$_4$ cell at 60°C, and (b) continuous charge and discharge curves of the core–shell at 60°C. Inset: Highlighted range between 4.5 and 5 V. (Adapted from [36].)

of which was provided by the upper 5-V plateau. The electrode had exceptional capacity retention upon cycling at 60°C, on the order of 95.2% over 100 cycles. In addition, the initial irreversible capacity, a deleterious phenomenon that affects most cathode materials, was confined to 2.1%. To our knowledge, such a high performance is rarely found in common lithium-ion battery chemistry, which is further evidence of the practical relevance of this core–shell structure. The outer LiNi$_{0.5}$Mn$_{1.5}$O$_4$ shell, in addition to the series of benefits discussed above, also confers overcharge protection by shuttling lithium ions in and out when the charge cutoff limit of a Li[(Li$_{0.05}$Mn$_{0.95}$)$_{0.8}$(Ni$_{0.25}$Mn$_{0.75}$)$_{0.2}$]$_2$O$_4$ core–shell electrode hits 4.7 V.

5 CONCLUSIONS

The development of electrode materials, especially cathode active materials, is important to satisfy requirements such as energy density, cost, and in particular, safety. The easiest route to cathode improvement is modification of the cathode surface. Here we suggest complete encapsulation of active material using other active materials that possess specific functions, such as good structural and thermal stability, good conductivity, and so on. These materials not only showed a very highly reversible capacity based on the particle bulk composition, but also excellent cycling and safety characteristics, which were attributed to the stability of the outer layer supported by the surface composition. It was especially noteworthy that in thermal-abuse tests the concentration-gradient material exhibited performance superior to that of bulk composition. This breakthrough is expected to contribute to practical applications in advanced lithium-ion batteries that meet the demanding performance and safety requirements of vehicles or energy storage systems.

Acknowledgments

This research was supported by a Human Resources Development of the Korea Institute of Energy Technology Evaluation and Planning grant funded by the Korean Government's Ministry of Knowledge Economy (No. 20124010203290).

REFERENCES

1. H.-J. Kweon, G.-B. Kim, and D.-G. Park, Korean Patent 0012005, 1998.
2. J. Cho, Y. Kim, and B. Park, *Chem. Mater*, **12**, 3788 (2000).
3. Z. X. Wang, C. Wu, L. Liu, F. Wu, L. Chen, and X. Huang, *J. Electrochem. Soc.*, **149**, A466 (2002).
4. E. Zhecheva, R. Stoyanova, G. Tyuliev, K. Tenchev, M. Mladenov, and S. Vassilev, *Solid State Sci.*, **5**, 711 (2003).
5. J. S. Gnanaraj, V. G. Pol, A. Gedanken, and D. Aurbach, *Electrochem. Commun.*, **11**, 940 (2003).
6. Y.-K. Sun, J.-M. Han, S.-T. Myung, S.-W. Lee, and K. Amine, *Electrochem. Commun.*, **8**, 821 (2006).
7. Y.-K. Sun, K.-J. Hong, J. Prakash, and K. Amine, *Electrochem. Commun.* **4**, 344 (2002).
8. Z. Chen and J. R. Dahn, *Electrochem. Solid-State Lett.*, **5**, A213 (2002).
9. S.-T. Myung, K. Izumi, S. Komaba, Y.-K. Sun, H. Yashiro, and N. Kumagai, *Chem. Mater.*, **17**, 3695 (2005).
10. S.-T. Myung, K. Izumi, S. Komaba, H. Yashiro, H. J. Bang, Y.-K. Sun, and N. Kumagai, *J. Phys. Chem. C*, **111**, 4061 (2007).
11. W.-S. Yoon, M. Balasubramanian, X.-Q. Yang, Z. Fu, D. A. Fischer, and J. McBreen, *J. Electrochem. Soc.*, **151**, A246 (2004).
12. S.-T. Myung, S. Komaba, K. Kurihara, K. Hosoya, N. Kumagai, Y.-K. Sun, I. Nakai, M. Yonemura, and T. Kamiyama, *Chem. Mater.*, **18**, 1658 (2006).
13. W. Schartl, *Adv. Mater.*, **12**, 1899 (2000).
14. F. Caruso, *Adv. Mater.*, **13**, 11–22 (2001).
15. T. K. Mandal, M. S. Fleming, and D. R. Walt, *Chem. Mater.*, **12**, 3481-3487 (2001).
16. V. Suryanarayanan, A. S. Nair, R. T. Tom, and T. Pradeep, *J. Mater. Chem.*, **14**, 2661-2666 (2004).
17. T. Nakashima and N. Kimizuka, *J. Am. Chem. Soc.*, **125**, 6386-6387 (2003).
18. J.-S. Hu, Y.-G. Guo, H.-P. Liang, L.-J. Wan, C.-L. Bai, and Y.-G. Wang, *J. Phys. Chem. B*, **108**, 9734-9738 (2004).
19. F. Caruso, M. Spasova, A. Susha, M. Giersig, and F. Caruso, *Chem. Mater.*, **13**, 109–118 (2001).
20. C.-W. Guo, Y. Cao, S.-H. Xie, W.-L. Dai, and K.-N. Fan, *Chem. Commun.* 700–701 (2003).
21. Y. Zhang, G. Li, and L. Zhang, *Inorg. Chem. Commun.*, **7**, 344–346 (2004).
22. Y.-K. Sun, S.-T. Myung, M.-H. Kim, J. Prakash, and K. Amine, *J. Am. Chem. Soc.*, **127**, 13411 (2005).
23. Y.-K. Sun, S.-T. Myung, B.-C. Park, and K. Amine, *Chem. Mater.*, **18**, 5159 (2006).
24. Y.-K. Sun, S.-T. Myung, H.-S. Shin, Y. C. Bae, and C. S. Yoon, *J. Phys. Chem. B*, **110**, 6810 (2006).
25. K.-S. Lee, S.-T. Myung, and Y.-K. Sun, *J. Power Sources*, **195**, 6043 (2010).
26. T. Ohzuku and Y. Makimura, *Chem. Lett.*, **30**, 744 (2001).
27. Y.-K. Sun, S.-T. Myung, M.-H. Kim, and J.-H. Kim, *Electrochem. Solid-State Lett.*, **9**, A171 (2006).
28. K.-S. Lee, S.-T. Myung. K. Amine, H. Yashiro, and Y.-K. Sun, *J. Electrochem. Soc.*, **154**, A971 (2007).
29. Y.-K. Sun, S.-T. Myung, H. J. Bang, S.-C. Park, S.-J. Park, and N.-Y. Sung, *J. Electrochem. Soc.*, **154**, A937 (2007).

30. S.-T. Myung, K. Amine, and Y.-K. Sun, *J. Mater. Chem.*, **20**, 7074 (2010).
31. Y.-K. Sun, D.-H. Kim, C. S. Yoon, S.-T. Myung, J. Prakash, and K. Amine, *Adv. Funct. Mater.*, **20**, 485 (2010).
32. Y.-K. Sun, S.-T. Myung, B.-C. Park, J. Prakash, I. Belharouak, and K. Amine, *Nat. Mater.*, **8**, 320 (2009).
33. Y.-K. Sun, B.-R. Lee, H.-J. Noh, H. Wu, S.-T. Myung, and K. Amine, *J. Mater. Chem.*, **21**, 10108 (2011).
34. J. Arai, T. Yamaki, S. Yamauchi, T. Yuasa, T. Maeshima, T. Sakai, M. Koseki, and T. Horiba, *J. Power Sources*, **146**, 788–792 (2005).
35. R.J. Gummow, A. de Kock, and M. M. Thackeray, *Solid State Ionics*, **69**, 59–67 (1994).
36. S.-T. Myung, K.-S. Lee, D.-W. Kim, B. Scrosati, and Y.-K. Sun, *Energy Environ. Sci.*, **4**, 935 (2011).

CHAPTER **6**

PROBLEMS AND EXPECTANCY IN LITHIUM BATTERY TECHNOLOGIES

K. Kanamura

1 INTRODUCTION

Energy cannot be seen by the human eye, although we can feel energy when we use it. Therefore, the management of energy is very difficult. To minimize energy consumption, we have to use energy very carefully.

Thermodynamics teaches us that energy does not disappear; this is the energy conservation law. Total energy consists of two parts: Gibbs free energy and an entropy

Lithium Batteries: Advanced Technologies and Applications, First Edition.
Edited by Bruno Scrosati, K. M. Abraham, Walter van Schalkwijk, and Jusef Hassoun.
© 2013 John Wiley & Sons, Inc. Published 2013 by John Wiley & Sons, Inc.

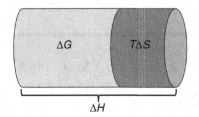

FIGURE 1 Energy content.

term, as shown in Figure 1. The entropy term is heat energy, which is utilized in electric power plants to produce a lot of electric energy at a low level of utilization of fuel energy, 40% or less. This is a problem. We need renewable energy, such as solar, wind, and geothermal energy.

In our homes, we sometimes heat water to use for coffee or a bath. In such cases, electricity is utilized to prepare the water. For example, in creating 40°C water from 20°C water, we lose 97% of the energy. This is a huge loss. Another example is that of a car with an internal combustion engine. When we drive a car, we make kinetic energy from fuel energy. Then, when we stop a car, we use a braking system. The braking system is a type of energy conversion device, changing from kinetic energy to heat energy. We lose all the energy when we stop the car. Every day, we produce a lot of heat energy as the result of human activities. We must recover such heat energy.

2 IMPORTANCE OF ENERGY STORAGE

Figure 2 shows a strategy for energy conversion using rechargeable batteries as energy storage devices. Natural energy is usually not as stable, depending on the weather. To use natural energy in an electric power system, electric energy should be accumulated in a buffer system. The most promising storage system is a rechargeable battery.

In a car, the braking system is altered to a regenerative form. We can produce electric energy when we stop a car. This electricity should be stored in rechargeable batteries and then used for to drive a car after it is started. This is on-site utilization of wasted energy by a hybrid-car system. In both cases, the storage of electric energy is very important when using natural energy and wasted energy. To save energy, high-performance rechargeable batteries must be developed and employed in these applications.

Figure 3 shows the energy density of rechargeable batteries that are used in a variety of human activities. The lead–acid battery is one of the batteries that have been used in cars as starter electric power. The energy density of this battery is roughly 50 W h/kg, which is much lower than that of gasoline. We can use this battery to start an engine, but it cannot be used as power for driving the car. For natural energy backup applications, we may use a lead–acid battery, but the energy storage system may need to be huge. We need more space for battery modules, which is a problem. Therefore, we need rechargeable batteries with higher energy density

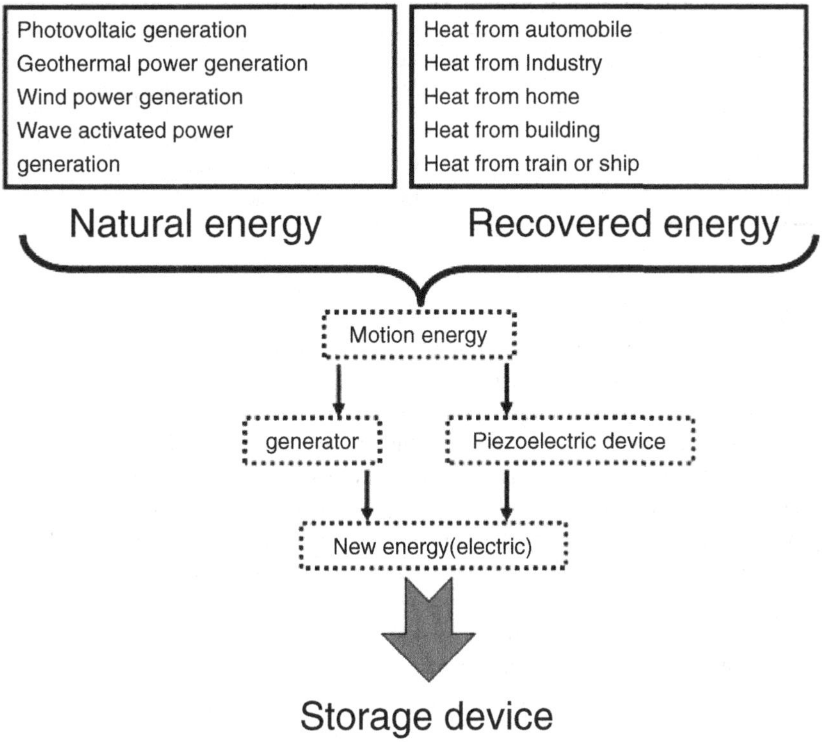

Photovoltaic generation
Geothermal power generation
Wind power generation
Wave activated power generation
Heat from automobile
Heat from Industry
Heat from home
Heat from building
Heat from train or ship
Natural energy
Recovered energy
Motion energy
generator
Piezoelectric device
New energy(electric)
Storage device

FIGURE 2 Energy conversion strategy.

Energy density (W h l-1)
Energy density (W h kg-1)
Smaller size
Lighter weight
LIB
Ni-MH
Ni-Cd
Lead Acid
0
100
200
300
400
500
600
700
0
50
100
150
200
250
300

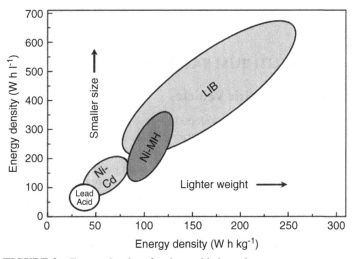

FIGURE 3 Energy density of rechargeable batteries.

TABLE 1   Energy and Power Density Required for Electric Vehicles

Type	Energy density (Wh/kg)	Power density (W/kg)	Technological problems
Plug-in hybrid car	200	2500	Improvement in current batteries
Hybrid car	70–100	2000	Current active materials
Electric vehicle	500–700	1000	New active materials

(Fig. 3). At present rechargeable lithium-ion batteries (LIBs) have the largest energy density among practical batteries: around 100 to 150 W h/kg.

Although this is a very attractive energy density value, it is not enough for electric cars and other energy storage applications.

The highest energy density value is that of gasoline, which is in theory, 12,000 Wh/kg. When gasoline is used in real devices, a tank and other elements are needed, corresponding to the cell case for a battery. In practice, the energy density of gasoline is estimated to be 1000 Wh/kg or less. If a rechargeable battery has an energy density of 500 to 800 Wh/kg, we can use a battery instead of gasoline.

Electric vehicles require batteries of varying energy densities. The energy densities for various applications are provided in Table 1. Commercialized lithium-ion batteries may satisfy the demands of plug-in hybrid and hybrid applications. However, a car cannot be driven for a long distance, such as a few hundred kilometers per one-time charge using a current lithium-ion battery.

For mobile use (especially for a smart phone), current lithium-ion batteries are not powerful enough, due to their higher energy consumption, and so require development of improved rechargeable batteries.

3   DEVELOPMENT OF LITHIUM BATTERIES

3.1   Lithium Batteries for Electric Vehicles

Three types of electric vehicles have been developed and commercialized: (1) hybrid electric vehicles (HEVs), (2) plug-in hybrid electric vehicles (PHEVs), and (3) pure electric vehicles (PEVs). The energy density for each type of vehicle is different due to their differing driving ranges and uses. For HEVs, the main power is gasoline, so that small batteries are enough, such as 100 to 150 Wh/kg. For PHEVs, a similar battery can be used, although more energy density is probably, needed to drive longer distance. Therefore, the energy density of batteries for PHEVs may be 150 to 200 Wh/kg. In contrast, PEVs need extremely high energy battery density. A driving range of PEVs may be 300 to 500 km per charge. PEVs can drive 1 km per 1 Wh/kg of energy density and so require rechargeable batteries valued at 300 to 500 Wh/kg. We cannot realize such high energy battery density simply by improving current rechargeable lithium-ion batteries.

3.2   Lithium Batteries for Mobile Applications

Smart phones have become ever more popular among consumers. They are not simply cellular phones, but resemble small computers and therefore consume a lot of electric energy during use. They require the development of batteries of higher energy density. Current rechargeable lithium-ion batteries consist of a graphite anode and a $LiCoO_2$ cathode. These materials limit the energy density of batteries so we have to develop new materials.

4   DEVELOPMENT OF MATERIALS FOR RECHARGEABLE LITHIUM BATTERIES

4.1   Safety

The safety issues for rechargeable lithium batteries are very important in their commercialization. From this point of view, the development of a new electrolyte system is required to suppress battery combustion. The electrolytes used in current rechargeable lithium-ion batteries are esters containing Li salt, such as $LiPF_6$. For example, a mixed solvent of ethylene carbonate and dimethyl carbonate has been used in practical cells. These solvents are basically volatile and flammable. New nonflammable electrolytes should be developed to solve safety issues related to rechargeable lithium-ion batteries. Some possible new electrolyte systems have been studied extensively by many research groups around the world: (1) ionic liquids, (2) polymer electrolytes, and (3) ceramic electrolytes. However, these new materials should be improved for use in practical batteries. Other approaches to achieving a high level of safety of rechargeable lithium-ion batteries are focused on additives that have a retardant function for organic electrolytes. There are many types of additives for rechargeable lithium-ion batteries.

Batteries are sometimes overcharged, resulting in explosions. This is due to an internal short circuit through some foreign materials or dendrite of lithium metal. Foreign materials can be removed by more precise control of the manufacturing process for lithium-ion batteries. On the other hand, the suppression of lithium dendrite formation is very critical. New technologies or new materials should be developed. The safety of rechargeable lithium-ion batteries can be enhanced by using more stable electrolytes or additives for flame retardance.

4.2   Lifetime

For batteries used in portable applications, the lifetime of batteries is less than five years, due to the lifetime of electronic devices. But for batteries used in electric vehicles and smart grid systems, lifetimes of at least 10 to 15 years have to be realized. The lifetime of batteries depends on degradation of the materials used in batteries. For example, the degradation of anode and cathode materials takes place during discharge and charge cycles. In other cases, the separator or binder is decomposed. Because the degradation of materials can be specified in interface and bulk problems, it has to be analyzed carefully.

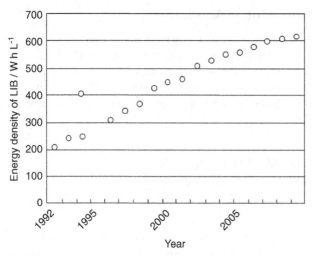

FIGURE 4 Developments in energy density of rechargeable lithium-ion batteries from 2000 to 2009. (From [1].)

4.3 High Energy Density

Today's rechargeable lithium-ion batteries have an energy density of 100 to 150 Wh/kg—not so low compared with those of conventional batteries such as lead–acid and nickel-metal hydride batteries. However, new applications, especially electric vehicle applications, require a higher energy density. When an $LiCoO_2$ cathode and a graphite anode are used in rechargeable lithium-ion batteries, the energy density cannot reach 250 Wh/kg. To realize a high energy density such as 300 Wh/kg, new cathode and anode materials have to be used. Figure 4 shows the development of the energy density of rechargeable lithium-ion batteries from 2000 to 2009 [1]. The increase in energy density is due to progress in battery construction technologies, not to material development. Therefore, within a few years the improvement in energy density will be saturated. High-capacity density materials should be utilized in practical batteries.

4.4 Cathode Materials

Transition metal oxides including Li, primarily $LiCoO_2$, have been used as a cathode material. Recently, foreign metal ions such as Mn and Ni have been substituted for Co. $LiNi_{0.8}Co_{0.2}O_2$ and $LiMn_{1/3}Ni_{1/3}Co_{1/3}O_2$ are promising cathode materials that can deliver higher capacity than that of $LiCoO_2$ [2–5]. Figure 5 shows discharge and charge curves for $LiNi_{0.8}Co_{0.15}Al_{0.05}O_2$ and $LiNi_{1/3}Mn_{1/3}Co_{1/3}O_2$. The capacities obtained are 190 and 150 mAh/g, respectively, which are larger than that of $LiCoO_2$. The charge potential of substituted layered transition metal oxide compounds are slightly lower than that of $LiCoO_2$, and their phase transitions during the extraction of Li^+ ion are different from those of $LiCoO_2$. These characteristics result in a larger discharge capacity for substituted layered transition metal oxide compounds.

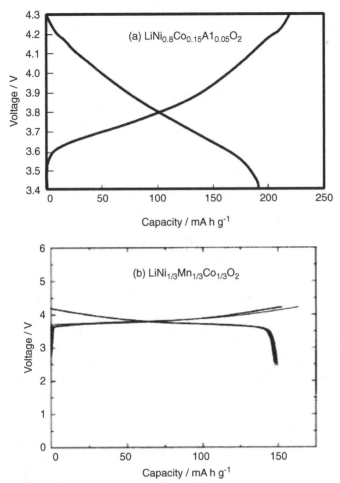

FIGURE 5   Charge and discharge curves of (a) $LiNi_{0.8}Co_{0.15}Al_{0.05}O_2$ and (b) $LiNi_{1/3}Mn_{1/3}Co_{1/3}O_2$. (From [2,4].)

To obtain a larger discharge capacity for cathode materials, more Li^+ ions should be utilized in the host matrix.

Recently, based on this strategy, Li_2MO_3, Li_2MSiO_4, and Li_2MPO_4F have been proposed [6–11]. These materials involve two Li^+ ions for one transition metal ion. When two electrons can be utilized, two Li^+ ions can be extracted or inserted during charge and discharge, respectively. In the case of Li_2MSiO_4 and Li_2MPO_4F, the synthesis of materials is so difficult but one or more Li^+ ions can be used, which corresponds to 200 mAh/g. On the other hand, Li_2MO_3 has been investigated as a high-capacity material. The theoretical capacity is larger than 300 mAh/g. Li_2MO_3 is not electrochemically active. To increase its electrochemical reactivity, a solid solution between Li_2MO_3 and $LiMO_2$ has been studied. Figure 6 shows the discharge and charge curves of $0.3Li_2MnO_3$–$0.7LiNi_{0.5}Mn_{0.5}O_2$ [12]. A reversible capacity

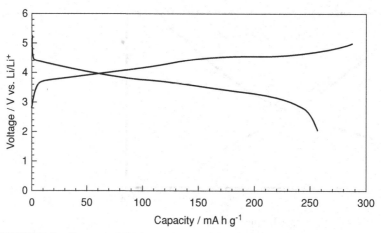

FIGURE 6 Charge and discharge curves of $0.3Li_2MnO_3-0.7LiNi_{0.5}Mn_{0.5}O_2$. (From [6].)

greater than 250 mAh/g was obtained successfully after the first charging process. At the first charge process, some changes in this cathode material took place, leading to the generation of an activated state for $0.3Li_2MnO_3-0.7LiNi_{0.5}Mn_{0.5}O_2$.

Another aspect of research on cathode materials is the development of safe cathode materials, such as $LiMn_2O_4$, $LiFePO_4$, and $LiMnPO_4$ [13–15]. $LiMn_2O_4$ and $LiFePO_4$ have already been used in practical batteries for electric vehicle and smart grid applications. The discharge capacities of $LiMn_2O_4$ and $LiFePO_4$ are 120 and 165 mAh/g, respectively. Of course, transition metal silicates and fluorophosphates are safe materials.

The discharge capacity of cathode materials required for mobile application is around 150 to 170 mAh/g; that for electric vehicles is greater than 250 mAh/g, and that for a smart grid (stationary application) is 100 to 200 mAh/g. Figure 7 shows the energy density and operation voltage and theoretical capacity density of active materials when using a lithium metal anode. The left circle indicates the current status of cathode materials. In the future, the materials existing in the right circle should be developed.

4.5 Anode Materials

The most popular anode material is graphite, which has a theoretical discharge capacity of 372 mAh/g. Another carbon material has also been developed based on the noncrystalline phase. Such noncrystalline material sometimes delivers a capacity of more than 372 mAh/g. However, the cyclability of noncrystalline carbon is not so good and the irreversible capacity is relatively larger. These problems must be solved before the material can be used in practical batteries. Silicon-based materials such as SiO and Si–O–C have been studied [16–20]. The problems with these materials are the same as those of noncrystalline carbon. Recently, Si–C–O-based anode material was reported. Figure 8 shows the discharge and charge curves of Si–C–O based anode material. This anode exhibited 600 to 800 mAh/g, and its cyclability is not too bad.

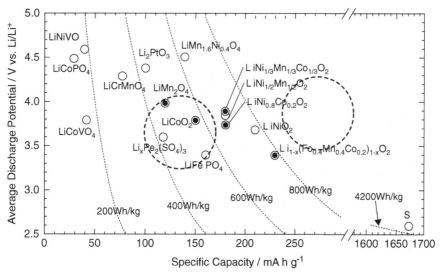

FIGURE 7   Energy density, operation voltage, and theoretical capacity density of active materials.

At the first discharge and charge cycle, an irreversible capacity was observed. To use this material in practical batteries, this irreversible capacity should be diminished.

Some metals can form an alloy with lithium metal. Si and Sn are well-known alloy materials with lithium metal. These materials are also promising candidates for an anode of high-energy-density batteries. An Si-alloy anode provides a capacity of more than 2000 mAh/g, and Sn delivers more than 600 mAh/g [21–25].

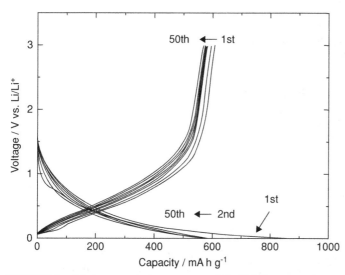

FIGURE 8   Discharge and charge curves of Si–C–O-based anode material. (From [17].)

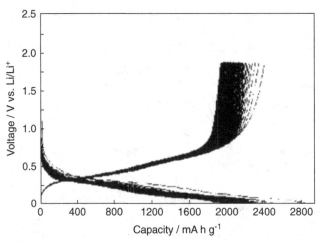

FIGURE 9 Discharge and charge curves of an Si anode. (From [21].)

Figure 9 shows the discharge and charge curves of an Si anode. The cyclability of an Si anode depends on the microstructure of an electrode to suppress the degradation of Si material. The degradation is caused by volume expansion during the lithium alloying process. The expansion of Si is 400%. The same problem was observed for other materials.

Figure 10 shows a unique Sn–Ni structure prepared by a template method combined with MEMS technology [26]. This type of microstructure is effective in suppressing the degradation of alloy materials. The highest capacity density is realized by a lithium-metal anode, whose theoretical capacity is 3861 mAh/g. In

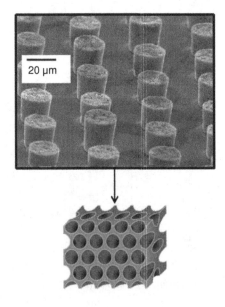

FIGURE 10 Unique structure of Sn–Ni prepared by a template method combined with MEMS technology. (From [26].)

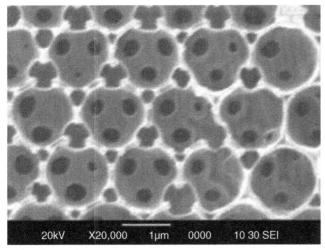

FIGURE 11   New separator with three dimensionally ordered macroporous structure. (From [27].)

the course of charging, lithium dendrite is formed, leading to a loss of capacity of lithium metal during discharge and charge cycles. Simultaneously, the surface area of lithium metal increases extremely with dendrite formation, resulting in the explosion of lithium batteries. Therefore, the suppression of lithium dendrite formation is necessary for production of lithium-metal batteries. Recently, a new separator has been developed and used in lithium-metal batteries. Three-dimensionally ordered macroporous (3DOM) structure was used in this new separator, as shown in Figure 11 [27]. Even after 3000 cycles, no lithium dendrite formation was confirmed using a symmetrical coin-type cell. This technology will be applied to lithium-metal batteries in the near future.

Various materials could be used as anodes to achieve higher energy density. However, only a few materials can be used as anodes for batteries used in electric vehicles designed to have a 300 to 500 km driving distance. These would include lithium metal and silicon, due to their high-capacity densities. In addition, the new type of cellular phone called a smart phone requires higher energy density. In this case, anode materials that have a capacity of 600 to 800 mAh/g are suitable. Sn alloy and Si–O–C anodes are attractive for this application. Figure 12 summarizes the of capacity density of various anode materials.

4.6   Electrolytes

Nonprotonic organic solvents containing lithium salt have been used in rechargeable lithium-ion batteries. Ethylene carbonate, diethyl carbonate, dimethyl carbonate, and ethylmethyl carbonate are popular nonprotonic organic solvents. Their ionic conductivities are around 10^{-2} to 10^{-3} S/cm. The transference number of Li^+ ions is about 0.3 to 0.4. To achieve high ionic conductivity, the viscosity of the electrolyte must

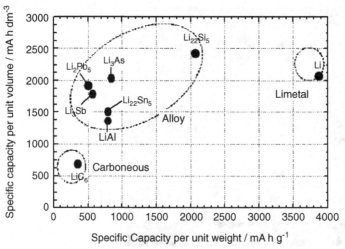

FIGURE 12 Summary of capacity density of anode materials.

be adjusted. Therefore, mixed solvents between ethylene carbonate and other esters have been used in practical cells. The organic electrolytes have wide electrochemical windows, leading to the high operational voltage of rechargeable lithium-ion batteries. On the other hand, organic electrolytes are usually flammable and volatile. These properties are not suitable, for reasons of battery safety. In fact, safety becomes more and more of an issue with the increasing energy density of rechargeable lithium-ion batteries.

As noted in Section 4.1, three types of new electrolytes have been developed to realize greater safety in rechargeable lithium-ion batteries.

1. *Ionic liquid:* an electrolyte with no vapor pressure and a reasonable ionic conductivity.

2. *Polymer electrolyte:* a solid but flexible electrolyte with a low ionic conductivity and a low Li^+ ion transference number.

3. *Ceramic electrolyte:* a solid electrolyte with reasonable ionic conductivity and an Li^+ ion transference number of 1.

These electrolytes are attractive in terms of the safety of rechargeable lithium batteries. However, each has problems. In the case of solid electrolytes, constructing a cell is relatively difficult.

Figure 13 shows the chemical composition of typical ionic liquids. This electrolyte consists of anion and cation only. The melting point of an ionic salt is lower than room temperature, and the chemistry and electrochemical properties of ionic liquids are not well understood. Ionic liquids have several disadvantages: (1) the cost of materials; (2) the low ionic conductivity compared with those of organic electrolytes; and (3) is a low transference number of Li^+ ions.

Figure 14 shows the nanostructure of a polymer electrolyte with a nanophase separation of polystyrene and polyether [28]. This type of phase-separated polymer

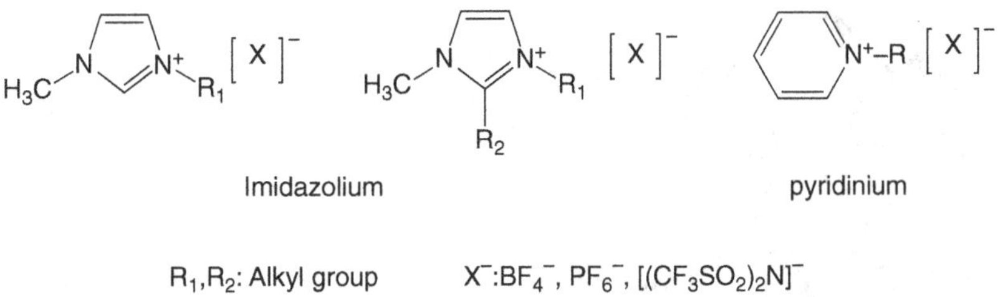

Imidazolium   pyridinium

R$_1$,R$_2$: Alkyl group   X$^-$:BF$_4^-$, PF$_6^-$, [(CF$_3$SO$_2$)$_2$N]$^-$

FIGURE 13   Chemical composition of typical ionic liquids.

has been developed by some groups. The polymer shown has 10^{-4} S/cm at room temperature. The mechanical strength of this polymer is sustained by a polystyrene block.

Figure 15 shows the crystal structure of Li$_{0.35}$La$_{0.55}$TiO$_3$, which has Li$^+$ ion conductivity [29]. The ionic conductivity is close to 10^{-3} S/cm, and the transference number of Li$^+$ ion is 1. The ceramics are so rigid that battery fabrication is very difficult, due to the poor contact between active materials and solid electrolytes.

A few different types of ceramic solid electrolytes have been studied, such as sulfide, oxide, and phosphate. In the case of sulfide solid electrolytes, batteries can be fabricated using a powder pressing method, because the soft nature of sulfide materials provides a good contact between active materials and solid electrolytes [30]. In fact, good discharge and charge behavior have been confirmed using a sulfide solid electrolyte, as shown in Figure 16.

On the other hand, the fabrication of all-solid-state batteries using oxide materials is not so easy, due to their hard nature. Some new design of all-solid-state batteries

Nano-phase separated polymer electrolyte

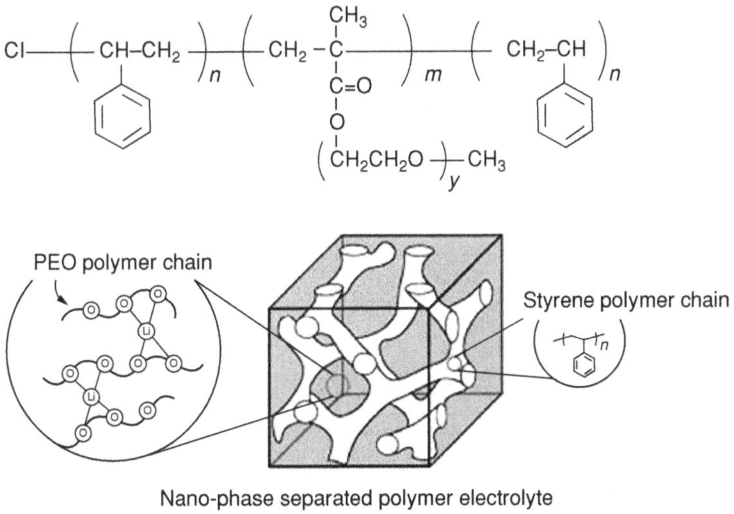

FIGURE 14   Nanostructure of a polymer electrolyte with a nanophase separation of polystyrene and polyether. (From [28].)

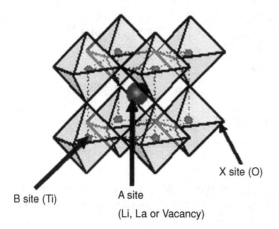

B site (Ti)   A site
                  (Li, La or Vacancy)

X site (O)

FIGURE 15 Crystal structure of $Li_{0.35}La_{0.55}TiO_3$.

should be developed. For example, a porous matrix of a solid electrolyte, the 3DOM structure cited in Section 4.5, has been studied to prepare all-solid-state batteries, as shown in Figure 17. When macropores are occupied by active materials such as $LiCoO_2$ or $Li_4Ti_5O_{12}$, an electrochemical system consisting of a solid electrolyte and an active material is established. This is one method of producing a large and uniform interfacial area between an electrolyte and an active material. In this way, the good contact between an electrolyte and an active material is a key factor in fabricating all-solid-state batteries.

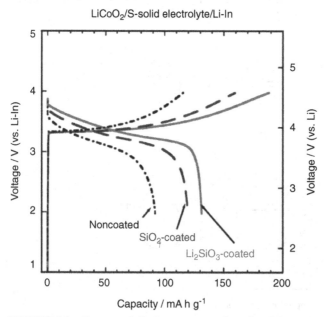

$LiCoO_2$/S-solid electrolyte/Li-In

Noncoated

SiO_2-coated

Li_2SiO_3-coated

FIGURE 16 Charge and discharge curves of an all-solid-state battery using a sulfide electrolyte. (From [30].)

Three Dimensionally Ordered
Macroporous Soild Electrolyte

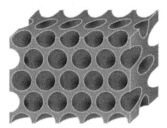 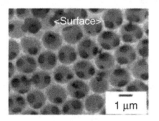

FIGURE 17   Porous matrix of a solid electrolyte for all-solid-state batteries. (From [31].)

5   PRODUCTION OF ELECTRODES FOR LITHIUM BATTERIES

5.1   Energy and Power Density

Figure 18 shows an electrochemical system consisting of powder of an active material and separator with a liquid electrolyte. The anode and cathode are both prepared using a coating process of ink involving active material powder, conductive material, and binding material. Figure 19 shows the microstructure of a porous electrode prepared using this coating process. The electrochemical kinetics is limited by the diffusion of Li^+ ions in this porous electrode. Therefore, the power density of an electrode

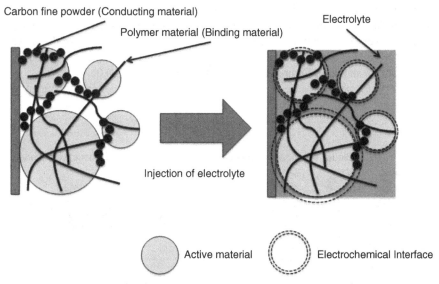

FIGURE 18   Electrochemical system consisting of active, conducting, and binding materials with a liquid electrolyte.

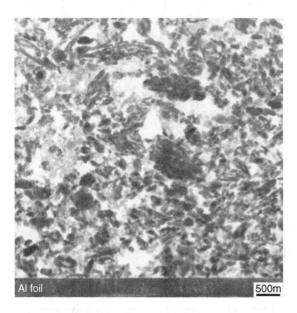

FIGURE 19 SEM image of the microstructure of a porous electrode prepared by a coating process.

(in other words, the rate capability) decreases with increasing electrode thickness. On the other hand, a thin electrode has a high rate capability, but a more powerful current collector and separator are required to obtain the same discharge capacity, leading to lower energy density.

Thus, the energy and power densities of batteries are always a trade-off. Therefore, the microstructure of electrodes, which depends on the particle nature of the active material, the binding material, and the preparation method used for a composite electrode, is very important. For example, the energy density of the electrode shown in Figure 20a is higher than that in Figure 20b. On the other hand, the power density of the electrode shown in Figure 20a is lower than that in Figure 20b.

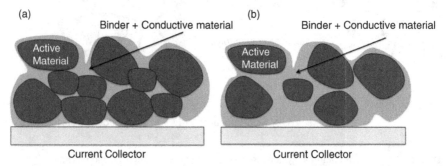

FIGURE 20 Microstructures of (a) dense and (b) sparse porous electrodes.

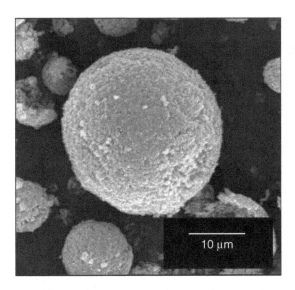

FIGURE 21   SEM image of LiFePO$_4$ secondary particles.

5.2 Particle Nature

The porous structure of a composite electrode depends on the size and shape of the active material particles. Nanosized particles are not suitable for the preparation of a composite electrode but microsized particles are appropriate. If nanoparticles have to be used as active materials, the formation of secondary particles is more important. By the way, the tap density of an electrode is sometimes determined by the particle shape, a spherical particle shape usually being preferable. This is very important in the realization of a higher electrode energy density. A higher energy density rate in an electrode results in a low rate capability. Particle size and shape both influence the process of coating a current collector with a composite consisting of active material, binding material, and conductive material. For example, LiFePO$_4$ cathode material should be nanosized; otherwise, it does not work well. In this case, the formation of secondary particles is a key factor in developing a high-performance electrode. Figure 21 shows an example of LiFePO$_4$ secondary particles.

5.3 Composite Electrodes

To produce a high-performance electrode, preparation of the slurry of active materials should be controlled to obtain low viscosity and high uniformity, so that material mixing is optimized before coating beings. The properties of the slurry depend on the active material, conducting material, and binding material that are used. Strong adhesion of an active material layer to a current collector is determined by the physical and chemical states of the slurry.

The drying process influences the microstructure of a composite electrode. In some cases, crack formation and peeling of an electrode layer from a current collector are observed, as shown in Figure 22. Therefore, the conditions for a drying process should be optimized very carefully. The preparation of slurry has been conducted

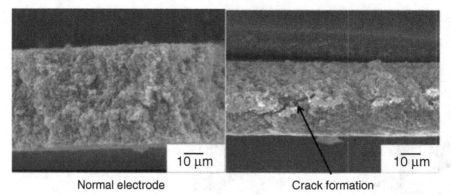

Normal electrode   Crack formation

FIGURE 22 Crack formation in an electrode.

by using water or NMP (*N*-methylpyrrolidone) solvents. For environmental reasons, water is more preferable than NMP organic solvent. SBR with carboxymethylcellulose has been utilized to both positive and negative electrodes. In composite electrodes the binding materials contact active materials directly, which influences the electrochemical kinetics of the active materials. Recently, some functional binders have been developed and used in rechargeable lithium-ion batteries to lower electrode polarization.

5.4 Current Collectors

Al and Cu foils have been used as current collectors for positive and negative electrodes, respectively. The surfaces of both current collectors are covered with oxides. The aluminum oxide layer on an Al current collector is stable when using $LiPF_6$ electrolyte salt. However, when the electrolyte contains imide salt, the surface film on the Al current collector becomes unstable, leading to corrosion of the collector. The selection of electrolyte salt is important in keeping the Al current collector stable. The surface of a Cu current collector is not stable when the electrolyte contains a small amount of HF, due to an acid–base reaction between the Cu oxide on a Cu current collector and HF impurity. Therefore, the concentration of HF impurity should be lessened. The weight of a Cu current collector is greater than that of an Al current collector, so that the energy density of rechargeable lithium-ion batteries can be increased by using a thin Cu current collector.

6 SUMMARY

Many types of materials are used in rechargeable lithium-ion batteries. To realize the high performance of these batteries, each material should be optimized. In addition, the process of preparation of a composite electrode has to be controlled. The electrochemical performance of rechargeable lithium-ion batteries does not depend only on the properties of materials, but also on the battery production process. Research on both battery materials and processes must be continued to realize batteries of high energy and high power density.

REFERENCES

1. Meeting information, Japan Chemical Innovation and Inspection Institute, unpublished.
2. M. Guilmard, C. Pouillerie, L. Croguennec, and C. Delmas, *Solid State Ionics*, **160**, 39–50 (2003).
3. C. Delmas and I. Saadoune, *Solid State Ionics*, **53–56**, 370–375 (1992).
4. T. Ohzuku and Y. Makimura, *Chem. Lett.*, **1**, 642 (2001).
5. T. Ohzuku, K. Ariyoshi, Y. Makimura, N, Yabuuchi, and K. Sawai, *Electrochemistry*, **73**, 2–11 (2005).
6. C. S. Johnson, J.-S. Kim, C. Lefief, N. Li, J. T. Vaughey, and M. M. Thackeray, *Electrochem. Commun.*, **6**, 1085 (2004).
7. M. E. Arroyo-de Dompablo, J. M. Gallardo-Amores, J. García-Martínez, E. Morán, J.-M. Tarascon, and M. Armand, *Solid State Ionics*, **179**, 1758–1762 (2008).
8. R. Dominko, M. Bele, M. Gaberšček, A. Meden, M. Remškar, and J. Jamnik, *Electrochem. Commun.*, **8**, 217–222 (2006).
9. T. N. Ramesh, Kyu Tae Lee, B. L. Ellis, and L. F. Nazar, *Electrochem. Solid-State Lett.*, **13**, A43–A47 (2010).
10. N. R. Khasanovaa, A. N. Gavrilova, E. V. Antipova, K. G. Bramnikb, and H. Hibstb, *J. Power Sources*, **196**, 355–360 (2011).
11. D. Wanga, J. Xiaoa, W. X., Z. Niea, C. Wangb, G. Graffa, and J.-G. Zhanga, *J. Power Sources*, **196**, 2241–2245 (2011).
12. S.-H. Kang, P. Kempgens, S. Greenbaum, A. J. Kropf, K. Amine, and M. M. Thackeray, *J. Mater. Chem.*, **17**, 2069–2077 (2007).
13. R. J. Gummow, A. de Kock, and M. M. Thackeray, *Solid State Ionics*, **69**, 59–67 (1994).
14. A. K. Padhi, K. S. Nanjundaswamy, and J. B. Goodenough, *J. Electrochem. Soc.*, **144**, 1188 (1997).
15. M. Yonemura, A. Yamada, Y. Takei, N. Sonoyama, and R. Kanno, *J. Electrochem. Soc.*, **151**, A1352 (2004).
16. T. Morita and N. Takami, *J. Electrochem. Soc.*, **153**, A425 (2006).
17. H. Fukui, H. Ohsuka, T. Hino, and K. Kanamura, *J. Power Sources*, **196**, 371–378 (2011).
18. W. Xing, A. M. Wilson, G. Zank, and J. R. Dahn, *Solid State Ionics*, **93**, 239 (1997).
19. A. M. Wilson, G. Zank, K. Eguchi, W. Xing, and J. R. Dahn, *J. Power Sources*, **68**, 195 (1997).
20. A. M. Wilson, W. Xing, G. Zank, B. Yates, and J. R. Dahn, *Solid State Ionics* **100**, 259 (1997).
21. T. Takamura, S. Ohara, M. Uehara, J. Suzuki, and K. Sekine, *J. Power Sources*, **129**, 96–100 (2004).
22. N. Tamura, M. Fujimoto, M. Kamino, and S. Fujitani, *Electrochim. Acta*, **49**, 1949 (2004).
23. H. Mukaibo, T. Momma, M. Mohamedi, and T. Osaka, *J. Electrochem. Soc.*, **152**, A560 (2005).
24. J. Hassoun, S. Panero, and B. Scrosati, *J. Power Sources*, **160**, 1336 (2006).
25. M. M. Thackeray, J. T. Vaughey, A. J. Kahaian, K. D. Kepler, and R. Benedek, *Electrochem. Commun.*, **1**, 111 (1999).
26. S.-W. Woo, N. Okada, M. Kotobuki, K. Sasajima, H. Munakata, K. Kajihara, and K. Kanamura, *Electrochim. Acta*, **55**, 8030–8035 (2010).
27. D. Yamamoto, H. Munakata, and K. Kanamura, *J. Electrochem. Soc.*, **155**, B303–B308 (2008).
28. T. Niitani, M. Shimada, K. Kawamura, and K. Kanamura, *J. Power Sources*, **146**, 386 (2005).
29. Y. Inaguma, L. Q. Chen, M. Itoh, T. Nakamura, T. Uchida, H. Ikuta, and M. Wakihara, *Solid State Commun.*, **86**, 689 (1993).
30. A. Sakuda, H. Kitaura, A. Hayashi, K. Tadanaga, and M. Tatsumisago, *J. Power Sources*, **189**, 527–530 (2009).
31. K. Kanamura, N. Akutagawa, and K. Dokko, *J. Power Sources*, **146**, 86–89 (2005).

CHAPTER 7

FLUORINE-BASED POLYANIONIC COMPOUNDS FOR HIGH-VOLTAGE ELECTRODE MATERIALS

P. Barpanda and J.-M. Tarascon

1 INTRODUCTION

Energy production and storage together with global warming associated with CO_2 emissions are popular topics in today's energy-conscious society. To cope with the foreseen population increase coupled with human beings' lifestyle improvement, we must double our present rate of 14 TW energy production by 2050 while preserving our CO_2 content at today's level, bearing in mind that fossil fuels are finite [1]. The

Lithium Batteries: Advanced Technologies and Applications, First Edition.
Edited by Bruno Scrosati, K. M. Abraham, Walter van Schalkwijk, and Jusef Hassoun.
© 2013 John Wiley & Sons, Inc. Published 2013 by John Wiley & Sons, Inc.

need to produce an additional 14 TW of CO_2-free energy conspires to make advanced renewable energy technology a worldwide imperative. Exploiting inherently intermittent renewable energies sources (e.g., wind, solar, biomass, sea-wave, geothermal), and switching from thermal to electric transportation to limit environmental pollution, which requires energy on board, conspire to make energy storage one of the greatest technological challenges of the twenty-first century.

In response to the needs of modern society and emerging ecological concerns, it is now mandatory that new and environmentally friendly energy storage sources be found: hence, the fast-developing research in that field involving, among others, primary and rechargeable batteries, supercapacitors, and so on. For more than three decades, intensive efforts have yielded several generations of highly successful batteries starting with primary lithium batteries, followed by rechargeable batteries, and finally the development of lithium-ion batteries. Lithium-ion batteries are now essential to our society, as they power a wide spectrum of devices, ranging from portable electronic devices to electric automobiles, and stand as a serious contender for grid application [2–4]. However, we are currently facing significant limitations in their storage capacity and power. To address this issue, present approaches are divided between (1) exploring other attractive technologies, such as Li–air, able to provide a three- to fourfold increase in energy density compared to lithium-ion [5], on the condition that we can master the complex Li–air chemistry, and (2) improving the present lithium-ion technology via material innovation, providing greater energy density while maintaining low cost and increasing safety.

The 1980s and 1990s recorded many successful discoveries and the commercialization of oxide-based cathode materials such as the layered compounds $LiCoO_2$, $LiNiO_2$, spinel $LiMn_2O_4$, and mixed oxides (e.g., $LiMn_{1.5}Ni_{0.5}O_4$, $LiMn_{0.33}Co_{0.33}Ni_{0.33}O_2$, and $LiNi_{0.8}Co_{0.15}Al_{0.05}O_2$) [6–10]. Among the massive research carried out on oxide insertion compounds, in 1997, Padhi et al. introduced the concept of polyanionic compounds as an alternative class of cathodes, $LiFePO_4$ being the first example [11,12]. In the lithium-battery field, the inception of olivine $LiFePO_4$ as a competent 3.43-V Fe-based cathode (theoretical capacity of 170 mAh/g) was a major milestone. Although initially criticized for its low conductivity, the application of carbon coating and particle downsizing enabled $LiFePO_4$ to stand as a front-runner cathode for large-scale applications [13–17]. The success story of $LiFePO_4$ acted as a seed leading to the discovery of a myriad of polyanionic cathode insertion compounds in the last 15 years (1997 to 2011). Among them are Li_2MSiO_4 (silicates) [18], $LiMBO_3$ (borates) [19], and $Li_{2-y}MP_2O_7$ (pyrophosphates) [20], to name only a few. All of them benefit from the dual nanosizing/coating approach to reach optimum performances.

All these oxide and polyanionic compounds are based on the insertion reaction. In addition to metal oxides, a parallel story runs for metal fluorides. Based on conversion reactions [21], the fluoride cathodes outperform oxides in terms of energy density, owing to their higher average potential. This improved potential stems from the high ionic character of the Me–F bond, owing to the electronegativity of fluorine species. Exploiting this feature, various metal fluorides have been investigated, beginning in the 1960s [22,23]. A question arises: Is it possible to produce fluorine-based polyanionic cathode materials capable of high energy density (along with high potential)? Indeed it is. Similar to oxide-based polyanionic cathodes, many fluorine-based polyanionic compounds can be designed; the first example, $LiVPO_4F$, was demonstrated by Barker et al. [24,25]. The success of PO_4-based cathodes and the quest to increase their energy density by implementing highly electronegative fluorine species, led to the "fluorophosphate" class of insertion materials. Soon after, based on the inductive descriptor proposed by Padhi et al. [12], a novel group of "fluoro-sulfate" cathode materials was unveiled by Tarascon et al., among them $LiFeSO_4F$, which turned out to be the most attractive with respect to applications [26].

In this chapter we summarize the discovery and development of high-potential electrode materials based on fluorine-based polyanionic compounds. First, a brief history of the role of fluorine in batteries is captured, enlisting metal fluoride conversion reaction cathodes, nonaqueous electrolytes, and additives. Then the syntheses, crystal structure, and electrochemical performance of the recently discovered fluorine-based high-voltage polyanionic cathode compounds are presented and discussed. For the sake of clarity, the fluorophosphates group of compounds is described first, followed by the fluorosulfate group of compounds. Finally, perspectives linked to fluorine-based electrode materials in terms of future discoveries or applications, as perceived personally, are given.

2   BRIEF HISTORY OF FLUORINE-BASED CATHODE MATERIALS

Fluorine-based materials are integral parts of the lithium-ion battery technology, spanning from electrodes to electrolytes. Since the early 1990s, fluorine-based electrolytes (e.g., $LiPF_6$, $LiBF_4$) and binders [e.g., Poly(vinylidene fluoride)] have been used to realize stable electrochemical operations at high potential. The versatility of fluorine-based compounds has contributed positively to the development of lithium-ion batteries. While fluorinated salt-based electrolytes contribute to wide electrochemical potential window, and therefore to safe operation, the fluorinated binders facilitate the mechanical stability of electrode composites during battery operation [27,28].

Returning to electrodes, fluorine has been used as a dopant to produce various fluorine-doped intercalation cathodes. A few of these compounds are layered transition metal oxyfluorides ($Li_{1+x}Ni_{1-x}O_{2-y}F_y$) [29], spinel lithium manganese oxyfluorides ($Li_{1+x}Mn_{2-x}O_{4-y}F_y$) [30], and orthorhombic lithium manganese oxyfluorides ($Li_{1.07}Mn_{0.93}O_{1.92}F_{0.08}$) [31]. In all cases, fluorine substitution led to much improved cycling stability, rate capability, and thermal stability. Further, the transport properties and consequently the electrochemical performance of many oxide cathodes can be improved by surface fluorination (e.g., AlF_3-coated $LiCoO_2$), as the transport of both lithium-ions and electrons occurs at the surface of the cathode [32]. Another class of fluorinated cathodes is that of the carbon fluorides (CF_x), based on the conversion–displacement reaction [33]. A similar group of compounds, the metal fluorides (MF_x), has been investigated extensively as high-capacity (>700 mAh/g) cathode materials capable of multiple electron conversion reactions [23,34,35]. These metal fluoride cathodes are made possible by embedding nanocrystalline metal fluoride particles (10 to 20 nm) inside a conductive matrix. The resulting nanocomposites circumvent

TABLE 1   Redox Potential vs. Li and Electrochemical Capacity of Several Metal Fluoride Compounds Based on Conversion Reaction

Compound	Redox potential (V)	Capacity (mAh/g)	Energy density (Wh/kg)
MnF_2	1.92	577	1108
FeF_2	2.66	571	1519
CoF_2	2.854	553	1578
NiF_2	2.96	554	1640
CuF_2	3.55	528	1874
VF_3	1.86	745	1386
CrF_3	2.28	738	1683
MnF_3	2.65	719	1905
FeF_3	2.74	712	1951
BiF_3	3.13	302	945

the poor lithium-ion transport, thus realizing their full capacity [34]. Later, several metal oxyfluoride (e.g., BiOF, FeOF) cathodes were reported [36]. However, all these compounds suffer from low redox potentials. A few conversion reaction-based metal fluorides are listed in Table 1. Due to space restrictions, most of the compounds noted above will not be discussed further. In the rest of this chapter we dwell solely on fluorine-based polyanionic electrode materials.

At this stage it should be stressed that fluorine is not an easy element to quantify in inorganic compounds from either a structural point of view (as neutrons are helpless) or quantitatively (as the analytical methods reported are all very inaccurate). That's why a few results dealing with coatings or fluorine-content in mixed F/OH-based samples are subject to a lot of controversy.

In 2002–2003, Barker's group introduced a new class of fluorine-based polyanionic compounds now known as fluorophosphates, the first candidate being the tavorite structured compound $LiVPO_4F$, followed by $NaVPO_4F$ [24,37,38]. Since then, extensive research work has been performed by Barker's group on $LiVPO_4F$ as a 4-V cathode material [39–42]. Recently, taking advantage of both $V^{2+/3+}$ and $V^{3+/4+}$ redox centers, symmetric (−) $LiVPO_4F$ ‖ $LiVPO_4F$ (+) cells have been configured, capable of delivering sustained reversible capacities of about 130 mAh/g at about 2 V [43,44]. The next candidate, orthorhombic Li_2CoPO_4F, was introduced by Okada et al. in 2005 in an effort to produce a high-voltage ($\sim$5 V) Co-based cathode [45]. Then in 2007, Nazar's group discovered orthorhombic Na_2FePO_4F as a multifunctional cathode with redox potentials located at 3 and 3.5 V, the first ever electrochemically feasible Fe-based fluorophosphate compound [46]. Studying this compound and its chemical derivatives further, the Li_2FePO_4F phase was introduced in 2010 [47]. In the same year, our group worked on a lithium metal fluorophosphate system equipped with a novel low-temperature ionothermal synthesis method and reported two new tavorite structured compounds, $LiFePO_4F$ and $LiTiPO_4F$, having redox potentials of 2.8 and 2.5 V, respectively [48]. In an independent pursuit, tavorite $LiFePO_4F$ was also later synthesized by Nazar's group via a two-step solid-state synthesis [49]. The latest fluorophosphate compound to join the bandwagon is orthorhombic Li_2NiPO_4F, which shows the highest redox potential, at 5.3 V [50].

Let's now turn our attention to the fluorosulfates, which made a rather late entry into the battery arena. The only known phase was the electrochemically inactive $LiMgSO_4F$ first reported by Sebastian et al. [51]. In their seminal paper the authors stated that "Li-ion conduction in $LiMgSO_4F$ suggests that isostructural metal analogues $LiMSO_4F$ (M = Mn, Fe, Co) would be important for redox extraction/insertion of lithium involving M^{II}/M^{III} oxidation states." In light of such a remark, our group shifted its attention to fluorosulfates in an effort to make high-voltage cathode materials by exploiting the higher electronegativity of sulfur and fluorine. Using a mild heat treatment and nonaqueous reaction media, we could synthesize the tavorite-type $LiFeSO_4F$ phase, the first-ever fluorosulfate cathode materials capable of delivering about 140 mAh/g with a high redox potential of 3.6 V [26]. Soon after, using the same synthesis protocol, a range of Li- and Na-based metal fluorosulfate compounds [$AMSO_4F$; A = Li, Na; M = Mg, Mn, Fe, Co, Ni, Zn, Cu] were synthesized to the point that in less than two years (2010–2011), 15 novel fluorosulfate compounds were discovered [52–56]. They were shown to exhibit both a rich crystal chemistry and electrochemistry, depending on the nature of the alkali and metal ions and on the synthesis process. For example, triplite derivative fluorosulfates $LiFe_{1-x}Mn_xSO_4F$ ($0 < x < 1$) were shown to deliver reversible capacities of about 130 mAh/g at a potential of 3.9 V, which is the highest potential ever reported for an $Fe^{2+/3+}$ redox couple [57,58].

Next, we focus on the synthesis, crystal structure, and electrochemical properties of these fluorine-based polyanionic compounds. To maintain the clarity and lineage of compounds, fluorophosphate compounds are treated first, followed by fluorosulfate polyanionic compounds.

3   ALKALI METAL FLUOROPHOSPHATES

Layered oxide $LiCoO_2$ material has pioneered the evolution and commercial success of rechargeable lithium-ion batteries, due to its low molecular weight and high operating voltage (4 V). Therefore, its attractive theoretical capacity of 274 mAh/g could not be reached, due to the structure collapse associated with a significant repulsion of the negatively charged CoO_2 layers when more than 50% of the Li is removed. In contrast, polyanionic compounds having interconnected polyanionic units [e.g., $(XO_4)^{n-}$] and transition metal polyhedra [e.g., $(MO_x)^{n-}$] creating robust structural frameworks delimiting lithium-ion migration channels, such as $LiFePO_4$, do not suffer from structural collapse, but show less theoretical capacity, due to the heavier polyanionic $(XO_4)^{n-}$ groups. Nevertheless, $AMXO_4$-type compounds are widely studied as cathode materials for their structural diversity, chemical and thermal stability, and easy tunability of the $M^{n+/(n+1)+}$ redox potential, through inductive effect, as proposed originally by Padhi et al. [12]. In polyanionic compounds, the relative electronegativity of X in $(XO_4)^{n-}$ units alters the ionicity of M–O bonds, which in turn modifies the redox potential. A careful selection of M and X can deliver cathode compounds with high redox potential. Additionally, a higher redox potential can be realized by the implementation of fluoride-based electrodes, owing to the larger ionicity of M–F bonds vis-à-vis M–O bonds. Coupling the rigidity of polyanionic frameworks with the electronegativity of fluorine, chemists have attempted to synthesize many fluorine-based polyanionic cathode materials. Table 2 enlists the structural

TABLE 2 Crystallographic Lattice Parameters and Cathode Properties of Li- and Na-Based Fluorophosphate Compounds that Have Potential Applications in Rechargeable Batteries

Compound	Crystal structure	Space group	a (Å)	b (Å)	c (Å)	α(deg)	β(deg)	γ(deg)	Volume (Å³)	Redox potential (V)	Capacity (mAh/g)	Ref.
					Lithium Metal Fluorophosphates ($Li_y T_m PO_4 F$)							
LiVPO$_4$F	Triclinic	P-1	5.1730(8)	5.3090(6)	7.2500(3)	72.479(4)	107.767(7)	81.375(7)	174.35(0)	4.20	115	24
LiVPO$_4$F	Triclinic	P-1	5.276(5)	5.361(3)	7.388(6)	74.42(6)	70.70(7)	98.87(8)	183.06(6)	4.20	110	25
LiVPO$_4$F	Triclinic	P-1	5.3094(1)	7.4993(6)	5.1688(8)	112.933(0)	81.664(0)	113.125(0)	174.30(6)	4.25/1.8	288	43
LiTiPO$_4$F	Triclinic	P-1	5.1991(2)	5.3139(2)	7.2428(3)	106.975(3)	108.262(4)	97.655(4)	176.10(2)	2.90/1.70	150	48
LiFePO$_4$F	Triclinic	P-1	5.1551(3)	5.3044(3)	7.2612(4)	107.357(5)	107.855(6)	98.618(5)	173.91(2)	2.80	130	48
LiFePO$_4$F	Triclinic	P-1	5.30022(2)	7.2601(2)	5.1516(2)	107.880(3)	98.559(3)	107.343(3)	173.67(6)	2.80	145	49
LiFePO$_4$F	Orthorhombic	Pbcn	5.0550(2)	13.5610(2)	11.0520(3)	90.000	90.000	90.000	757.62(1)	—	—	59
Li$_2$VPO$_4$F	Monoclinic	C2/c	7.2255(1)	7.9450(1)	7.3075(1)	90.000	116.771(1)	90.000	374.53(7)	4.25/1.8	288	43
Li$_2$FePO$_4$F	Triclinic	P-1	5.3746(3)	7.4437(3)	5.3256(4)	109.038(2)	94.423(6)	108.259(9)	189.03(4)	2.80	145	49
Li$_2$FePO$_4$F	Orthorhombic	Pbcn	5.0550(2)	13.5610(2)	11.0520(3)	90.000	90.000	90.000	757.62(1)	—	—	46
Li$_2$CoPO$_4$F	Orthorhombic	Pnma	10.4520(2)	6.3911(8)	10.8740(2)	90.000	90.000	90.000	726.40(3)	5.00	—	45
Li$_2$NiPO$_4$F	Orthorhombic	Pnma	10.4730(3)	6.2887(8)	10.8460(1)	90.000	90.000	90.000	714.33(2)	—	—	60
Li$_2$NiPO$_4$F	Orthorhombic	Pnma	10.4730(2)	6.2890(8)	10.8460(2)	90.000	90.000	90.000	714.37(2)	5.30	—	61
VPO$_4$F	Monochnic	C2/c	7.1553(2)	7.1014(1)	7.1160(2)	90.000	118.089(1)	90.000	319.00(8)	4.25/1.8	288	43
					Sodium Metal Fluorophosphates ($Na_y T_m PO_4 F$)							
NaVPO$_4$F	Tetragonal	I4/mmm	6.387(2)	6.387(2)	10.734(3)	90.000	90.000	90.000	437.88(1)	3.88/4.43	110	38
NaVPO$_4$F	Monoclinic	C2/c	12.730	6.380	8.950	90.000	102.31	90.000	710.18(3)	3.40/4.20	87.7	62
NaFePO$_4$F	Orthorhombic	Pbcn	5.1047(2)	14.1326(6)	11.3655(5)	90.000	90.000	90.000	820.00(1)	—	—	46
NaFePO$_4$F	Orthorhombic	Pbcn	5.1018(7)	14.1224(6)	11.3600(1)	90.000	90.000	90.000	818.48(0)	—	—	47
Na$_{1.25}$FePO$_4$F	Orthorhombic	Pbcn	5.2040	13.9100	11.4780	90.000	90.000	90.000	830.87	—	—	46
Na$_{1.5}$FePO$_4$F	Monoclinic	P2/c	13.9296(2)	5.2009(7)	11.5147(5)	90.000	91.220	90.000	834.03(1)	—	—	46
NaLiFePO$_4$F	Orthorhombic	Pbcn	5.0750	13.6600	11.1350	90.000	90.000	90.000	771.92	—	—	46
Na$_{1.75}$FePO$_4$F	Orthorhombic	Pbcn	5.2190	13.8930	11.6020	90.000	90.000	90.000	841.23	—	—	47
Na$_2$FePO$_4$F	Orthorhombic	Pbcn	5.2200(2)	13.8540(6)	11.7792(5)	90.000	90.000	90.000	851.85(1)	3.00/3.50	135	46
Na$_2$CoPO$_4$F	Orthorhombic	Pbcn	5.2475(9)	13.795(2)	11.689(2)	90.000	90.000	90.000	846.2(3)	4.80	—	47
Na$_2$NiPO$_4$F	Orthorhombic	Pbcn	—	—	—	90.000	90.000	90.000	823.4	—	—	47
Na$_2$MnPO$_4$F	Monoclinic	P2$_1$/n	13.694(14)	5.3054(4)	13.708(17)	90.000	119.727(6)	90.000	864.87(16)	—	—	63

and cathode properties of the existing fluorophosphate compounds. These fluorinated polyanionic compounds are described below, focusing on their crystal structure and electrochemical properties.

3.1 $Li_{1\pm x}VPO_4F$

The era of fluorophosphates began with the introduction of $LiVPO_4F$ as a positive insertion material by Barker et al. in 2003 [24]. $LiVPO_4F$ was synthesized by a two-step carbothermal reduction method using particulate carbon as a reducing agent [64]. First, a stoichiometric mixture of V_2O_5, $NH_4H_2PO_4$, and carbon was annealed at 750°C for 4 h to obtain VPO_4 as an intermediate product, which was afterward reacted with equimolar LiF at 750°C for 0.25 to 1 h (in Ar flow) to produce single-phase $LiVPO_4F$. The latter reaction between VPO_4 and LiF could also be completed by a hydrothermal reaction involving autogenous pressure (~1800 psi) created at 250°C for 48 h [25]. A single-step carbothermal reaction using all precursors (LiF + V_2O_5 + $NH_4H_2PO_4$ + C) was also implemented successfully. The use of carbon served multiple purposes: (1) to reduce the transition metal $V^{5+} \rightarrow V^{3+}$ by oxidizing $C \rightarrow CO$, (2) to act as a nucleation site for product formation while inhibiting extensive grain growth, and (3) to facilitate subsequent electrode processing and final electrochemical performance. Moreover, $LiVPO_4F$ can also be obtained by an ion exchange of the $NaVPO_4F$ counterpart with excess LiBr carried out at 70°C for 10 h [25]. $LiVPO_4F$ powders were made of agglomerates ranging from about 10 μm for solid-state synthesis to 1 to 3 μm for hydrothermal synthesis.

$LiVPO_4F$ was found to be isostructural with the naturally occurring mineral tavorite, $LiFePO_4 \cdot OH$ [65]. It crystallizes in a triclinic unit cell (space group P-1), and its structure consists of $V^{3+}O_4F_2$ octahedra linked together by fluorine vertices in the trans position, forming chains along the c-axis (Fig. 1a). These parallel $(V^{3+}O_4F_2)_\infty$ chains are in turn connected by corner-sharing PO_4 tetrahedra to build a spacious three-dimensional framework with wide tunnels running along the (100), (010), and (001) directions accommodating the Li atoms in two sites parted by about 0.79 Å. While the Li1 site adopts a five-coordinate geometry having low occupancy (~18%), the Li2 site adopts a six-coordinate geometry with high occupancy (~82%).

Turning to its electrochemical performance, $LiVPO_4F$ was shown by Barker et al. to react with Li. Although earlier work on $LiVPO_4F$ focused primarily on the one-electron $V^{3+/4+}$ redox activity around 4 V, vanadium-based compounds offer the possibility of realizing multiple redox reactions using the multivalent nature of V (V^{2+} to V^{5+}). Barker et al. first reported the presence of multivalent reactions in $LiVPO_4F$ located at 1.8 V ($V^{3+/2+}$) and 4.2 V ($V^{3+/4+}$). Similar to the $V^{3+/4+}$ redox at 4.2 V, the $V^{3+/2+}$ redox reaction was found to be a two-phase reaction capable of reversibly (de)inserting about 0.9 Li, corresponding to a capacity of 140 mAh/g (Fig. 1c) [66].

Turning to its electrochemical performance, Barker et al. initially showed that $LiVPO_4F$ can reversibly react with 0.9 Li^+ at a $V^{3+/4+}$ redox potential of 4.19 V, leading, after electrode optimization, to a sustainable reversible capacity of 140 mAh/g (i.e., 92% of the theoretical capacity of 155 mAh/g) at a rate of C/5 (at 23°C). Further scrutinizing the voltage profile revealed a small inflection near $x = 0.3$

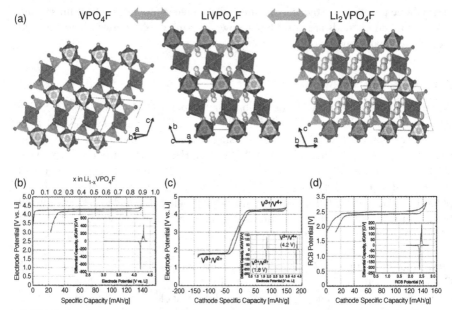

FIGURE 1 (a) Crystal structure of $Li_{1\pm x}VPO_4F$ compounds: VPO_4F (left), $LiVPO_4F$ (center) and Li_2VPO_4F (right). Each constitutes parallel chains of corner-shared VO_6 octahedra (grey color), PO_4 tetrahedra (light grey color) and statistically occupied two Li sites. (b) Galvanostatic voltage profile of carbothermally prepared $LiVPO_4F$ cathode with a redox potential of about 4.2 V. Inset: Differential capacity plot. (c) Two independent two-phase redox plateaus shown involving $V^{3+/4+}$ and $V^{3+/2+}$ located at 4.2 and 1.8 V, respectively (inset). (d) Galvanostatic voltage profile of symmetric (–) $LiVPO_4F \parallel LiVPO_4F$ (+) cell with a potential window of 2.4 V and reversible capacity of 130 mAh/g. The inset shows the differential capacity plot illustrating the redox reaction at 2.4 V.

in the charging profile and two distinct differential capacity peaks located at 4.25 and 4.29 V in the charging dx/dV derivative curves. It revealed two insertion–deinsertion domains, which were assigned to the progressive extraction of Li^+ from two energetically distinct crystallographic sites (Li1 and Li2). In contrast, a single asymmetric peak located near 4.19 V was observed on the differential discharge capacity curve (Fig. 1b), and to our knowledge this unusual behavior has not yet been explained satisfactorily. Conventional $LiVPO_4F$/graphite lithium-ion cells capable of delivering reversible capacity of 130 mAh/g of $LiVPO_4F$ at an average discharge voltage of 4.1 V with excellent capacity retention (90% of the initial capacity after 500 cycles) were also demonstrated.

In an effort to modify the underlying structure and electrochemical properties of pristine $LiVPO_4F$, Barker's group attempted a partial substitution of V for Al. Al was chosen particularly as a substituent as (1) the ionic radius of Al^{3+} (53.5 pm–AlO_6) is close to that of V^{3+} (64 pm–VO_6), and (2) the tavorite $LiVPO_4F$, is isostructural to the mineral amblygonite, $LiAlPO_4F$. Various $Li(V_{1-x}Al_x)PO_4F$ ($x = 0$ to 0.5) compositions were prepared successfully by the two-step carbothermal reduction method via making $(V_{1-x}Al_x)PO_4$ intermediate phases. The presence of Al was shown to benefit the cycling performances with (1) a lower irreversibility and polarization, (2) a gradual upshift in the $V^{3+/4+}$ redox peak from 4.19 to 4.28 V, and (3) improved cycling stability [42].

Overall, this optimization work demonstrates that $LiVPO_4F$ or its Al derivatives could be attractive electrodes for lithium-ion batteries, as they show an average discharge voltage of 4.1 V (i.e., 0.3 V higher than $LiCoO_2$), a reversible capacity nearing 140 mAh/g, good rate capability, and excellent cycling retention with, in addition, a better thermal stability than that of most oxide-based cathodes. This material caught the battery community's attention but never materialized into a real product, probably because of toxicity issues linked to the use of V-based compounds, although we must recall that early Li–polymer batteries used LiV_3O_8 as a positive electrode.

Although earlier work on $LiVPO_4F$ focused primarily on the one-electron $V^{3+/4+}$ redox activity around 4 V, vanadium-based compounds offer the possibility of realizing multiple redox reactions using the multivalent nature of V (V^{2+} to V^{5+}). Barker et al. first reported the presence of multivalent reactions in $LiVPO_4F$ located at 1.8 V ($V^{3+/2+}$) and 4.2 V ($V^{3+/4+}$). Similar to the $V^{3+/4+}$ redox at 4.2 V, the $V^{3+/2+}$ redox reaction was found to be a two-phase reaction capable of reversibly (de)inserting about 0.9 Li, corresponding to a capacity of 140 mAh/g [66] (Fig. 1c). Taking advantage of nearly a one-electron reaction by both oxidation (at 4.2 V) and reduction (at 1.8 V) of the $LiVPO_4F$ parent phase, a symmetric (–) $LiVPO_4F \parallel LiVPO_4F$ (+) cell could be configured. It was the first example of using a symmetric cell having one compound acting as a cathode as well as an anode. This symmetric cell, based on 3d-metal fluorophosphate, has a potential window of 2.4 V with a possible reversible capacity of 130 mAh/g (Fig. 1d). Although limited by lower energy density and cycling stability, it was a novel demonstration of building a safe full lithium-ion cell. Since then, further structural and electrochemical analyses of $LiVPO_4F$-based symmetric cells have been carried out by Nazar's group [43] and by Okada's group [44]. The latter group achieved low capacity in the symmetric cell using a conventional organic electrolyte, owing to high acidity and possible vanadium dissolution. Instead, they used nonflammable room-temperature ionic liquid electrolytes (1 M $LiBF_4/EMIBF_4$) to obtain stable, safe, and highly reversible symmetric cells at both room and high temperatures (80°C).

Although the electrochemical properties of $LiVPO_4F$ were obtained with both cathodic and anodic sweeps, the crystal structure of the resulting end phases (i.e., VPO_4F and Li_2VPO_4F) was not well understood. Recently, Nazar's group investigated these end members combining neutron and x-ray diffraction [43]. VPO_4F and Li_2VPO_4F were prepared by the chemical oxidation – reduction of the parent $LiVPO_4F$ compound. The crystal structure of these compounds is illustrated in Figure 1a and Table 2. Although very similar to $LiVPO_4F$, Li_2VPO_4F undergoes a slight change in symmetry to adopt a monoclinic framework (space group $C2/c$) made of chains of $[V^{2+}O_4F_2]$ oxyfluoride octahedra, interconnected by PO_4 tetrahedra units. It has two Li sites (Li1 and Li2), which are filled with equal probability and can be observed clearly using solid-state nuclear magnetic resonance (NMR) spectroscopy [67]. The delithiated phase VPO_4F also adopts a monoclinic $C2/c$ structure with corner-shared $[V^{4+}O_4F_2]$ oxyfluoride octahedra chains interconnected by PO_4 groups. The volume strain was determined to be 7.4% for the $Li_2VPO_4F \rightarrow LiVPO_4F$ transition and 8.5% for the $LiVPO_4F \rightarrow VPO_4F$ transition, respectively; this is in accordance with the standard two-phase redox behavior.

3.2 NaVPO$_4$F

The research on tavorite $LiVPO_4F$ naturally led Barker's group to dwell with its sodium counterpart (i.e., $NaVPO_4F$). Extending the carbothermal reduction route to the sodium system, $NaVPO_4F$ was first synthesized in 2003 by reacting NaF with an intermediate product, VPO_4, at 750°C for 1 h (in Ar flow) [38]. The final $NaVPO_4F$ phase is not isostructural with $LiVPO_4F$ but has a tetragonal structure (space group $I4/mmm$) related to that of the sodium aluminum fluorophosphate α-$Na_3Al_2(PO_4)_3F_2$ [68]. This Al-based compound was reported to have good ionic conductivity with facile Na^+ diffusion inside an open framework consisting of $[Al^{3+}O_4F_2]$ octahedra and PO_4 tetrahedra having large cavities accepting the constituent Na species. Nevertheless, the structure of $NaVPO_4F$ is yet to be solved completely. $NaVPO_4F$ has a theoretical capacity of 143 mAh/g with a $V^{3+/4+}$ redox activity. When cycled vs. Li, about 0.71 Na^+ could be extracted in two steps, located at 3.8 and 4.3 V vs. Li. Subsequently, Li^+ could be inserted reversibly in a two-step process around 4.25 and 3.75 V vs. Li. The reversibility of Li^+ (de)insertion process showed evidence of the structural robustness during electrochemical cycling. Subsequently, a full cell was prepared, with mass-balanced $NaVPO_4F$ as the cathode and hard carbon as the anode, which delivered a reversible capacity of 82 mAh/g with an average discharge voltage of 3.7 V [38]. However, it was marred by a high-capacity loss upon cycling.

In 2006, Zhuo et al. [62] produced Cr-substituted $Na(V_{1-x}Cr_x)PO_4F$ compounds and in the process discovered a monoclinic polymorph (space group $C2/c$) of $NaVPO_4F$, which is isostructural to the $Na_3Al_2(PO_4)_2F_2$ compound [69]. Again, the details of this monoclinic polymorph have not yet been solved. The $Na(V_{1-x}Cr_x)PO_4F$ compositions form a complete solid solution, as shown by the gradual decrease in their lattice parameters, owing to the smaller ionic radius of Cr^{3+} (61 pm) vis-à-vis V^{3+} (64 pm). A small amount of Cr doping was found to improve the cycling stability while delivering a reversible capacity of about 80 mAh/g. Recently, Zhao et al. successfully synthesized both the tetragonal and monoclinic polymorphs of $NaVPO_4F$ via sol–gel synthesis. While the monoclinic phase was stabilized at 700°C, it was transformed into the tetragonal phase at 750°C [70]. The resulting cathode was found to deliver 3.6 V redox activity (vs. Li) with a capacity exceeding 117 mAh/g with very good cycling stability.

3.3   (Na$_{2-x}$Li$_x$)MPO$_4$F

Although fluorophosphates started with vanadium-based compounds showing excellent electrochemical properties, the high cost and toxicity associated with vanadium naturally forced the battery community to design novel fluorophosphate cathodes that are safe and economical. Naturally, the Fe-based compound was the obvious choice. Pursuing this direction, in 2007, Nazar's group discovered a sodium–lithium iron fluorophosphate A_2FePO_4F (A = Na/ Li), which could be synthesized by either sol–gel or solid-state synthesis. This was done by annealing the constituent precursor mixtures of NaF, $NaCH_3COO$, $Fe(CH_3COO)_2$, and H_3PO_4 or, alternatively, NaF, $NaHCO_3$, $FeC_2O_4 \cdot 2H_2O$, and $(NH_4)H_2PO_4$ at a final temperature of 525 to 625°C for 6 h [46]. Additionally, the hydrothermal method could also be used to prepare Na_2FePO_4F powder made of nanoscale (75 to 200 nm) particles by using a precursor mixture of NaF, NaOH, $(NH_4)_2Fe(SO_4)_2$, and H_3PO_4 at a lower reaction temperature of 170 to 220°C [47].

Na_2FePO_4F was found to be isostructural with known minerals such as Na_2MgPO_4F, Na_2FePO_4OH, and Na_2CoPO_4F, which crystallizes in an orthorhombic structure (e.g., *Pbcn*). It offers a two-dimensional diffusion path for Na, with its structure consisting of $Fe^{3+}O_4F_2$ octahedra building blocks, which uniquely form $Fe_2^{2+}O_6F_3$ bioctahedral units made by face-sharing of two adjacent $[Fe^{3+}O_4F_2]$ units (Fig. 2a). In turn, these $Fe_2^{2+}O_6F_3$ bioctahedra share corners via F atoms to form

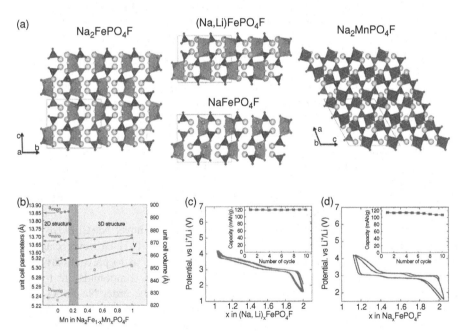

FIGURE 2 (a) Crystal structure of Na_2MPO_4F compounds having a two-dimensional structure. (left) Na_2FePO_4F and (center) section of $(Na, Li)FePO_4F$ prepared by ion exchange and $NaFePO_4F$ prepared by chemical oxidation. All these compounds have an orthorhombic structure with varied amount of Na/Li. The FeO_4F_2 octahedra are shown in grey and PO_4 tetrahedra are shown in black, with the sodium atoms presented as spheres. (Right) Monoclinic Na_2MnPO_4F having a two-dimensional structure is illustrated. The MnO_4F_2 octahedra are shown in black. (b) Evolution of unit cell parameters in $(Na_{2-x}Mn_x)PO_4F$ compositions, showing a sudden change of about $x = 0.2$. Galvanostatic voltage profiles and cycling stability (inset) of ionothermally synthesized Na_2FePO_4F, cycled vs. Li (c) and vs. Na (d) counter electrodes.

parallel chains along the a-axis. These parallel chains are connected to each other by PO_4 tetrahedra along the c-axis to form $(Fe^{2+}PO_4F)_\infty$ infinite slabs. The constituent Na atoms occupy two distinct crystallographic sites. The Na1 site adopts $6 + 1$ coordination, being surrounded by four neighboring O atoms, two neighboring F atoms, and a slightly far distant O atom. The Na2 site is very similar in nature to the Na1 site, but slightly smaller.

When electrochemically tested vs. Li, Na_2FePO_4F shows a sloping voltage–composition profile upon oxidation. Although this compound can theoretically involve a two-electron redox activity ($Fe^{2+} \rightarrow Fe^{3+} \rightarrow Fe^{4+}$), the highly unstable Fe^{4+} limits it to one-electron ($Fe^{2+} \rightarrow Fe^{3+}$) redox reaction with extraction of only one Na atom. The sloping voltage profile suggests that the removal of one Na^+ ($Na_2FePO_4F \rightarrow NaFePO_4F$) occurs via a single-phase process. Nevertheless, after a few cycles the voltage–composition curve is modified completely with the appearance of two equal and distinct sloping composition domains which are completely reversible upon cycling. Simply, once Na^+ has been removed from the structure via an (in situ) electrochemical ion-exchange reaction [46], Li^+ ions favorably get intercalated into the oxidized $NaFePO_4F$, forming an $(Na,Li)FePO_4F$ electrode, which during further cycling acts as a Li-intercalation cathode solely involving Li (de)intercalation, with Na^+ acting as a spectator ion. The resulting electrode delivered a reversible capacity of 120 to 130 mAh/g, operating with an average voltage of 3.3 V vs. Li.

The overall framework of the parent Na_2FePO_4F is retained perfectly upon Na^+ extraction and subsequent Li^+ intercalation with only 3.7% volume strain; therefore, the structure of $NaFePO_4F$ and $(NaLi)FePO_4F$ remains identical to that of the parent Na_2FePO_4F phase with a slight change in the Na–Li positions [46] (Fig. 2a). This was confirmed by the structural analysis of both end members prepared by chemical oxidation–reduction and ion-exchange reaction. However, halfway to chemical oxidation, the $Na_{1.5}FePO_4F$ deviates from the orthorhombic *Pbcn* structure with a slight increase in β from $90°$ to $91.22°$, which can be an effect of the structural ordering of mixed oxidation sites (Fe^{2+} and Fe^{3+}).

Along with Na_2FePO_4F, Nazar's group also reported novel Na_2CoPO_4F and Na_2NiPO_4F obtained for the first time by solid-state synthesis [47]. However, caution has to be exercised to avoid any carbonaceous precursor so as to prevent carbothermal reduction from forming metallic Co or Ni. All these compounds are found to be isostructural with their lattice parameters following the order $Na_2FePO_4F >$ $Na_2CoPO_4F > Na_2NiPO_4F$. The Co ($\sim$4.8 V) and Ni (>5 V) homologues involve redox activity at a much higher voltage, making it difficult to see realizable cathode performance. Density functional theory (DFT) calculations have recently been performed on these fluorophosphates, asserting high voltages for Co and Ni compounds [59,71].

Although Na_2FePO_4F was electrochemically promising and easy to synthesize, its thermal stability was found to be less than that of other phosphate-based cathodes, such as $LiFePO_4$, as it decomposes at $T > 650°C$, hence preventing the use of high-temperature carbon coatings for improving the electrode kinetics. To bypass this kinetic issue, our group successfully prepared highly divided particles delivering excellent electrochemical capacity using ionothermal synthesis [62]. More specifically, a stoichiometric precursor mixture of Na_3PO_4, FeF_2, and $FeCl_2$

was (1) loaded into a 25-mL Teflon-lined Parr reactor inside an Ar-filled glove box, (2) immersed in a minimal amount of C2-protected 1,2-dimethyl-3-butylimidazolium bis(trifluoromethanesulfonyl imide) ionic liquid medium,(3) heated to 270°C (at a rate of 5°C/min) for 48-h annealing, and (4) washed with organic solvent to remove the ionic liquid as well as the NaCl by-product. Following the same method, Na_2MnPO_4F as well as mixed $Na_2(Fe_{1-x}Mn_x)PO_4F$ ($x = 0$ to 1) phases were synthesized (Fig. 2b). In all cases, the ionothermal synthesis led to highly divided nanometric and well-dispersed particles (25 to 80 nm). Such highly divided Na_2FePO_4F powders show an excellent reversible capacity of 120 mAh/g ($x \sim 1$), with low irreversible capacity loss and negligible polarization when cycled vs. Li (Fig. 2c). Moreover, when tested in a full sodium battery configuration vs. Na they also show a discharge capacity of 115 mAh/g ($x \sim 0.8$) with excellent cycling stability (Fig. 2d).

Surprisingly, while the Fe, Co, and Ni homologs of the Na_2MPO_4F phases are isostructural with a layered two-dimensional orthorhombic *Pbcn* structure, Na_2MnPO_4F has a tunnel, three-dimensional monoclinic $P2_1/n$ structure. In the latter case, $Mn^{3+}O_4F_2$ octahedra are corner-shared and connected by an F vertex to form $(Mn_2O_8F_2)_\infty$ along the *b*-axis. These chains are in turn linked by PO_4 tetrahedra along the *a*- and *c*-axes, forming a three-dimensional structure accommodating Na ions into their tunnels. Unlike Na_2FePO_4F with two F atoms in the cis position, in the case of Na_2MnPO_4F the F atoms are located in the trans position. It changes from two-dimensional packing of the MO_4F_2 octahedra (face-sharing) to three-dimensional packing (corner-sharing octahedra). The corner-sharing monoclinic three-dimensional structure has a longer M–M distance accommodating the larger Mn species. Through a study of the $Na_2(Fe_{1-x}Mn_x)PO_4F$ series, the two-dimensional → three-dimensional phase transition was shown to occur near $x = 0.2$ (Fig. 2b).

Na_2MnPO_4F was found to be electrochemically inactive vs. Li^0/Li^+. In contrast, for the mixed $Na_2(Fe_{1-x}Mn_x)PO_4F$ phases the capacity dropped rapidly with increased Mn content and totally disappeared for $x = 0.2$ (e.g., as we moved away from the two-dimensional structure). The underlying Jahn–Teller effect associated with Mn^{3+} and the poor conductivity could be the reasons behind this poor performance. Although the tunnel structure should favor easy Na diffusion, it is practically impossible to take Na out of the structure.

Overall, owing to its easy synthesis, open structure, and excellent electrochemical performance vs. both Li and Na, Na_2FePO_4F truly forms an attractive polyanionic cathode to develop lithium-ion as well as sodium-ion batteries. Recently, carbon-coated Na_2FePO_4F with excellent capacity retention has been reported for sodium-ion batteries [72].

3.4 Li_2MPO_4F (M = Fe, Co, Ni)

Following the successful investigation of Na_2FePO_4F, Nazar's group first reported its lithium-counterpart (i.e., Li_2FePO_4F) [46]. Phase-pure Li_2FePO_4F was prepared by taking Na_2FePO_4F as seed material and implementing the complete ion exchange of Na by Li without altering the structure. The resulting Li_2FePO_4F assumed an orthorhombic, two-dimensional layered structure. Its galvanostatic voltage profile was very similar to that of Na_2FePO_4F after a few cycles, involving a solid-solution reaction to deliver about 120 mAh/g capacity at an average voltage of 3.5 V.

On another note, Li_2CoPO_4F was introduced by Okada's group (in 2005) via a two-step solid-state synthesis by reacting the equimolar mixture of LiF and $LiCoPO_4$ in a vacuum quartz tube at 780°C for 78 h [45]. The reaction product presented an orthorhombic structure with *Pnma* symmetry. Li_2CoPO_4F can also be prepared by a complete ion exchange of the Na_2CoPO_4F phase, which retains the structure of the parent phase (i.e., the orthorhombic *Pbcn* structure [45]. However, it is metastable and quickly converts to the above-mentioned structure upon annealing at 580°C. When cycled vs. Li, Li_2CoPO_4F was found to involve a $Co^{2+/3+}$ redox reaction centered at 5 V. The exact crystal structure of Li_2CoPO_4F was recently illustrated by Hadermann et al. using precession electron diffraction [73,74]. It consists of parallel chains of edge-sharing CoO_4F_2 octahedral units along the *b*-axis. These chains are interlinked by PO_4 tetrahedra, thus forming tunnels along the *b*-axis which accommodate the Li atoms in three distinct sites with full occupancy. Whereas two Li sites (Li1 and Li2) are five-coordinated sites, the third one (Li3) is a distorted six-coordinated site. This compound was found to undergo an irreversible structural transformation during the first oxidation segment by rotation (distortion) of CoO_4F_2 octahedra and PO_4 tetrahedra units, resulting in a slight expansion of the unit cell. This expansion ($\sim$3.5%) makes the structure more open, so as to facilitate the lithium-mobility during subsequent cycling. Consequently, a reversible discharge capacity of 60 mAh/g (i.e., 0.4 Li) was obtained with a 5-V redox reaction. Isostructural to Li_2CoPO_4F, Li_2NiPO_4F has been reported as early as 1999 [60]. Recently, Li_2NiPO_4F was found to be electrochemically active at a high voltage of 5.3 V (vs. Li) by using a highly stable 1 M $LiBF_4$EC/DMC/sebaconitrile electrolyte [61]. While these Li_2MPO_4F compounds are interesting for structural analysis, they all invariably suffer from poor electrochemical properties.

3.5   LiFePO$_4$F

Following the discovery of $LiVPO_4F$, Barker's group attempted to synthesize many other $LiMPO_4F$ (M = Ti, Cr, Fe, Co) homologues. Although all of them were claimed to be isostructural to $LiVPO_4F$ (i.e., triclinic tavorite structure), their detailed crystal structure and electrochemical activity were not elucidated [75]. Based on our success-ful experience with the synthesis of Na_2MPO_4F (M = Fe, Mn) phases, we decided after all to synthesize novel $LiMPO_4F$ phases by either solid-state or ionothermal syn-theses [48]. For the solid-state synthesis, an equimolar mixture of Li_3PO_4 and FeF_3 sealed inside a Pt tube was annealed at 700°C for 24 h, resulting in $LiFePO_4F$ with the reaction $Li_3PO_4 + FeF_3 \rightarrow LiFePO_4F + 2LiF$ driven by the strong lattice energy of LiF. The reaction product was rinsed rapidly with cold water and acetone to get rid of LiF by-product and was oven-dried at 60°C. For the ionothermal synthesis, the same precursor mixture was used, but for stability issues we started with the classical 1-ethyl-3-methylimidazolium bis(trifluoromethanesulfonyl)imide (EMI-TFSI) ionic liquid and then used 1-butyl-3-ethylimidazolium trifluoromethanesulfonate (triflate), which is not reduced by Fe^{3+}. Contrary to the solid-state synthesis requiring high temperature, the ionothermal synthesis could be realized at 260°C for a reaction time of 48 h, with the reaction product being washed with dichloromethane (CH_2Cl_2) and distilled water and then oven-dried at 60°C. Apart from being less energy demanding (400°C lower than for the solid-state synthesis), the ionothermal synthesis led to nanoscale ($\sim$20 nm) particles as compared to micrometric (2 to 5 μm) particles obtained by the solid-state route. In an independent effort, Nazar's group synthesized $LiFePO_4F$ either by reacting an equimolar mixture of LiF and $FePO_4$ at 575°C for 75 min or by reacting stoichiometric amounts of Li_2CO_3, $FePO_4$ and NH_4F at 575°C for 90 min under an Ar atmosphere. Both cases led to the formation of micrometric particles (1 μm) [49]. Further, $LiFePO_4F$ was reduced chemically with the help of $LiAlH_4$ in tetrahydrofuran medium (at 25°C for 2 days) to obtain Li_2FePO_4F. Like $LiVPO_4F$, both $LiFePO_4F$ and Li_2FePO_4F are isostructural to the tavorite $LiFePO_4 \cdot OH$ and crystallize in a triclinic cell (space group P-1) with the following lattice parameters: $a = 5.3002(2)$ Å, $b = 7.2601(2)$ Å, $c = 5.1516(2)$ Å, $\alpha = 107.880(3)°$, $\beta = 98.559(3)°$, $\gamma = 107.343(3)°$, and $V = 173.67(6)$ Å^3 for $LiFe^{3+}PO_4F$, and $a = 5.3746(3)$ Å, $b = 7.4437(4)$ Å, $c = 5.3256(4)$ Å, $\alpha = 109.038(2)°$, $\beta = 94.423(6)°$, $\gamma = 108.259(9)°$, and $V = 189.03(4)$ Å^3 for $Li_2Fe^{2+}PO_4F$. Note that the insertion of an extra Li into the parent $LiFePO_4F$ results in 7.9% volumetric expansion, owing, among other things to the larger size of Fe^{2+} compared to Fe^{3+}.

More specifically, the $LiFePO_4F$ structure consists of parallel chains of corner-sharing $[Fe^{3+}O_4F_2]$ oxyfluoride octahedra linked together by PO_4 tetrahedra and forming a three-dimensional framework delimiting tunnels, running along (100), (010), and (101) directions, which host Li ions and are responsible for efficient Li diffusion in tavorite. In addition, the Fe adopts two independent crystallographic sites (Fe1 and Fe2), resulting in a Fe1–F–Fe2 angle = 130.8° (Fig. 3).

When tested in half-cell configuration, $LiFePO_4F$ delivers a reversible capacity of about 145 mAh/g (theoretical capacity $\sim$ 152 mAh/g) with excellent cycling retention, with the average $Fe^{3+/2+}$ redox potential located at 2.8 V. Such high-degree performance was obtained with limited electrode optimization (e.g., carbon coating), attesting to efficient Li diffusion (Fig. 3). Nevertheless, its relatively low redox potential comes as a drawback for applications in terms of poor energy density. Also, $LiFePO_4F$ cannot act as a Li reservoir, as it cannot be oxidized, hence limiting its use to lithium-metal polymer batteries.

3.6 LiTiPO₄F

Apart from $LiVPO_4F$ and $LiFePO_4F$, the only other reported tavorite compound is $LiTiPO_4F$, which was successfully synthesized by our group, according to the reaction $Li_3PO_4 + TiF_3 \rightarrow LiTiPO_4F + 2LiF$, using both solid-state and ionothermal routes [48]. The solid reaction was conducted in a stainless-steel container and reacted at 700°C for 24 h, with the resulting LiF being washed off in cold water. In contrast, the ionothermal reaction was carried out in a Teflon liner steel bomb in the presence of 3 cm^3 of 1,2-dimethyl-3-(3-hydroxypropyl)imidazolium bis(trifluoromethanesulfonyl)imide ionic liquid. Single-phase $LiTiPO_4F$ powders were obtained at temperatures as low as 260°C for a reaction time of 48 h. This tavorite $LiTiPO_4F$ phase is built from two crystallographically independent and slightly distorted titanium-based TiO_4F_2 octahedra linked by fluorines in the trans position so

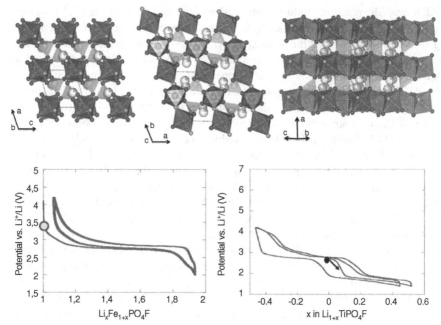

FIGURE 3 (Top) Crystal structure of tavorite structured $LiMPO_4F$ (M = Fe, Ti) compounds comprising interconnected chains of MO_4F_2 octahedra and PO_4 tetrahedra. The crystallographic view along [010] (left), [001] (center), and [011] (right) directions are illustrated. The MO_4F_2 octahedra are shown in grey and PO_4 tetrahedra are shown in light grey, with the lithium atoms presented in spheres. (Bottom) Galvanostatic potential-composition profiles of tavorite $LiFePO_4F$ (left) and $LiTiPO_4F$ (right) are shown vs. Li as counter electrodes. Circles represent the starting points of electrochemical cycling.

as to form one-dimensional chains interlinked by PO_4 tetrahedra (Fig. 3). It has the following lattice parameters: $a = 5.1991(2)$ Å, $b = 5.3139(2)$ Å, $c = 7.2428(3)$ Å, $\alpha = 106.975(3)°$, $\beta = 108.262(4)°$, $\gamma = 97.655(4)°$, and $V = 176.10(2)$ Å^3.

Unexpectedly, $LiTiPO_4F$ shows a distinct staircase voltage–composition profile with two reversible pseudoplateaus centered near 2.9 and 1.7 V, respectively. Combining these two processes leads to a reversible capacity approaching the theoretical one of 150 mAh/g (Fig. 3). By analogy to $LiVPO_4F$ (which shows two voltage plateaus corresponding to the ($V^{4+} \rightarrow V^{3+}$ and $V^{3+} \rightarrow V^{2+}$) redox couples, it will be tempting to relate the 2.9- and 1.7-V plateaus in Ti-based compounds to the Ti^{4+}/Ti^{3+} and Ti^{3+}/Ti^{2+} redox couples, respectively. For such a hypothesis to be valid, the degree of oxidation and reduction of the parent Ti^{3+} species should not be limited to 50% capacity. More intriguing is also the fact that the ratio of the 2.9- and 1.7-V plateaus was found to be sensitive to mild changes in the sample synthesis and recovery procedures, raising questions about the real oxidation state of Ti in our starting material. Whatever the exact reasons of this staircase profile, the presence of multiple redox reactions enlisting low voltages (below 3 V) makes $LiTiPO_4F$ poorly attractive for commercial applications.

4   ALKALI METAL FLUOROSULFATES

Although fluorophosphates offer a variety of polyanionic cathode systems, the Fe-based fluorophosphate, which is the most commercially relevant, suffers from a low redox potential of 2.8 V. A more useful Fe-containing fluorine-based polyanionic cathode can be designed by simple substitution of $(PO_4)^{3-}$ for $(SO_4)^{2-}$. Initially conceptualized by Goodenough as an inductive effect, the isostructural X: P $\rightarrow$ S transition in any NASICON $Li_xT_3{}^M(XO_4)_3$ compound can increase its redox potential by 0.8 V, independent of the $3d$ metal, owing to the larger electronegativity of sulfur with respect to phosphorus. So if the $LiFeSO_4F$ compound could be synthesized, it should have a redox potential of 3.6 V. Pursuing this direction, a variety of fluorosulfate phases have been unveiled in recent years. This is illustrated below showing their structural and electrochemical properties. A few of these structural and cathode properties of such fluorosulfate compounds are summarized in Table 3.

4.1   LiFeSO₄F

Unlike the umpteenth number of $(PO_4)^{3-}$-based cathodes (e.g., phosphates, pyrophosphates, fluorophosphates), there have been very few $(SO_4)^{2-}$-based electroactive compounds, and until 2010, no electroactive fluorosulfate compound existed. The only phase reported, with Li showing some mobility, was $LiMgSO_4F$ [51]. It was then tempting to make the $3d$ metal (e.g., Fe^{2+}, Co^{2+}, Ni^{2+}) equivalent to fluorosulfates, hoping for redox activity toward Li [26,77]. Of course, the Fe-based homologue ($LiFeSO_4F$) comes as a first choice. But the synthesis of such a compound was a tricky issue since, unlike $(PO_4)^{3-}$ compounds, the $(SO_4)^{2-}$ compounds are prone to (1) dissolution in water, hence eliminating aqueous syntheses, and (2) decomposition at moderately high temperatures ($T > 400°C$), thus ruling out the conventional solid-state synthesis using aggressive heat treatment. It is a synthesis riddle that pushes synthetic chemists into using nonaqueous routes involving moderate temperatures. To circumvent this issue, our group, which had earlier developed the ionothermal synthesis of a variety of cathode materials (oxides, phosphates, fluorophosphates, silicates) [77,78], applied the same ionothermal synthesis to produce fluorosulfates. Ionic liquids, which are nonaqueous media decomposing at temperatures near 300°C, offer partial precursor solubility, facilitating a nucleation and grain growth (Ostwald ripening) mechanism [78,79] if provided with the right choice of precursors. At this juncture, our group attempted the fluorosulfate synthesis by a clever choice of mono-hydrate precursors ($LiF + FeSO_4·H_2O$). The $FeSO_4·H_2O$ monohydrate precursor adopts a Szomolnokite structure with a striking structural similarity to the previously known $LiMgSO_4F$ tavorite fluorosulfate (Fig. 4a). Owing to this structural relation, a topotactic reaction enlisting the dual ion exchange (H^+ by Li^+ and OH^- by F^-) could be envisioned by reacting a mixture of LiF and $FeSO_4·H_2O$, without the need for complete decomposition of the initial precursor.

Following this soft chemistry ion-exchange mechanism, our group first tried the (nonaqueous) ionothermal synthesis of $LiFeSO_4F$ in 2010 [26]. The $FeSO_4·H_2O$ monohydrate precursor was made in situ by annealing the commercially obtained $FeSO_4·7H_2O$ heptahydrate (1) in an ionic liquid bath at 90 to 150°C or (2) inside

TABLE 3 Crystallographic Lattice Parameters and Cathode Properties of Li- and Na-Based Fluorosulfate Compounds that Have Potential Applications in Rechargeable Batteries

Compound	Crystal structure	Space group	a (Å)	b (Å)	c (Å)	α(deg)	β(deg)	γ(deg)	Volume (Å^3)	Redox potential (V)	Capacity (mAh/g)	Ref.
\multicolumn{13}{c}{Lithium Metal Fluorosulfates ($Li_y T_M SO_4 F$)}
$LiMgSO_4F$	Triclinic	$P\text{-}1$	5.1623(7)	5.388(1)	7.073(1)	106.68(1)	107.40(1)	97.50(1)	174.72(5)	—	—	51
$LiFeSO_4F$	Triclinic	$P\text{-}1$	5.1747(3)	5.4943(3)	7.2224(3)	106.522(3)	107.210(3)	97.791(3)	182.559(16)	3.6	140	26
$LiFeSO_4F$	Monoclinic	$C2/c$	13.0238(6)	6.3957(3)	9.8341(5)	90.000	119.68(5)	90.000	711.64(1)	3.9	85	76
$LiCoSO_4F$	Triclinic	$P\text{-}1$	5.1721(7)	5.4219(7)	7.1842(8)	106.859(6)	107.788(6)	97.986(5)	177.80(4)	—	—	52
$LiNiSO_4F$	Triclinic	$P\text{-}1$	5.1430(6)	5.3232(7)	7.1404(7)	106.802(9)	107.512(8)	98.395(6)	172.56(4)	—	—	52
$LiMnSO_4F$	Monoclinic	$C2/c$	13.2701(5)	6.4162(2)	10.0393(4)	90.000	120.586(2)	90.000	735.85(5)	—	—	57
$LiZnSO_4F$	Orthorhombic	$Pnma$	7.40357(9)	6.32995(7)	7.42016(9)	90.000	90.000	90.000	347.740(7)	—	—	54
$FeSO_4F$	Triclinic	$P\text{-}1$	5.0735(2)	5.0816(3)	7.3363(4)	110.975(4)	111.189(4)	88.157(3)	163.640(12)	3.6	140	26
$CoSO_4F$	Triclinic	$P\text{-}1$	5.0395	5.1304	7.0227	107.313	108.413	97.673	159.197	—	—	1
$NiSO_4F$	Triclinic	$P\text{-}1$	5.2162	5.0635	7.1662	108.266	109.317	94.787	165.881	—	—	1
$LiFe_{0.9}Zn_{0.1}SO_4F$	Monoclinic	$C2/c$	13.0109(2)	6.3950(9)	9.8295(2)	90.000	119.591(6)	90.000	177.80(2)	3.9	82.5	58
$LiFe_{0.85}Zn_{0.15}SO_4F$	Orthorhombic	$Pnma$	7.5421(3)	6.4751(3)	7.3094(3)	90.000	90.000	90.000	178.48(1)	3.6	97.5	58
$LiFe_{0.90}Mn_{0.10}SO_4F$	Monoclinic	$C2/c$	13.05166(3)	6.39591(1)	9.85550(2)	90.000	119.7876(1)	90.000	714.007(3)	3.9	125.0	57
\multicolumn{13}{c}{Sodium Metal Fluorosulfates ($Na_y T_M SO_4 F$)}
$NaMgSO_4F$	Monoclinic	$C2/c$	6.66958(10)	8.58747(12)	7.05209(10)	90.00	114.090(1)	90.00	368.729(9)	—	—	56
$NaFeSO_4F$	Monoclinic	$C2/c$	6.68015(19)	8.7062(2)	7.19131(18)	90.000	113.520(2)	90.00	383.491(18)	3.6	6	56
$NaCoSO_4F$	Monoclinic	$C2/c$	6.66685(18)	8.6216(2)	7.14386(19)	90.000	114.335(2)	90.00	374.138(17)	—	—	56
$NaZnSO_4F$	Monoclinic	$C2/c$	6.6888(2)	8.6320(2)	7.0865(2)	90.000	113.940(2)	90.00	373.959(16)	—	—	56
$NaCuSO_4F$	Monoclinic	$C2/c$	6.8231(2)	8.5246(3)	6.8778(2)	90.000	110.745(2)	90.00	374.11(2)	—	—	56
$NaMnSC_4F$	Monoclinic	$I2/c$	12.1635(5)	6.6364(2)	10.3488(4)	90.000	105.06(2)	90.00	806.65(5)	—	—	53
$NaFeSO_4F\cdot2H_2O$	Monoclinic	$P2_1/m$	5.75959(5)	7.38273(5)	7.25044(7)	90.000	113.3225(6)	90.000	283.109(11)	—	—	55
$NaCoSO_4F\cdot2H_2O$	Monoclinic	$P2_1/m$	5.73364(2)	7.314981(17)	7.18640(2)	90.000	113.5028(2)	90.000	276.40(9)	—	—	55
$NaNiSO_4F\cdot2H_2O$	Monoclinic	$P2_1/m$	5.70118(4)	7.27603(3)	7.15634(3)	90.000	113.8883(2)	90.000	271.429(15)	—	—	55
$NaCoSO_4F\cdot2D_2O$	Monoclinic	$P2_1/m$	5.73991(6)	7.32250(7)	7.19288(8)	90.000	113.512(1)	90.000	277.221(5)	—	—	1

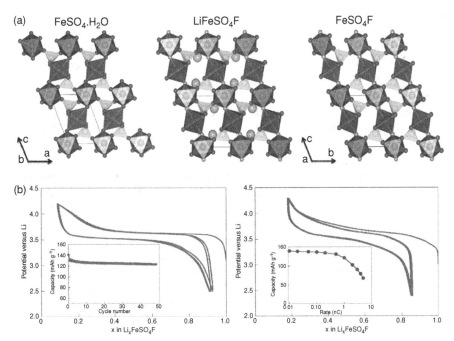

FIGURE 4   (a) Crystal structure of $FeSO_4 \cdot H_2O$ monohydrate precursor (left), tavorite $LiFeSO_4F$ (center), and delithiated $FeSO_4F$ (right) phases with striking structural similarity. In each case, the FeO_6/FeO_4F_2 octahedra are shown in grey and the SO_4 tetrahedra are shown in light grey, with Li atoms presented as spheres. (b) (Left) Galvanostatic cycling of ionothermal synthesized $LiFeSO_4F$ vs. Li at C/10 rate, showing a reversible capacity close to 130 mAh/g involving the $Fe^{2+/3+}$ redox reaction centered at 3.6 V. The cycling stability is illustrated in the inset. (Right) Galvanostatic cycling at the faster C/2 rate still shows appreciable reversible capacity with low polarization. The inset shows the rate capability study of tavorite $LiFeSO_4F$.

a primary vacuum pump at 110 to 150°C. Afterward the ionothermal synthesis was carried out by reacting an equimolar mixture of LiF and $FeSO_4 \cdot H_2O$ immersed inside 5 mL of ionic liquid media at a temperature of 275 to 300°C for 4 to 24 h. A slow heating rate of 1 to 2°C/min was used to avoid quicker dehydration of the precursor mixture with, additionally, a slight excess of LiF (~10%) was added to ensure the reaction completion. In a typical protocol, commercially available EMI-TFSI ionic liquid was used, although a variety of ionic liquid media can be used. The final reaction product could be isolated from ionic liquid media by repeated washings with organic solvents (e.g., dichloromethane or ethyl acetate) followed by oven drying at 60°C. The x-ray diffraction powder pattern of the recovered powder, greenish in appearance and made of 1- to 3-μm particulates, was similar to that of $LiMgSO_4F$, indicating the presence of single-phase $LiFeSO_4F$, whose elemental composition was confirmed by energy-dispersive x-ray measurements. We experienced the need to use the Fe^{3+}-free $FeSO_4 \cdot H_2O$ precursor for the reaction to succeed. In addition, the reaction mixture should be slowly heated to the final temperature so as to balance the kinetics of two competing steps: H_2O removal and LiF ingression. A faster heating rate can trigger quicker H_2O removal while not giving sufficient time for LiF to react, leading to incomplete reaction with the anhydrous $FeSO_4$ impurity phase. In this context, ionic liquids form an ideal reacting medium slowing down the H_2O dehydration, owing to their inherent hydrophobicity while favoring easy dissociation and reactivity of LiF. Keeping all these shades in mind, the synthesis of $LiFeSO_4F$ was achieved successfully.

Although the ionothermal synthesis led to the first synthetic production of $LiFeSO_4F$, it nevertheless involves rather expensive ionic liquid media, even though they can be recovered and reused for subsequent syntheses: hence, the need for alternative synthesis processes. Based on the same precursor mixture (LiF + $FeSO_4 \cdot H_2O$), the solid-state (dry) synthesis was carried out at a low temperature of 300°C by our group, the main difficulty being to balance the kinetics of H_2O removal with LiF ingression steps [80]. This was achieved by carrying the reaction in a Teflon-lined Parr reactor enclosing the precursor mixture (in the shape of pressed pellets) under an Ar atmosphere. The reactor was slowly heated to 290°C (at 0.2°C/ min) to avoid the quick dehydration of precursors. During such a treatment, the autogenous pressure of water, which can come close to 4.5 bar, inhibits the water dehydration kinetics, thus giving LiF adequate time to react and form $LiFeSO_4F$. Nevertheless, unlike the ionothermal method, the dry synthesis involving the diffusion of the chemical elements in the solid rather than the liquid state required a longer annealing time of 45 to 50 h.

In yet another successful alternative synthesis of $LiFeSO_4F$, to alleviate the cost issue and sluggish kinetics associated with the respective ionothermal and solid-state processes, our group pioneered the polymer-assisted synthesis, using powdered polymeric compounds such as poly(ethylene glycol) as reacting media [81]. Such polymers are powdered at room temperature, then melted above 60°C to form liquid media, which are thermally stable until temperatures exceed 300°C. Similarly to ionothermal synthesis, this polymer-assisted synthesis, which can be controlled by using polymeric media of different natures (viscosity, polarity, molecular weight, etc.), was conducted in a Teflon-lined Parr reactor. Thermal treatment at 290°C for 24 h was shown to be sufficient to produce phase-pure $LiFeSO_4F$ compound. Later the same year, Nazar's group independently reported a similar polymer-based synthesis using hydrophilic tetraethylene glycol as a reacting medium [82]. While they reduced the reaction temperature to 220°C, the final product was recovered after a longer period of 60 h.

As speculated, $LiFeSO_4F$ was found to be isostructural with the parent $LiMgSO_4F$ fluorosulfate phase. It crystallizes in a triclinic unit cell (P-1 space group) with lattice constants $a = 5.1747(3)$ Å, $b = 5.4943(3)$ Å, $c = 7.2224(3)$ Å, $\alpha = 106.522(3)°$, $\beta = 107.210(3)°$, $\gamma = 97.791(3)°$, and $V = 182.56(6)$ Å^3. Much like all the previous tavorite compounds, such as $LiFePO_4F$, the $LiFeSO_4F$ crystal structure is made of two slightly distorted crystallographically independent $Fe^{2+}O_4F_2$ oxyfluoride octahedra linked together by F vertices in the trans position, thus forming chains along the c-axis (Fig. 4a). These $(Fe^{2+}O_4F_2)_\infty$ chains are interwoven with isolated SO_4 tetrahedra, forming a three-dimensional structure delimiting three tunnels along the (100), (010), and (101) directions, where the Li^+ ions are located [26,83].

Contrary to early reports, synchrotron study revealed a well-defined (and no split) site for Li (with full occupancy) [83]. The delithiated version $FeSO_4F$ was easily derived from pristine $LiFeSO_4F$ by chemical oxidation with NO_2BF_4. It crystallizes into a monoclinic structure ($C2/c$ space group) with lattice constants: $a = 7.3037(1)$ Å, $b = 7.0753(1)$ Å, $c = 7.3117(1)$ Å, $\beta = 119.758(2)°$, and $V = 164.05(8)$ Å^3 [82]. Comparing $LiFe^{2+}SO_4F$ to $Fe^{3+}SO_4F$, delithiation is seen to improve the crystal symmetry along with a very important ($\sim10\%$) volume strain. This Li-driven structural modification, which results in $\Delta V/V$ of 10%, is accomplished by the rotation of FeO_4F_2 octahedra with an increase in the Fe–F–Fe bond angle from $129°$ (for $LiFe^{2+}SO_4F$) to $145°$ (for $Fe^{3+}SO_4F$).

When tested in standard lithium-half-cell architecture, $LiFeSO_4F$ depicts a voltage–composition profile having an $Fe^{2+/3+}$ redox activity located at 3.6 V [26] (Fig. 4b). This compound could reversibly insert up to 0.85 Li, via a two-phase process, giving rise to a capacity of 130 to 140 mAh/g (theoretical capacity = 151 mAh/g) at a $C/10$ rate with good cycling stability and rate capability. It is worth noting that no such performances were achieved with either highly divided powders or carbon coating. Such a difference was shown to be rooted in the existence of a three-dimensional diffusion for Li^+ in $LiFeSO_4F$ as opposed to one-dimension in $LiFePO_4$; this translates, for example, in (1) an ionic conductivity at 150°C, which is three-orders higher in magnitude for $LiFeSO_4F$ (10^{-6} S/cm) [26] than that of $LiFePO_4$ (10^{-9} S/cm), and (2) a faster Li^+ diffusion coefficient ($\sim10^{-14}$ cm^2/s) for $LiFeSO_4F$ [84]. Such experimental works were supported further by atomistic modeling, which revealed the absence of any Li–Fe antisite defects and low energy barriers (~0.4 eV) to Li migration in fluorosulfates, unlike olivines [85].

In addition to the electrochemical performance, another figure of merit regarding cathode applications deals with their water and thermal stability. The $LiFeSO_4F/FeSO_4F$ system was found to be stable up to 350°C, well beyond what is required for most battery use. In contrast, $LiFeSO_4F$ decomposes in water, implying the need to use nonaqueous solvents to prepare $LiFeSO_4F$-based electrodes. Overall, despite the need for careful handling, $LiFeSO_4F$ was found to be a workable Fe-based polyanionic cathode with an energy density (543 Wh/kg) comparable to that of $LiFePO_4$ (581 Wh/kg). Additionally, $LiFeSO_4F$ has a huge cost advantage, since it can be (1) prepared at low temperature and (2) made of abundant $FeSO_4 \cdot nH_2O$ precursor. Bearing in mind that within two years of its discovery it has stood out as a possible future cathode for large-scale applications, optimism must prevail regarding the future of this material.

4.2   LiMSO$_4$F (M = Co, Ni, Mn, Zn)

With the discovery of $LiFeSO_4F$, our group unveiled a new branch of fluorosulfate chemistry. This nascent family of compounds offered the possibility of designing other $3d$-metal (M = Mn, Co, Ni, etc.)-based fluorosulfates, preferably with higher redox potential while retaining a similar theoretical capacity. This work received great resonance from the battery community, and several reports appeared using density functional theory methods predicting higher redox potentials for novel $LiMSO_4F$

compounds (e.g., $Co^{2+/3+}$ redox at 4.9 V and $Ni^{2+/3+}$ redox at 5.4 V) [86–89]. This was an impetus to implement the freshly gained expertise in $LiFeSO_4F$ synthesis to that of other $3d$ homologs as summarized in Table 3. First, $LiMnSO_4F$, $LiCoSO_4F$, and $LiNiSO_4F$ were prepared using an equimolar reaction mixture of $MSO_4 \cdot H_2O$ (M = Co, Ni) and LiF by ionothermal synthesis at 290–300°C for 24 h [52]. Then, $LiZnSO_4F$ was reported using a $ZnSO_4 \cdot H_2O$ precursor [54]. Similar to $LiFeSO_4F$, all these $LiMSO_4F$ phases can be prepared successfully by low-temperature syntheses ($T < 300°C$) such as solid-state (dry) syntheses and polymer-assisted syntheses [80, 81], provided that $MSO_4 \cdot H_2O$ monohydrate precursors adopting the Szomolnokite structure ($C2/c$ space group) are used. In addition, although each phase involves the conversion of constituent octahedra from $M^{2+}O_6$ in the precursor to $[M^{2+}O_4F_2]$ in the final product, the size and local charge density of $3d$ metal (M) influence the size and relative degree of tilting or distortion of $M^{2+}O_4F_2$ octahedral units, which in turn can stabilize diverse $LiMSO_4F$ structures (Fig. 5a), as described below.

As expected, $LiCoSO_4F$ and $LiNiSO_4F$ were found to be isostructural to tavorite $LiFeSO_4F$, therefore with some slight difference in tilting of the $M^{2+}O_4F_2$ units [52,83]. More specifically, the M–F–M bond angle gradually rises from 129.47°

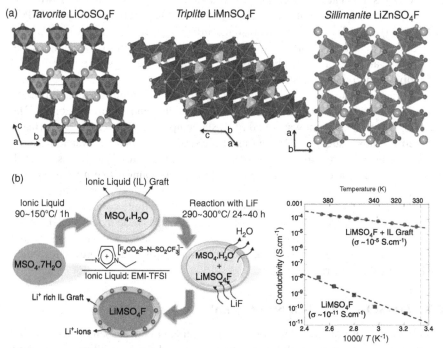

FIGURE 5   (a) Rich structural diversity in the fluorosulfate family of compounds shown for tavorite (P-1) $LiCoSO_4F$ (left), triplite ($C2/c$) $LiMnSO_4F$ (center), and sillimanite ($Pnma$) $LiZnSO_4F$ (right) compounds. (b) The mechanism of formation of ionic liquid graft during the synthesis of $LiMSO_4F$ compounds is illustrated schematically (left). Consequently, ionic liquid grafting leads to a sixfold increment in the ionic conductivity of $LiMSO_4F$ compounds, independent of their crystal structure (right).

(for Fe–F–Fe) to 130.28° (for Co–F–Co) to 131.38° (for Ni–F–Ni). When tested as possible cathode compounds vs. Li, these new fluorosulfates showed no sign of electrochemical activity up to 5 V. Future developments in cathode optimization and stable electrolytes should make it possible to tap into the nearly 4.9-V redox activity of the $LiMSO_4F$ (M = Co, Ni) phases. Using reaction conditions similar to those used earlier, the greater surprise came when the synthesis of $LiMnSO_4F$ was attempted, as it led to a completely different structure. The $LiMnSO_4F$ homologue adopted a monoclinic structure with $C2/c$ symmetry and was found to be isostructural to the naturally occurring mineral triplite $(Mg^{2+}, Ca^{2+}, Mn^{2+}, Fe^{2+})_2(PO_4)(F, OH)$ [90]. Similar to tavorite $LiMSO_4F$ phases, the $LiMnSO_4F$ structure consists of chains of $Mn^{2+}O_4F_2$ octahedra, interlinked with SO_4 tetrahedra (Fig. 5a). Nevertheless, the structure has two major differences from its tavorite $LiFeSO_4F$ counterpart. Unlike $LiFeSO_4F$, in the case of $LiMnSO_4F$, both Li and Mn atoms share the same crystallographic sites, each with a relative occupancy of about 50%. Moreover, the F atoms in the triplite $LiMnSO_4F$ occupy the cis position vis-à-vis the trans position in tavorite $LiFeSO_4F$, with the overall result being the formation of two interpenetrated chains of octahedra sharing either two O or F atoms. Overall, the triplite, which can be viewed as a disordered analog of tavorite structure, is quite dense and free of channels for Li^+ diffusion; this is a possible reason for the inactivity of $LiMnSO_4F$ toward Li. The larger size of Mn^{2+} compared to that of Fe^{2+}, Co^{2+}, and Ni^{2+} is believed to be at the origin of such a structural difference.

To gauge further the effect of metal cation size, the smallest $3d$ metal (i.e., the Zn homolog) was synthesized. Further structural diversity was unveiled as $LiZnSO_4F$, prepared at 300°C from an equimolar mixture of $ZnSO_4\cdot H_2O$ and LiF precursors, was found to crystallize in an orthorhombic framework with $Pnma$ symmetry with unit cell parameters $a = 7.40357(9)$ Å, $b = 6.32995(7)$ Å, $c = 7.42016(9)$ Å, and $V = 347.740(7)$ Å^3 [54]. Isostructural to the naturally occurring sillimanite $LiTiOPO_4$ mineral, $LiZnSO_4F$ is made up of chains of $Zn^{2+}O_4F_2$ octahedra sharing common F atoms and interconnected with SO_4 tetrahedra (Fig. 5a). Although structurally similar to the tavorite $LiMSO_4F$ discussed previously, the Zn homolog was found to offer a one-dimensional diffusion path for Li^+ (i.e., along the a-axis).

Besides their rich crystal chemistry, these $3d$-metal fluorosulfates display drastically different electrochemical properties. Except for $LiFeSO_4F$, they are all electrochemically inactive. Although research on these fluorosulfates could not yield potential cathodes, it led to an accidental finding of the concept of ionic liquid grafting as a way to improve the transport properties of inherently insulating ceramics [54,91]. The overall mechanism is illustrated schematically in Figure 5b. As described earlier, the synthesis of all fluorosulfates is accomplished by using in situ homemade $MSO_4\cdot H_2O$ precursors from corresponding $MSO_4\cdot nH_2O$ materials.

When the polyhedral materials are annealed inside ionic liquid media, parallel to the dehydration process, a thin layer of ionic liquid gets adsorbed to the surface of precursor particulates. This grafting was found to be bonded permanently to solid particles that could not be removed despite several vigorous washings in organic solvents or mild heat treatments up to 150°C. The subsequent reaction with LiF led to the simultaneous departure of structural H_2O as well as ingression of LiF through the ionic liquid layer, which dissolves some interfacial Li^+ ion during the synthesis, thus creating a Li^+-ion-rich ionic liquid graft surrounding $LiMSO_4F$ grains (Fig. 5b) [92]. The presence of a few nanometers (8 to 15 nm) thick ionic liquid grafts were noted clearly by probing with solid-state (^{19}F) nuclear magnetic resonance (NMR) and x-ray photoelectron spectroscopy (XPS) [54]. These materials were also tested for conductivity analysis by preparing pellets with ionically blocking electrodes. Conductivities as high as 10^{-5} S/cm (Fig. 5b) were obtained for ionothermally made $LiZnSO_4F$ compared solely to 10^{-11} S/cm for $LiZnSO_4F$ samples made by the solid-state process. Dc polarization measurements were conducted on both ionothermally and solid-state synthesized $LiZnSO_4F$ pellets, and a high current arising solely from Li^+ mobility was observed for the ionothermally made sample. Although the fluorosulfates have a rather low ionic conductivity, the ionic liquid graft forms a percolation network surrounding them, which favors easy Li^+ migration. This phenomenon was recorded for many $LiMSO_4F$ (M = Co, Mn, Zn) phases independent of their crystal structure [91]. It led to a generic concept that ionic liquid grafting can be an alternative way to improve the conductivity of otherwise insulating ceramic materials (e.g., oxides, fluorosulfates, phosphates) so as to enable their use as solid electrolytes in all-solid-state batteries.

4.3   Li(Fe₁₋ₓMₓ)SO₄F (M = Mn, Zn)

The fluorosulfates show a rich structural diversity, with their electrochemical properties being strongly correlated to their underlying structure. The Fe-based fluorosulfate adopts the tavorite structure with a redox activity at 3.6 V, while Mn- or Zn-based fluorosulfates crystallize, respectively, in triplite- and sillimanite-type structures with no redox activity. To throw some light on the origin of these tavorite–triplite or tavorite–sillimanite phase transitions and on their influence on electrochemical properties, their solid-solution phases were explored. $Li(Fe_{1-x}M_x)SO_4F$ (M = Mn, Zn) solid-solutions were prepared by reacting in-house fabricated mixed-metal $(Fe_{1-x}M_x) \cdot H_2O$ monohydrate precursors along with equimolar LiF at 300°C for 24 to 40 h by either dry solid-state or wet ionothermal/polymer-assisted synthesis [79–81]. Surprisingly, the structure of the resulting solid solution members was found to depend strongly on the reaction conditions (solid-state vs. ionothermal), the reason that the Mn and Zn substitutions for Fe are treated separately.

A nearly complete family of mixed-metal $Li(Fe_{1-x}Mn_x)SO_4F$ (0.03 < x < 1) with triplite structure was obtained by the dry synthesis route, whereas with wet synthesis the tavorite structure was retained until x = 0.2, after which it led to the formation of mixed phases, with anhydrous $MnSO_4$ as the main impurity (Fig. 6). The stabilization of two distinct polymorphs, despite using the same precursors and reaction temperatures, can be due to different reaction mechanisms in dry and wet routes.

Fe-rich $Li(Fe_{1-x}Mn_x)SO_4F$ (0 < x < 0.2) tavorite polymorphs tested in half-cell architecture revealed redox activity at 3.6 V much like that of $LiFeSO_4F$, with a standard two-phase voltage–composition curve with little polarization (Fig. 6) [57]. Surprisingly, members of $Li(Fe_{1-x}Mn_x)SO_4F$ triplite solid-solution polymorphs show a redox activity at 3.9 V. For example, the Fe-rich triplite $Li(Fe_{0.9}Mn_{0.1})SO_4F$ phase shows a reversible capacity close to 120 mAh/g (i.e., 0.7 to 0.8 Li^+) at a C/20 rate (Fig. 6). Such a redox potential was ascribed unambiguously to the Fe^{2+}/Fe^{3+}

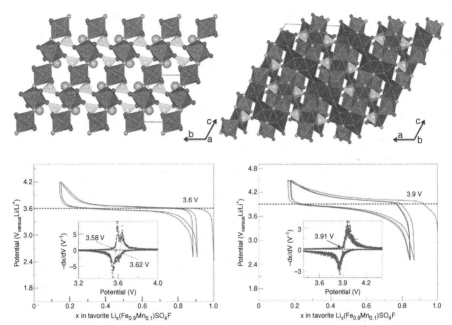

FIGURE 6 (Top) Mixed-metal $Li(Fe_{0.9}Mn_{0.1})SO_4F$ fluorosulfate demonstrating polymorphism by forming tavorite (left) and triplite (right) structure. (Bottom) While the tavorite structure gives rise to 3.6 V $Fe^{2+/3+}$ redox voltage (left), the triplite structure shows an $Fe^{2+/3+}$ redox voltage at 3.9 V (right). Both phases offer respectable reversible capacity and cycling stability. The inset diagrams show the differential capacity plots that point out the $Fe^{2+/3+}$ redox potentials exactly.

redox potential as confirmed by complementary in situ Mössbauer and XANES measurements. This constitutes the highest Fe^{2+}/Fe^{3+} redox potential ever reported for any compound. The decrease in capacity with increasing x for the triplite samples further confirms the electrochemical inactivity of Mn. Also, the greater polarization observed for the triplite phase compared to the tavorite phase can be explained simply by the Li/Mn site mixing, deduced by both synchrotron and Mössbauer measurements. Although at first glance the triplite structure seems to have no favorable Li^+ diffusion paths, Li^+ ions undergo a zigzag movement to avoid blocking transition metal sites. Although pure triplite structure does not offer favorable Li^+ diffusion paths (zigzag chains blocking Li^+ diffusion), we believe that the activity observed at low Mn content for the triplite phase is due to the preferential Li/Fe site mixing. With increasing Mn content, this preferential site mixing vanishes and no electrochemical activity is observed beyond $x = 0.5$. Another positive attribute of $Li(Fe_{0.9}Mn_{0.1})SO_4F$ triplite resides in the fact that Li insertion–removal is associated solely with a $\Delta V/V$ of 0.6% compared to 10% for the tavorite $LiFeSO_4F$ phase and 7.5% for olivine $LiFePO_4$. Such a small $\Delta V/V$ change inflicts minimum strain on the electrode upon cycling, hence enabling electrodes to display a long cycle life. Overall, the advent of the 3.9-V Fe-based triplite cathode is a major milestone for cathode research, and it calls for a synergistic investigation combining further experiments and atomistic modeling.

Similarly, our group studied the effect of wet vs. dry synthesis conditions on the synthesis of mixed-metal $Li(Fe_{1-x}Zn_x)SO_4F$ phases. In wet synthesis, the initial tavorite $LiFeSO_4F$ retained its structure up to 10% Zn substitution, after which it transformed into a sillimanite structure [i.e., tavorite for $x = 0$ to 0.1 and sillimanite for $x = 0.1$ to 1 $Li(Fe_{1-x}Zn_x)SO_4F$]. In contrast, the dry route led to the transition of tavorite to sillimanite through a brief zone of triplite [i.e., triplite for $x = 0.05$ to 0.15 and sillimanite for $x = 0.15$ to 1 in $Li(Fe_{1-x}Zn_x)SO_4F$] [58]. Thus, a thermodynamic trend of tavorite $\rightarrow$ triplite $\rightarrow$ sillimanite is marked. When tested electrochemically vs. Li, this sillimanite $Li(Fe_{0.9}Zn_{0.1})SO_4F$ delivered behavior identical to that of the parent tavorite $LiFeSO_4F$, delivering a reversible capacity of about 100 mAh/g, with the $Fe^{2+/3+}$ redox plateau located at 3.6 V. However, like the Mn-substituted triplite phases, the Zn-substituted intermediate $Li(Fe_{1-x}Zn_x)SO_4F$ ($x = 0.05$ to 0.15) triplite phases delivered a 3.9-V redox plateau with high polarization. The Fe–Zn binary systems further attest to the rich structural diversity and redox tunability in fluorosulfates.

The feasibility of triggering the transition from tavorite to sillimanite and/or triplite structure with a minimal (3 to 10%) degree of Mn or Zn substitution was an impetus to attempt the stabilization of $LiFeSO_4F$ in its triplite form, provided that the appropriate synthesis conditions could be determined. From a survey of several synthesis parameters and more specifically by using fast heating ($\sim6°C/min$) of the LiF and $FeSO_4 \cdot H_2O$ precursor mixture together with annealing at 300°C for 72 h [76], our group recently succeeded in preparing $LiFeSO_4F$ in its triplite form. This new $LiFeSO_4F$ polymorph shows a 3.9-V $Fe^{2+/3+}$ redox plateau vs. Li^+/Li^0 with a low capacity of about 85 mAh/g, large polarization and slow insertion–deinsertion kinetics. Further optimization at the synthesis level to favor Li/Fe site ordering via various annealing treatments and at the electrode configuration level is presently being pursued to approach its theoretical 150-mAh/g reversible capacity so as to turn this new 3.9-V electrode into an alternative candidate to $LiFePO_4$ for large-volume applications. Triplite $LiFeSO_4F$ could have an energy density similar to that of $LiFePO_4$ while providing cost advantages associated with its greener synthesis process, involving low temperature.

At this stage, a legitimate question deals with the origin of the 3.9-V potential specific to the triplite structure independent of Mn or Zn substitution. This increased potential can, in the first place, be rationalized on the basis of the ionocovalent character of the metal bonds. From synchrotron data, the average metal bond lengths were found to be 2.1034, 2.1062, and 2.1508 Å for the sillimanite, tavorite, and triplite phases, respectively [58,93]. The longest bond length reflects the greatest ionic character, and therefore, based on the inductive effect descriptor, it comes as no surprise that the Fe^{2+}/Fe^{3+} redox couple is the highest in triplite (3.9 V) and essentially of the same value in tavorite and sillimanite (3.6 V) since the corresponding bond length is nearly equal.

Now, turning back to the synthesis, we demonstrate that ceramic processes using fast or slow heating produce triplite and tavorite polymorphs. We believe that the origin of such a difference is rooted primarily in the crucial role of water, which dictates the reaction path. Slow heating rates delay the water departure and enable a topotactic reaction between $FeSO_4 \cdot H_2O$ and LiF, leading to the tavorite structure. In contrast, faster heating rates short-circuit the topotactic reaction by favoring a quick departure of water that leads to water-free sulfate precursors reacting with LiF. The complexity resides in the close thermodynamic stability of the two phases, which differs in energy by 2 kJ/mol, as deduced from preliminary calorimetric measurements [76].

Overall, this research on lithium metal fluorosulfates displays a rich crystal chemistry, which makes it possible to fine-tune the $Fe^{2+/3+}$ redox voltage. Further, it imparts a sublime message that cathode compounds on the verge of structural stability can be vital in realizing superior electrochemical performances.

4.4   NaMSO₄F (M = Fe, Co, Ni, Mn, Mg, Zn, Cu)

It took seven years (2003 to 2010) to discover Li-based fluorophosphates and solve their structure. Progress is faster with Li-based fluorosulfates, which were unveiled in 18 months and now count more than five Li-based compounds, with some of them showing attractive electrochemical ($LiFeSO_4F$) and ionic ($LiZnSO_4F$) performance. To further enlarge the fluorosulfate family, we considered in parallel the feasibility of preparing various $3d$ metals containing sodium fluorosulfates by reacting equimolar mixtures of NaF and homemade $MSO_4 \cdot H_2O$ monohydrate precursors at 300°C by either ionothermal, polymer-assisted, or solid-state syntheses [53,56]. Such a strategy, based on a topotactic reaction involving dual ion-exchange reactions, led to the discovery of numerous novel $NaMSO_4F$ (M = Fe, Co, Ni, Mn, Cu) phases. Alternatively, most of these phases could be prepared using a two-step route consisting of synthesizing dihydrated $NaMSO_4F \cdot 2H_2O$ phases via a dissolution–precipitation mechanism (as described in Section 4.5), followed by their dehydration at 200 to 300°C in either ionic liquid media or in a primary vacuum [55]. The $NaMgSO_4F$ and $NaZnSO_4F$ phases were synthesized by the latter route, as the former topotactic reaction ($MSO_4 \cdot H_2O$ + NaF) invariably led to mixed phases.

These new phases crystallized in a monoclinic framework (space group $C2/c$), isostructural to the naturally occurring mineral maxwellite [94]. Similar to tavorite and sillimanite $LiMSO_4F$ phases, the $NaMSO_4F$ compounds possess chains of corner-sharing $[M^{3+}O_4F_2]$ octahedra along the (001) direction, having F atoms in the trans position, and connected by SO_4 tetrahedra to form a three-dimensional framework offering large cavities to host Na atoms (Fig. 7a). Due mainly to the larger ionic radii of Na^+ compared to Li^+, the $NaMSO_4F$ phases show a larger unit cell volume (5 to 8%) than that of their $LiMSO_4F$ counterparts as well as higher symmetry (triclinic→ monoclinic). Comparing these $NaMSO_4F$ homologues, the unit cell volume gradually increases with increasing ionic radii of constituent $3d$ metals (Fig. 7b). The unit cell parameters of these $NaM^{2+}SO_4F$ homologues, follow the trend of M^{2+} ionic radii, with the exceptional case of $NaCuSO_4F$. In that case, due mainly to Jahn–Teller instability of Cu^{2+} ion in an octahedral environment, the $Cu^{3+}O_4F_2$ octahedra are highly distorted (e.g., elongated along the Cu–F direction) with Cu^{2+} presenting distinct square-planar coordination. Also, worth mentioning is the fact that $NaMnSO_4F$ behaves differently and crystallizes in a monoclinic triplite structure, similar to $LiMnSO_4F$ [53].

All these novel $NaMSO_4F$ phases were found to be electrochemically inactive except for $NaFeSO_4F$, which when tested vs. Na led to a meager 7% of its theoretical

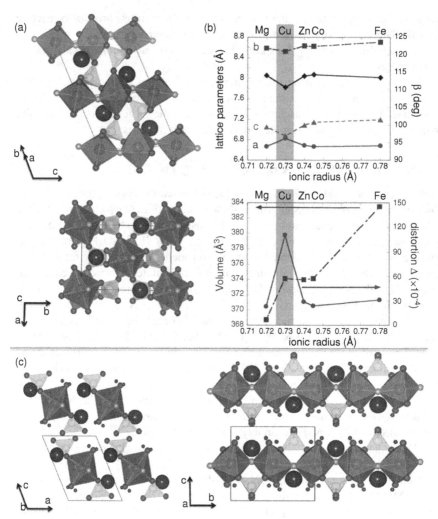

FIGURE 7 (a) Crystal structure of $NaMSO_4F$ phases showing the tavorite cages (top) and tilting of neighboring distorted octahedra (bottom). The MO_4F_2 octahedra, SO_4 tetrahedra, and Na atoms are shown in grey, light grey and black colors, respectively. (b) Structural evolution in $NaMSO_4F$ compounds. The variation of lattice parameters (top) and octahedral distortion (bottom) are shown to highlight Jahn–Teller distortion in the case of $NaCuSO_4F$. (c) Structure of $NaMSO_4F \cdot 2H_2O$ phases shown perpendicular (left) and parallel (right) to the MO_4F_2 octahedra chains.

capacity (137 mAh/g) with a visible $Fe^{2+/3+}$ redox plateau located at 3.6 V [53]. There was little success in chemical oxidation by NO_2BF_4 as well as in ion exchange by $LiCl/LiBr$. This reluctant Na^+ activity may be due to the large volume contraction ($\sim16\%$) involved with $NaFe^{2+}SO_4F/Fe^{3+}SO_4F$ redox reaction, which is energetically not feasible despite the lattice energy associated with the formation of $NaCl$ or $NaBr$. Further atomistic modeling study hints at a one-dimensional zigzag Na^+-ion motion with a high activation energy barrier ($\sim$0.9 eV). Thus, the electrochemical activity of NaMSO$_4$F phases is structurally limited [85].

4.5   NaMSO$_4$F·2H$_2$O (M = Fe, Co, Ni, Zn)

Pursuing the moisture stability of the NaMSO$_4$F systems further, in 2010, our group unveiled a series of novel bihydrated NaMSO$_4$F·2H$_2$O compounds via solvothermal processing [55,95]. Overall, stoichiometric mixtures of NaF and MSO$_4$·nH$_2$O precursors were dissolved in distilled water and the solution was heated at 80°C under stirring so as to evaporate about 90% of the water. The resulting saturated solution was treated with a large excess of ethanol to obtain the NaMSO$_4$F·2H$_2$O (M = Co, Ni) precipitates, which were recovered and then vacuum dried. For NaFeSO$_4$F·2H$_2$O, the same process was used with special care to prevent Fe$^{2+/3+}$ oxidation by handling the aqueous solution and precipitation under steady N$_2$ flow. In the case of NaZnSO$_4$F·2H$_2$O, an extra step of washing with methanol was used to remove any impurities [56].

These novel bihydrated phases were found to be isostructural to the uklonskovite mineral (NaMgSO$_4$F·2H$_2$O) with a monoclinic structure (space group $P2_1/m$) [96]. Their structure is built from parallel chains of M^{2+}O$_4$F$_2$ octahedra (along the b-axis) linked together with F atoms in the trans position. Of the four equatorial O atoms, two O atoms are linked to 2 H atoms each (forming H$_2$O units) while the remaining two O atoms are part of the SO$_4$ tetrahedra (Fig. 7c). Na atoms occupy the cavities forming NaO$_3$F (pseudo) tetrahedra units. In fact, these NaO$_3$F tetrahedra are highly distorted and act as bridging units between neighboring (M^{2+}O$_4$F$_2$)$_\infty$ chains to form a three-dimensional structure. Interestingly, these bihydrated compounds can be dehydrated in ionic liquid media or in primary vacuum at temperatures of 200°C, and this water removal was shown to be associated with a structural rearrangement to form respective NaMSO$_4$F products [55]. Similar to NaMSO$_4$F compounds, these bihydrated phases showed no activity when examined for (electro)chemical oxidation–reduction or ion exchange. At this juncture it is worth mentioning that the NaMSO$_4$F phases obtained from temperature-driven water removal can convert back to their precursor bihydrated phase upon moisture exposure, whereas NaMSO$_4$F made directly from a topotactic reaction (MSO$_4$·H$_2$O + NaF) phase does not uptake water. Besides the existence of a memory effect, which could exist for the former, we believe that such a difference is the result of a surfacial effect probably due to the presence of a hydrophobic ionic liquid monolayer at the surface of the particles, grown during the preparation of the precursor MSO$_4$·H$_2$O phase in ionic liquid media. This hydrophobic layer provides the enhanced stability of these materials toward water reactivity.

5   PERSPECTIVES AND SUMMARY

Rechargeable batteries have steadily become an indispensable part of modern life rich in portable electronics and communications. The continuous demands for sleeker, more efficient, more cost-friendly devices, and more demand for eco-friendly automotives challenge electrochemists to build better batteries to cater to these needs [97]. In this case, polyanionic framework compounds offer chemists a vast platform from which to design newer, more attractive cathodes, $LiFePO_4$ being the first example. Moving ahead of $LiFePO_4$, alternative cathodes can be found in fluorine-based polyanionic compounds which cleverly combine the higher electronegativity of F with the inductive effect of polyanion to tune the redox potential of the final material.

Such a strategy was initiated earlier to produce $LiVPO_4F$ compounds showing redox activity at 4.1 V and, more recently, the Fe-based fluorosulfate $LiFeSO_4F$ showing the highest potential ever reported for a Fe^{2+}/Fe^{3+} redox couple, also with the early feasibility to reach the Fe^{3+}/Fe^{4+} redox couple in $LiFePO_4F$ electrochemically. Similarly, numerous Ni- and Co-based fluorophosphates and fluorosulfates with a potential greater than 4.5 V, as deduced by experimental calculations or predicted by DFT calculations, could be achieved. Therefore, the practical difficulty here resides in tapping into the attractive theoretical capacity of these high-voltage materials, owing to the lack of electrolytes capable of sustaining proper operation at voltages greater than 4.5 V without progressively falling apart. No doubt, pursuing research toward high-voltage materials is a great thing, but for such effort to have a practical meaning it must be coupled with similar activities toward designing electrolytes that are highly resistant to oxidation. Thus, there is a need for the battery community to beef up its activity in this sector by looking at both liquid and inorganic solid electrolytes.

F-based compounds usually display higher redox potential with respect to their oxide counterparts, owing to the M–F bond's larger ionicity compared to the M–O bond, but a counter effect of this benefit with respect to practical applications is that F-based compounds are poorer electronic conductors than their oxide counterparts, and therefore usually show sluggish electrode kinetics. To bypass this issue, greater efforts must be devoted toward improving electrode wiring, which can be achieved via the use of dual-carbon-coating nanosizing approaches, provided that the F-based materials studied are thermally stable up to temperatures of 600°C under a reducing environment.

To date, fluoride-based materials have been made primarily by solid-state reactions using several bake-and-shake methods. Through this review we have shown how innovative low-temperature synthesis enlisting well-known hydro(solvo)thermal reactions or newly developed ionothermal approaches were used (1) to prepare known Li- or Na-based polyanionic electrodes such as $LiFePO_4$, $LiFePO_4F$, $LiTiPO_4F$, and Na_2FePO_4F at temperatures 400 to 500°C lower than what is necessary for solid-state reactions or (2) to unravel new metastable and attractive electrode materials such as the fluorosulfate $LiFeSO_4F$. Through this finding we also recalled the importance of understanding the reaction mechanism governing the formation of a new compound. Indeed, having identified the structural relationship between the precursor phase (mother phase) and daughter phase underlying the topotactic reaction, we could in less than 18 months synthesize more than 18 new fluorophosphate phases. This highlights the importance for solid-state chemists of spotting a homothetic family of compounds in order to hunt for new materials. Regarding these novel fluorosulfate phases, we systematically found a different structure and electrochemical behavior for Mn-based compounds, a repeat of what has also been encountered with F(OH)-based phosphates. We hope that such constancy in the peculiar behavior of Mn-based polyanion compounds will attract the theoreticians' attention so that we soon have a rational explanation. Pursuing high-throughput theoretical calculations to predict and spot new materials is certainly needed, and new formalism must be developed to help solid-state chemists find new materials even though success so far has been limited. Nevertheless, although less elegant, sorting out all the reported inorganic compounds by family, on the basis of structural relation, would be of great help to solid-state chemists during this mid-term period, and let's hope that this call will receive answers.

Besides being prepared by eco-efficiency processes, $LiFePO_4$ and $LiFeSO_4F$ rely on the use of abundant resources to meet the conditions of sustainability. Indeed, iron phosphates and sulfates are both sources of cheap Fe salt. $LiFePO_4$, and especially, $FeSO_4 \cdot H_2O$ are abundant minerals known as triphyllite and szomolnokite, respectively, the latter being the product of pyrite FeS_2 weathering. So developing such electrode materials constitutes a first step toward the development of sustainable electrodes, and efforts in these directions must be pursued. Searching for new phosphates or sulfate-based electrode materials is quite attractive, although such a chemistry is quite complex. The reason for this is that sulfates are known for their water solubility, owing to the fact that, in solution, M–O bonds are equally likely to form with either H_2O or $(SO_4)^{2-}$ as the bonding valence for the oxygen within these two species is ≈ 0.17 v.u. This solubility problem is less crucial for the phosphate $(PO_4)^{3-}$ species, due to their greater oxygen bond valences (≈ 0.25 v.u.). So, to eschew some issues it is important to work with nonaqueous media: hence the interest in ionic liquids or other solvents, such as dimethyl fluoride or dimethyl sulfoxide. Also, stabilizing Fe^{2+}-based compounds is always troublesome, owing to the ease with which Fe^{2+} gets oxidized in Fe^{3+}. Based on such remarks and on our own experience, we advise those willing to explore sulfate chemistry to start by stabilizing new phases with Co, as Co^{2+} is very stable in these media. Then, if a Li–Co-based sulfate or fluorosulfate does form, it should be an impetus to prepare the Fe homologue, provided that we find the right protocol to maintain Fe in its Fe^{2+} state.

Finally, depending on the synthesis conditions, we have shown and illustrated with $LiFeSO_4F$ the feasibility of preparing two polymorphs showing either 3.6 or 3.9 V for the Fe^{2+}/Fe^{3+} redox couple. We hope through this example to have conveyed the message that the electrochemical property that a compound possesses can never be attributed to a single factor and is, instead, a harmonious integration of the atomic structure, the elemental composition, and the material processing. This opens a myriad of opportunities for discovering new materials by mastering the synthesis, such as the possibility of transforming known tavorite structures into new triplite, or stabilizing Na-based triplite, to name but a few.

In summary, F-based fluorophosphates and fluorosulfates form a major class of polyanionic cathode materials, offering suitable crystal structure and competent electrochemical capacity. The constituent alkali and 3d-metal cations play a key role in forming a rich structural diversity, which in turn offers redox activity tunability. The current chapter is meant to capture the decade-long (2003 to 2012) journey of conception and evolution of the rich world of F-based polyanionic cathode compounds, elaborating their syntheses, crystal structure, and electrochemical performance. In particular, Na_2FePO_4F (3.3 V) and $LiFeSO_4F$ polymorphs (3.6 to 3.9 V) are commercially feasible for large-scale application. This novel group of materials, still in its infancy, forms an alternative platform for the discovery and development of the next generation of cathode materials. Further, low-temperature synthesis and abundant precursors make them attractive for commercial ventures. Currently, the F-based polyanionic compounds are expected to realize commercial battery production in the coming 5 to 10 years.

Acknowledgments

The authors thank the group members of LRCS and Alistore-ERI for their scientific assistance, for stimulating many technical discussions, and G. Rousse for her expertise in crystallography. We especially acknowledge Michèle Nelson for her careful proofreading and semantic improvement of the manuscript. Financial support from ALISTORE-ERI and the French Ministry of Science is gratefully acknowledged. The first author (P.B.) is grateful to the Japan Society for the Promotion of Sciences for a fellowship at the University of Tokyo (Japan) and to Professor Atsuo Yamada for his kind support.

REFERENCES

1. R. E. Smalley, *MRS Bull.*, **30**, 412 (2005).
2. M. Armand and J.-M. Tarascon, *Nature (London)*, **451**, 652 (2008).
3. B. Dunn, H. Kamath, and J.-M. Tarascon, *Science*, **334**, 928 (2011).
4. J.-M. Tarascon, *Philos. Trans. A*, **368**, 3227 (2010).
5. P. G. Bruce, S. A. Freunberger, L. J. Hardwick, and J.-M. Tarascon, *Nat. Mater.*, **11**, 19 (2012).
6. K. Mizushima, P. C. Jones, P. J. Wiseman, and J. B. Goodenough, *Mater. Res. Bull.*, **15**, 783 (1980).
7. T. Ohzuku, M. Kitagawa, and T. Hirai, *J. Electrochem. Soc.*, **137**, 769 (1990).
8. K. Amine, H. Tukamoto, H. Yasuda, and Y. Fujita, *J. Power Sources*, **68**, 604 (1997).
9. T. Ohzuku and Y. Makimura, *Chem. Lett.*, **7**, 642 (2001).
10. M. Guilmard, C. Pouillerie, L. Croguennec, and C. Delmas, *Solid State Ionics*, **160**, 39 (2003).
11. A. K. Padhi, K. S. Nanjundaswamy, and J. B. Goodenough, *J. Electrochem. Soc.*, **144**, 1188 (1997).
12. A. K. Padhi, K. S. Nanjundaswamy, C. Masquelier, S. Okada, and J. B. Goodenough, *J. Electrochem. Soc.*, **144**, 1609 (1997).
13. N. Ravet, Y. Chouinard, J. F. Magnan, S. Besner, M. Gauthier, and M. Armand, *J. Power Sources*, **97**, 503 (2001).
14. A. Yamada, S. C. Chung, and K. Hinokuma, *J. Electrochem. Soc.*, **148**, 224 (2001).
15. A. Yamada and S. C. Chung, *J. Electrochem. Soc.*, **148**, 960 (2001).
16. A. Yamada, Y. Kudo, and K. Y. Liu, *J. Electrochem. Soc.*, **148**, A1153 (2001).
17. S. Y. Chung, J. T. Blocking, and Y. M. Chiang, *Nat. Mater.*, **1**, 123 (2002).
18. A. Nyten, A. Abouimrane, M. Armand, T. Gustafsson, and J. O. Thomas, *Electrochem. Commun.*, **7**, 156 (2005).
19. A. Yamada, I. Iwane, Y. Harada, S. Nishimura, Y. Koyama, and I. Tanaka, *Adv. Mater.*, **22**, 3583 (2010).
20. S. Nishimura, M. Nakamura, R. Natsui, and A. Yamada, *J. Am. Chem. Soc.*, **132**, 13596 (2010).
21. P. Poizot, S. Laruelle, S. Grugeon, L. Dupont, J.-M. Tarascon, *Nature (London)*, **407**, 496(2001).
22. H. F. Bauman, *Proceedings of the 20th Annual Power Sources Conference*, 1996 PSC Pub., p. *73*.
23. G. G. Amatucci and N. Pereira, *J. Fluorine Chem.*, **128**, 243 (2007).
24. J. Barker, M. Y. Saidi, and J. L. Swoyer, *J. Electrochem. Soc.*, **150**, 1394 (2003).
25. J. Barker, M. Y. Saidi, and J. L. Swoyer, *J. Electrochem. Soc.*, **151**, 1670 (2004).

26. N. Recham, J.-N. Chotard, L. Dupont, C. Delacourt, W. Walker, M. Armand, and J.-M. Tarascon, *Nat. Mater.*, **9**, 68 (2010).
27. K. M. Abraham, *J. Power Sources*, **14**, 179 (1985).
28. T. Nakajima, and H. Groult, Eds., *Fluorinated Materials for Energy Conversion*, Elsevier, New York, 2005.
29. K. Kubo, M. Fujiwara, S. Yamada, S. Arai, and M. Kanda, *J. Power Sources*, **68**, 553 (1997).
30. G. G. Amatucci, A. Blyr, C. Schmutz, and J.-M. Tarascon, *Prog. Batteries Battery Mater.*, **16**, 1 (1997).
31. T. J. Kim, D. Son, J. Cho, and B. Park, *J. Power Sources*, **154**, 268 (2006).
32. Y. K. Sun, J.-M. Han, S. T. Myung, S. W. Lee, and K. Amine, *Electrochem. Commun.*, **8**, 821 (2006).
33. R. Yazami and A. Hamwi, *Solid State Ionics*, **28**, 1756 (1988).
34. F. Badway, F. Cosandey, N. Pereira, and G. G. Amatucci, *J. Electrochem. Soc.*, **150**, 1318 (2003).
35. H. Li, G. Ritcher, and J. Maier, *Adv. Mater.*, **15**, 736 (2003).
36. N. Pereira, F. Badway, M. Wartelsky, S. Gunn, and G. G. Amatucci, *J. Electrochem. Soc.*, **156**, 407 (2009).
37. J. Barker, M. Y. Saidi, and J. Swoyer, U.S. Patent 6,387,568, 2002.
38. J. Barker, M. Y. Saidi, and J. L. Swoyer, *Electrochem. Solid-State Lett.*, **6**, 1 (2003).
39. J. Barker, R. K. B. Grover, P. Burns, A. Bryan, M. Y. Saidi, and J. L. Swoyer, *J. Electrochem. Soc.*, **152**, 1776 (2005).
40. J. Barker, R. K. B. Grover, P. Burns, M. Y. Saidi, and J. L. Swoyer, *J. Power Sources*, **146**, 516 (2005).
41. R. K. B. Grover, P. Burns, A. Bryan, M. Y. Saidi, J. L. Swoyer, and J. Barker, *Solid State Ionics*, **177**, 2635 (2006).
42. J. Barker, M. Y. Saidi, R. K. B. Grover, P. Burns, and A. Bryan, *J. Power Sources*, **174**, 927 (2007).
43. B. L. Ellis, T. N. Ramesh, L. J.-M. Davis, G. R. Goward, and L. F. Nazar, *Chem. Mater.*, **23**, 5138 (2011).
44. L. S. Plashnitsa, E. Kobayashi, S. Okada, and J. Yamaki, *Electrochim. Acta*, **56**, 1344 (2011).
45. S. Okada, M. Ueno, Y. Uebou, and J. Yamaki, *J. Power Sources*, **146**, 565 (2005).
46. B. L. Ellis, W. R. M. Makahnouk, Y. Makimura, K. Toghill, and L. F. Nazar, *Nat. Mater.*, **6**, 749 (2007).
47. B. L. Ellis, W. R. M. Makahnouk, W. N. Rowan-Weetaluktuk, D. H. Ryan, and L. F. Nazar, *Chem. Mater.*, **22**, 1059 (2010).
48. N. Recham, J.-N. Chotard, J. C. Jumas, L. Laffont, M. Armand, and J.-M. Tarascon, *Chem. Mater.*, **22**, 1142 (2010).
49. T. Ramesh, K. T. Lee, B. L. Ellis, and L. F. Nazar, *Electrochem. Solid-State Lett.*, **13**, 43 (2010).
50. M. Nagahama, N. Hasegawa, and S. Okada, *J. Electrochem. Soc.*, **157**, 748 (2010).
51. L. Sebastian, J. Gopalakrishnan, and Y. Piffard, *J. Mater. Chem.*, **12**, 374 (2002).
52. P. Barpanda, N. Recham, J.-N. Chotard, K. Djellab, W. Walker, M. Armand, and J.-M. Tarascon, *J. Mater. Chem.*, **20**, 1659 (2010).
53. P. Barpanda, J.-N. Chotard, N. Recham, C. Delacourt, M. Ati, L. Dupont, M. Armand, and J.-M. Tarascon, *Inorg. Chem.*, **49**, 7401 (2010).
54. P. Barpanda, J.-N. Chotard, C. Delacourt, M. Reynaud, Y. Filinchuk, M. Armand, M. Deschamps, and J.-M. Tarascon, *Angew. Chem. Int. Ed.*, **50**, 2526 (2011).
55. M. Ati, L. Dupont, N. Recham, J.-N. Chotard, W. T. Walker, C. Davoisne, P. Barpanda, V. Sarou-Kanian, M. Armand, and J.-M. Tarascon, *Chem. Mater.*, **22**, 4062 (2010).
56. M. Reynaud, P. Barpanda, G. Rousse, J.-N. Chotard, B. C. Melot, N. Recham, and J.-M. Tarascon, *Solid State Sci.*, **14**, 15 (2012).
57. P. Barpanda, M. Ati, B. C. Melot, G. Rousse, J.-N. Chotard, M. L. Doublet, M. T. Sougrati, S. A. Corr, J. C. Jumas, and J.-M. Tarascon, *Nat. Mater.*, **10**, 772 (2011).
58. M. Ati, B. C. Melot, G. Rousse, J.-N. Chotard, P. Barpanda, and J.-M. Tarascon, *Angew. Chem. Int. Ed.*, **50**, 10574 (2011).
59. M. Ramzan, S. Lebegue, P. Larsson, and R. Ahuja, *J. Appl. Phys.*, **106**, 043510 (2009).
60. M. Dutreilh, C. Chevalier, M. El-Ghozzi, and D. Avignant, *J. Solid State Chem.*, **142**, 1 (1999).
61. M. Nagahama, N. Hasegawa, and S. Okada, *J. Electrochem. Soc.*, **157**, A748 (2010).

62. H. Zhuo, X. Wang, A. Tang, Z. Liu, S. Gambao, and P. J. Sebastian, *J. Power Sources*, **160**, 698 (2006).
63. N. Recham, J.-N. Chotard, L. Dupont, K. Djellab, M. Armand, and J.-M. Tarascon, *J. Electrochem. Soc.*, **156**, 993 (2009).
64. H. T. Ellingham, *J. Soc. Chem. Ind.*, **63**, 125 (1944).
65. O. V. Yakubovich and V. S. Urusov, *Geokhimiya*, **7**, 720 (1997).
66. J. Barker, R. K. B. Gover, P. Burns, and A. Bryan, *Electrochem. Solid-State Lett.*, **8**, A285 (2005).
67. L. J.-M. Davis, B. L. Ellis, T. N. Ramesh, L. F. Nazar, A. D. Bain, and G. R. Goward, *J. Phys. Chem. C*, **115**, 22603 (2011).
68. J.-M. Le Meins, M.-P. Cronier-Lopez, A. Hemon-Ribaud, and G. Courbion, *Solid State Ionics*, **111**, 67 (1998).
69. J.-M. Le Meins, M.-P. Cronier-Lopez, A. Hemon-Ribaud, and G. Courbion, *J. Solid State Chem.* **148**, 260 (1999).
70. J. Zhao, J. He, X. Ding, J. Zhuo, Y. Ma, S. Wu, and R. Huang, *J. Power Sources*, **195**, 6854 (2010).
71. J. Yu, K. M. Rosso, J. G. Zhang, and J. Liu, *J. Mater. Chem.*, **21**, 12054 (2011).
72. Y. Kawabe, N. Yabuuchi, M. Kajiyama, N. Fukuhara, T. Inamasu, R. Okuyama, I. Nakai, and S. Komaba, *Electrochem. Commun.*, **13**, 1225 (2011).
73. J. Hadermann, A. M. Abakumov, S. Turner, Z. Hafideddine, N. R. Khasanova, E. V. Antipov, and G. Van Tendeloo, *Chem. Mater.*, **23**, 3540 (2011).
74. N. R. Khasanova, A. N. Gavrilov, E. V. Antipov, K. G. Bramnik, and H. Hibst, *J. Power Sources*, **196**, 355 (2011).
75. J. Barker, M. Y. Saidi, and J. L. Swoyer, *US Patent*, US2003013019-A1.
76. M. Ati, B. C. Melot, J.-N. Chotard, G. Rousse, M. Reynaud, and J.-M. Tarascon, *Electrochem. Commun.*, **13**, 1280 (2011).
77. N. Recham, J.-M. Tarascon, and M. Armand, *World Patent*, WO2010046610-A1.
78. N. Recham, L. Dupont, M. Courty, K. Djellab, D. Larcher, M. Armand, and J.-M. Tarascon, *Chem. Mater.*, **21**, 1096 (2010).
79. J.-M. Tarascon, N. Recham, M. Armand, J.-N. Chotard, P. Barpanda, W. Walker, and L. Dupont, *Chem. Mater.*, **22**, 724 (2010).
80. M. Ati, M. T. Sougrati, N. Recham, P. Barpanda, J. B. Leriche, M. Courty, M. Armand, J. C. Jumas, and J.-M. Tarascon, *J. Electrochem. Soc.*, **157**, 1007 (2010).
81. M. Ati, W. T. Walker, K. Djellab, M. Armand, N. Recham, and J.-M. Tarascon, *Electrochem. Solid-State Lett.*, **13**, 150 (2010).
82. R. Tripathi, T. N. Ramesh, B. L. Ellis, and L. F. Nazar, *Angew. Chem. Int. Ed.*, **49**, 8738 (2010).
83. B. C. Melot, G. Rousse, J.-N. Chotard, M. Ati, J. Rodriguez-Carvajal, M. C. Kemei, and J.-M. Tarascon, *Chem. Mater.*, **23**, 2922 (2011).
84. C. Delacourt, M. Ati, and J.-M. Tarascon, *J. Electrochem. Soc.*, **158**, 741 (2011).
85. R. Tripathi, G. R. Gardiner, M. S. Islam, and L. F. Nazar, *Chem. Mater.*, **23**, 2278 (2011).
86. M. Ramzan, S. Lebegue, and R. Ahuja, *Phys. Rev. B*, **82**, 125101 (2010).
87. C. Frayret, A. Villesuzanne, N. Spaldin, E. Bousquet, J.-N. Chotard, N. Recham, and J.-M. Tarascon, *Phys. Chem. Chem. Phys.*, **12**, 15512 (2010).
88. Y. Cai, G. Chen, X. Xu, F. Du, Z. Li, X. Meng, C. Wang, and Y. Wei, *J. Phys. Chem. C*, **115**, 7032 (2011).
89. T. Muller, G. Hautier, A. Jain, and G. Ceder, *Chem. Mater.*, **23**, 3854 (2011).
90. J. R. Rea and E. Kostiner, *Acta Crystallogr. B*, **28**, 2525 (1972).
91. P. Barpanda, R. Dedryvère, M. Deschamps, C. Delacourt, M. Reynaud, A. Yamada, and J.-M. Tarascon, *J. Solid State Electrochem.*, **16**, 1743 (2012).
92. M. Armand, F. Endres, D. R. MacFarlane, H. Ohno, and B. Scrosati, *Nat. Mater.*, **8**, 621 (2009).
93. R. D. Shannon and C. T. Prewitt, *Acta Crystallogr. B*, **25**, 925 (1969).
94. M. A. Cooper and F. C. Hawthorne, *Neues Jahrb. Mineral.*, **3**, 97 (1995).
95. B. C. Melot, J.-N. Chotard, G. Rousse, M. Ati, M. Reynaud, and J.-M. Tarascon, *Inorg. Chem.*, **50**, 7662 (2011).
96. C. Sabelli, *Bull. Mineral.*, **108**, 133 (1985).
97. J.-M. Tarascon and M. Armand, *Nature (London)*, **414**, 359 (2001).

CHAPTER **8**

LITHIUM–AIR AND OTHER BATTERIES BEYOND LITHIUM-ION BATTERIES

K. M. Abraham

1 INTRODUCTION

The nonaqueous lithium–air battery, comprised of a lithium-metal anode and an oxygen cathode, is the highest-energy-density battery that can be developed into a practical system. It promises to overcome the energy density limitations of lithium-ion batteries, which have played an unequaled part in the modern electronic technology revolution. Lithium-ion batteries are indispensable in our everyday life as power sources for wireless telephones, laptop and tablet computers, music players, digital cameras, and many other consumer devices. The discovery of high-energy-density (energy stored per unit mass and volume) battery chemical couples based on Li intercalating negative and positive electrodes together with chemically and electrochemically compatible nonaqueous electrolytes, and engineering of the resulting Li-ion

Lithium Batteries: Advanced Technologies and Applications, First Edition.
Edited by Bruno Scrosati, K. M. Abraham, Walter van Schalkwijk, and Jusef Hassoun.
© 2013 John Wiley & Sons, Inc. Published 2013 by John Wiley & Sons, Inc.

cells to perform over a wide range of temperatures for hundreds of discharge–charge cycles, ushered in the modern era of rechargeable batteries. Today Li-ion batteries have replaced conventional aqueous batteries in most portable devices. Despite their huge success and ubiquitous presence, there is an ever-increasing need for batteries with significantly higher energy and power densities than those of Li-ions to meet the demands of new power-hungry consumer devices, electric automobiles, large-scale energy storage, aerospace technologies, power tools, and many other future applications.

The high degree of motivation for the development of Li–air batteries can be illustrated by examining the energy density requirements of rechargeable batteries for the all-electric automobiles capable of driving 250 to 300 miles on a single charge. It can be shown that the rechargeable battery pack for a 300-mile-range all-electric family car such as the Toyota Camry requires cells having a specific energy greater than 500 Wh/kg, as opposed to the 225-Wh/kg Li-ion cells available today [1]. In the Camry, the engine plus the gasoline tank contribute approximately 400 kg to the total 1500-kg mass of the vehicle. This is the weight available for the battery pack without altering the conveniences in the car. Assuming that 1 kWh of battery will provide 3 miles of driving, a 100-kWh battery pack is needed for driving 300 miles on a single charge. In practice, it is necessary to oversize the battery by about 40% for two purposes; (1) to have a 20% reserve capacity during normal operation of the car, and (2) to compensate for the 20% capacity fade during the life of the battery. Thus, the 300-mile-range car would be powered initially by a 140-kWh battery. The weight energy density of such a 400-kg battery pack is then 350 Wh/kg. If the efficiency for the conversion of cells to a battery pack is 60%, the energy density of the cells needed to build the 300-mile-range battery pack is about 580 Wh/kg, more than twice that of today's Li-ion cells. To build such batteries using Li-ion technology, positive electrode materials with significantly higher specific capacities than those presently available have to be developed. Today's highest-specific-capacity Li-ion positive electrode is $LiNi_{0.80}Co_{0.15}Al_{0.05}O_2$ (NCA), a variant of the original Li intercalating layered $LiCoO_2$ [2]. NCA has a reversible specific capacity of about 190 mAh/g, which is 70% of the theoretical one-electron reaction of this electrode material. Commercial 18650-size Li-ion cells built with NCA and a graphite negative electrode have a specific energy of 240 Wh/kg. The reversible capacity for a transition metal oxide positive electrode material capable of one-electron positive electrode reaction is about 270 mAh/g. A 3.6-V Li-ion cell built with such a metal oxide cathode and a graphite negative electrode would exhibit a specific energy of about 300 Wh/kg. A class of positive electrode materials potentially capable of such a high capacity is that of the lithium-rich layered manganese oxides (layered LMOs) [3], exemplified by the molecular formula $Li[Li_{0.2}Mn_{0.54}Ni_{0.13}Co_{0.13}]O_2$. This mixed metal oxide and others related to it have shown practical reversible capacities up to 250 mAh/g, albeit at low rates. The projected energy density of the corresponding 18650 Li-ion cells is 280 Wh/kg, increasing to approximately 320 Wh/kg if the graphite negative electrode is replaced by silicon.

The evolution of Li-ion battery technology is presented in Figure 1 using the highly engineered 18650 Li-ion cells as examples. Practically all of the 18650 cells manufactured until 2006 were based on the graphite/$LiCoO_2$ system, and the graphite/NCA cell was introduced in 2009. The 4000 mAh projected for the Si/layered

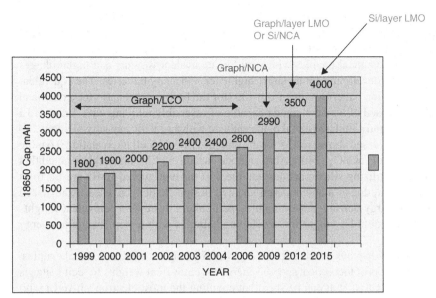

FIGURE 1   Past and projected capacity evolution of 18650 Li-ion cells. Until 2006, the cells were based on the graphite/LiCoO$_2$ couple. The graphite/NCA cell was introduced in 2009. The layer-LMO represents future cells based on Li-rich layered manganese oxides such as Li[Li$_{0.2}$Mn$_{0.54}$Ni$_{0.13}$Co$_{0.13}$]O$_2$. The 18650 cell with silicon anode and layered LMO is projected to have a 4000-mAh capacity and about 320 Wh/kg. Their approximate volumetric energy densities may be obtained by multiplying this value by 2.5.

LMO 18650 cells corresponds to a specific energy of 320 Wh/kg, which clearly falls short of the 580-Wh/kg cells required for 300-mile vehicles. However, the 150 to 200-mile-range all-electric vehicle possible with Si/layered LMO Li-ion technology is attractive if the battery can be made cost-effective. In this case, a 200-mile-range car would be powered by a 90-kWh battery pack, which at the best-case scenario price of \$400 kWh^{-1} will cost \$36,000, too high for an average family car. Clearly, significant changes in the energy density and cost of Li-ion batteries have to occur for them to be widely used for all-electric vehicle propulsion. Intensive research efforts are in place worldwide for the realization of new generations of Li-ion batteries. However, achieving a step change in battery energy density for long-range vehicle propulsion and other low-cost storage applications requires the discovery and development of new Li cell technologies that differ from Li-ion-based systems. Consequently, the search continues for superhigh-energy-density rechargeable batteries such as the Li–air battery.

2   ULTRAHIGH-ENERGY-DENSITY BATTERIES

The specific energy (expressed as Wh/kg) of an electrochemical couple is given by the formula

$$\text{energy density (Wh/kg)} = \frac{\text{cell voltage } (V) \times 1000 \text{ g} \times 26.8 \text{ Ah}}{(\text{kg}) \times \text{equiv.wt (g)}} \qquad (1)$$

where, V is the voltage of the electrochemical couple and 26.8 Ah is the theoretical capacity available from the reaction between 1 equivalent weight of the anode and cathode reactants in the battery cell. Equation (1) shows that identifying and developing electrochemical couples with very low equivalent weights and high voltages are the key to developing future super high-energy-density batteries. The equivalent weights of Li-ion batteries range from 130 (for the best-case scenario for a chemical couple composed of a silicon anode capable of reversibly alloying with 4 mol of Li per mol of Si and a lithiated metal oxide cathode with a specific capacity of about 250 mAh/g, as discussed above) to 235 for the graphite/$LiCoO_2$ cell intercalating 0.6 mole of Li per mol of $LiCoO_2$ (140 mAh/g cathode capacity) [4,5]. It should be noted that Li-ion cells utilizing silicon anodes and the 250-mAh/g layered LMO are far from being practical, due to many material and cell design challenges. Clearly, advanced superhigh-energy-density battery couples, particularly those with equivalent weights below 100, should be sought for developing batteries with a step change in energy densities.

Figure 2 displays a theoretical energy density map for electrochemical couples, depicted as plots of theoretical specific energy vs. equivalent weights for cell voltages ranging from 1 to 6 V. It can be seen that coupling the most electropositive Li with highly electronegative and light elements from groups VIB and VIIB of the periodic table would provide the highest-energy-density batteries. As presented in Figure 3, the Li–F_2 couple is the highest-energy-density battery theoretically possible, but it is an impractical system, due to the extreme chemical reactivity of F_2. Similarly, practical implementation of the Li–Cl_2 couple is difficult. This leaves the Li–O_2

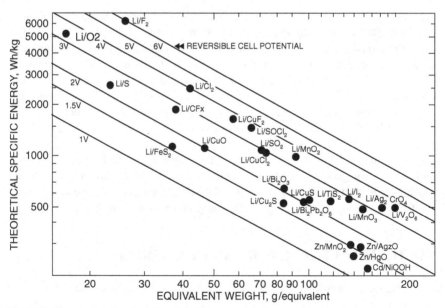

FIGURE 2 Energy density map depicting theoretical energy densities of battery couples vs. equivalent weights for various cell voltages.

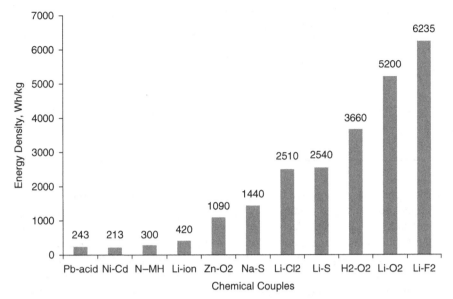

FIGURE 3   Theoretical energy densities of Li battery couples.

couple as candidate for developing the ultimate high-energy-density battery. The Li–S system, also being developed actively, occupies a distant second place in the chase for superhigh-energy-density batteries. The Li–O_2 couple has an equivalent weight of 15 and an open-circuit voltage of 2.91 V (calculated from the Gibbs free energy change (ΔG) of –268 kcal for the reaction between Li and O_2 to form Li_2O) [6]. The most important attribute of the LiO_2 battery, in addition to its very high energy density, is the low cost of the positive electrode material, which would make a practical Li–O_2 battery very attractive for cost-sensitive applications such as electric vehicle propulsion.

3   RECHARGEABLE LITHIUM–AIR BATTERIES

Metal–air batteries utilizing alkali and alkaline earth metal negative electrodes (anodes) and oxygen as the positive electrode (cathode) offer very high energy densities, as can be gleaned from the theoretical specific energies listed in Table 1. They are unique power sources because the cathode active material, oxygen, does not have to be stored in the battery but can be accessed from the environment. Of the various metal–air battery chemical couples, the Li–air battery is the system most highly sought after, since the cell discharge reaction between Li and oxygen to yield Li_2O, according to $4Li + O_2 = 2Li_2O$, has an open-circuit voltage of 2.91 V and a theoretical specific energy of 5200 Wh/kg. In practice, oxygen is not stored in the battery, and the theoretical specific energy excluding oxygen is 11,140 Wh/kg.

The Li–O_2 battery is popularly known as the Li–air battery because the fully developed system is expected to use oxygen from atmospheric air as the positive

TABLE 1   Theoretical Voltages and Specific Energies of Metal–Air Batteries

| Metal–air battery | Calculated OCV (V) | Theoretical specific energy (Wh/kg) | |
		Including oxygen	Excluding oxygen
Li–O_2	2.91	5,200	11,140
Na–O_2	1.94	1,677	2,260
Ca–O_2	3.12	2,990	4,180
Mg–O_2	2.93	2,789	6,462
Zn–O_2	1.65	1,090	1,350

electrode. The first rechargeable Li–air cell [6,7] comprised a Li foil anode, gaseous oxygen cathode (supplied continuously into a porous carbon electrode), and a Li-ion conducting gel-polymer electrolyte membrane that served as both the separator and the Li-ion transporting medium between the anode and the cathode (Fig. 4). The Li-ion conducting gel-polymer electrolytes used to construct those polymer Li–O_2 cells include polyacrylonitrile- [8] and poly(vinylidene fluoride)-based materials [9]. During cell discharge, O_2 gas, admitted to the porous carbon electrode through holes in the plastic cell, is reduced to form lithium oxide discharge products while generating electricity. Preliminary experiments with cells utilizing atmospheric air as the cathode active material showed open-circuit and discharge load voltages, and discharge capacities, similar to those operated with pure oxygen, suggesting that the oxygen in the air can be used as the cathode active material in a practical battery, albeit with appropriate cell engineering to keep moisture from entering the cell.

There are different versions of the Li–air battery based on the electrolyte used: (1) the nonaqueous Li–air battery utilizing organic liquid, polymer, or ionic liquid electrolytes; (2) the aqueous Li–air battery, which is more aptly called a Li–water battery since the cell discharge reactions involve the participation of both H_2O and O_2; and (3) the solid-state Li–air battery, fabricated with a Li^+-conducting inorganic

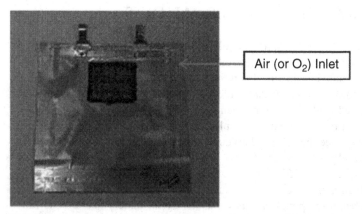

Air (or O_2) Inlet

FIGURE 4   Plastic pouch Li–air cell used to demonstrate Li–air technology.

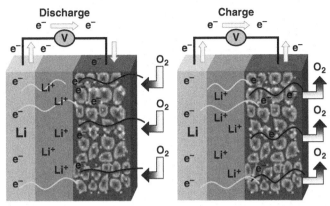

FIGURE 5   Schematic representation of a Li–air battery. (From [10], with permission of the *Journal of Physical Chemistry Letters*.)

solid electrolyte separator. All of these various forms of the Li–air battery can be represented by the general schematic diagram depicted in Figure 5. In the following sections these batteries are discussed with an emphasis on the electrochemistry of oxygen in nonaqueous organic and ionic liquid electrolytes. A summary of efforts to date to fabricate and test Li–air cells is also presented. The solid-state Li–air battery is discussed briefly. The aqueous Li–air battery is not covered in this chapter, as it is discussed in Chapter 9.

3.1   Nonaqueous Lithium–Air Battery

Several reviews of the Li–air battery have appeared recently [10–12], and as a result, in this chapter emphasis is on the fundamental electrochemistry of O_2 in nonaqueous electrolytes, including organic and ionic liquid electrolytes, and characterization of the discharge and charge reactions and products from Li–air cells. In the operation of a nonaqueous Li–air battery, oxygen gas from an external source enters the porous carbon cathode, where it is reduced to form lithium oxides as discharge products (see below). Since lithium oxides are insoluble in these electrolytes, they are stored in the carbon electrode pores, and the cell discharge ends when these pores are filled completely with the oxides. The amount of oxygen reacted is not measured, and as a result the cell capacity is expressed as Ah/g of the carbon in the porous electrode. This way of expressing capacity is also used in $Li/SOCl_2$ and Li/SO_2 primary liquid cathode cells, in which the discharge products are insoluble in the electrolytes. A drawback to expressing the specific capacity in this manner is that it can vary significantly depending on the nature of the carbon, particularly its surface area, and on the porosity and pore structure of the electrode used in the Li–air cell. As a result, widely different specific capacities have been reported for the Li–air battery by various authors [11,12].

The aforementioned plastic Li–air battery (Fig. 4) had an open-circuit voltage of about 3 V, and it discharged at 2.6 V, as depicted in Figure 6. The cell discharge

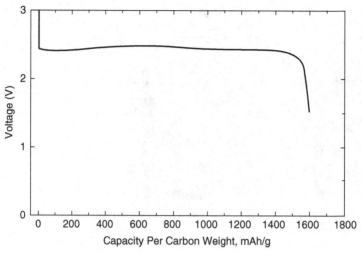

FIGURE 6 Discharge of the first nonaqueous Li/O_2 cell in a pouch cell. (From [6], with permission of the *Journal of the Electrochemical Society*.)

reaction that agreed with analytical results involves the formation of Li_2O_2 as the discharge product:

$$2Li + O_2 \rightleftharpoons Li_2O_2; \qquad \Delta G_0 = -145 \text{ kcal} \qquad (E_0 = 3.1 \text{ V}) \qquad (2)$$

Li_2O_2 was identified in the carbon electrodes of discharged cells by qualitative analysis [6,13], Raman spectroscopy [6,14] and more recently, from x-ray diffraction patterns [15]. The formation of Li_2O_2, the two-electron reduction product of O_2, as the principal discharge product is determined by kinetic factors. As discharge proceeds in a nonaqueous Li–air cell, the insoluble Li_2O_2 precipitates out in the pores of the porous carbon electrode. Under a steady supply of oxygen, the discharge reaction in equation (1) continues at 2.6 V until the pores of the carbon electrode are fully chocked by the Li_2O_2 deposited (Fig. 6). At this point the cathode becomes incapable of sustaining further O_2 reduction, and as a result, the discharge ends. The cell polarizes to lower voltages, but the carbon electrode is inactive for reducing Li_2O_2 to Li_2O:

$$Li_2O_2 + 2Li \rightarrow 2Li_2O \qquad (3)$$

An additional deterrent to this reaction is the electronic insulating property of Li_2O_2. The formation of Li_2O as the final reduction product of O_2 has been demonstrated in voltammetric studies [13, 15] and claimed in discharged Li–air cells [16], although its presence in Li–air cells has not yet been confirmed analytically. The theoretical energy density of the Li–air cell corresponding to the reaction in equation (2) involving the formation of Li_2O_2 is 3500 Wh/kg.

Interestingly, the reactions in both equation (2), involving two-electron O_2 reduction, and equation (4), corresponding to four-electron O_2 reduction, can explain the open-circuit potential of about 3 V observed for the Li–air cell since the differences between them are small [16,17].

$$4Li + O_2 \rightarrow 2Li_2O; \quad \Delta G_0 = -268 \text{ kcal} \quad (E_0 = 2.91 \text{ V}) \quad (4)$$

However, as discussed below, neither a one-step two-electron nor a direct four-electron reduction of O_2 occurs in the nonaqueous Li–air cell, even in the presence of catalysts. As discussed below in detail, O_2 reduction begins with a one-electron transfer and then continues in a stepwise fashion to additional reduction reactions to form the final product, Li_2O.

The Li–air battery can be recharged with and without the use of a catalyst in the carbon cathode, although the recharge efficiency appears to be improved by catalysts. Active catalysts demonstrated to be useful for recharging the Li–air cell include cobalt phthalocyanine (Co-Pc) [6], Pt/Au [18], and α-MnO$_2$ (14). Catalysts also help increase the discharge voltage slightly, although they have not been shown to increase the capacity of the Li–air cell significantly. Catalysts are discussed in detail later in the chapter.

3.2 Oxygen Reduction Reactions in Organic Electrolytes for the Lithium–Air Battery

There have been a number of studies dealing with the influence of organic electrolytes on the performance of the Li–O$_2$ battery [13,15,19–22]. Most of these were empirical studies with emphasis on the performance of primary Li–air cells. We have investigated O_2 electrochemistry in detail, correlating electrolyte properties with oxygen reduction reaction (ORR) kinetics, the ORR mechanism, and the nature of the reduction products formed at various reduction potentials. We performed these electrochemical investigations using Li^+-conducting electrolytes in selected organic solvents having a range of electron-donating properties, defined by the Gutmann donor numbers. We found [13,15] that the acid–base properties of the solvents and the solvated Li ions [Li^+(solvent)$_n$] present in the electrolyte, and the basicity of the oxygen reduction products formed strongly influence ORR chemistry. The properties of several solvents used in our investigations are listed in Table 2.

The ORR chemistry in nonaqueous electrolytes is described best from electrochemistry results obtained in dimethyl sulfoxide (DMSO)/LiPF$_6$ solutions [13].

TABLE 2 Solvent Properties

Solvent	ε 25°C	DN (kcal/mol)	Viscosity, η (cP)	Oxygen solubility (mM/cm^3)
Dimethyl sulfoxide	48.01	29.8	1.94	2.10
Acetonitrile	36.64	14.1	0.36	8.10
Tetraethylene glycol dimethyl ether	7.79	16.6	4.05	4.43
1,2-dimethoxyethane	7.20	20.0	0.46	9.57
Propylene carbonate	64.4	15.1	2.53	
Ethylene carbonate	89.6	16.4	1.85	1.55

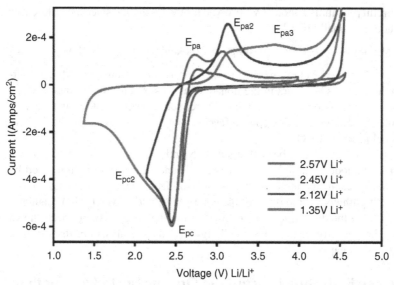

FIGURE 7 Cyclic voltammograms for the reduction of O_2 in 0.1 M $LiPF_6$/DMSO at various potential windows. All scans used a glassy carbon working electrode; scan rate of 100 mV/s. (From [15], with permission of the *Journal of Physical Chemistry*.)

Cyclic voltammetry data presented in Figure 7 provide evidence for the sequential reduction of O_2 as described in equations (5) to (8). The ORR process begins with the formation of superoxide, LiO_2 [equation (5)], which decomposes [equation (6)] or is reduced further electrochemically [equation (7)] to form Li_2O_2. Finally, Li_2O is formed [equation (8)] as the final reduction product when the electrode potential is swept lower.

$$O_2 + Li^+ + e^- = LiO_2 \qquad E_{pc} \qquad (5)$$
$$2LiO_2 = Li_2O_2 + O_2 \qquad \text{chemical} \qquad (6)$$
$$LiO_2 + Li^+ + e^- = Li_2O_2 \qquad E_{pc2} \qquad (7)$$
$$Li_2O_2 + 2Li^+ + 2e^- = 2Li_2O \qquad (8)$$

We have explained this ORR mechanism using the hard–soft acid–base (HSAB) properties of the solvated Li ions [$Li^+(\text{solvent})_n$] and DMSO present in the electrolyte and the basicity of the ORR products: O_2^-, O_2^{2-}, and O^{2-}.

Due to their small atomic radii ($r = 0.078$ nm), and the associated very high positive charge density, Lithium ions are hard Lewis acids. In electrolyte solutions, Li^+ are solvated by the solvents, usually by about four solvent molecules per Li^+, to form solvent-separated ion pairs [e.g., $Li^+(DMSO)_4PF_6^-$ in DMSO solutions]. According to the HSAB theory, the hard acid Li^+ likes to be associated with the hard Lewis base ions, peroxide (O_2^{2-}) and monoxide (O^{2-}), formed from the reduction of O_2. The initial one-electron O_2 reduction product, superoxide, O_2^-, formed according to equation [5], is a relatively soft base that cannot exist as stable LiO_2 by combining with Li^+. Consequently, the superoxide formed as the first ORR product of O_2 decomposes

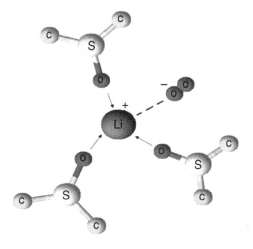

SCHEME 1 Ion pair between solvated Li^+ and O_2^-. (DMSO is shown as the example and the methyl hydrogen in DMSO is omitted in the structure.) (From [13].)

or undergoes a fast second reduction to form the hard base, O_2^{2-}, as depicted in equations (6) and (7). However, the acidity of Li^+ in the electrolytes is modulated by the electron-donating property of the solvent when it forms $Li^+(solvent)_n$; the higher the donor number of the of the solvent, the greater is its basicity and hence the greater is its ability to modulate (lower) Li^+ acidity. Thus, high-DN solvents such as DMSO make the solvated Li^+ a softer acid and, in turn, impart increased stability for the ion-pair complex, $[Li^+(solvent)_n–O_2^-]$, depicted in Scheme 1.

In such electrolytes, a distinct O_2/O_2^- reversible couple may be observed in the presence of Li^+, as shown in Figure 7 in the DMSO solution. Thus, high-donor-number solvents such as DMSO impart better stability to the initial ORR product LiO_2, although ultimately it decomposes to the stable product Li_2O_2. In solvents with low DN, the general tendency is for the LiO_2 to decompose quickly or undergo fast electrochemical reduction to Li_2O_2. In Li^+-containing electrolytes prepared with low-DN solvents, it may be possible to reduce O_2 fully to Li_2O. This was demonstrated in tertaethylene glycol dimethyl ether, dimethoxyethane, and acetonitrile electrolytes [13]. Thus, in nonaqueous organic electrolytes, although ORR proceeds through the initial product LiO_2, it will be stabilized to varying degrees depending on the basicity of the solvent before transforming to Li_2O_2 via chemical or electrochemical reactions. The multistep electrochemical reduction of O_2 in organic electrolytes can be depicted by the general reaction, shown in Scheme 2.

Cyclic voltammetric results showed that all three reduction products of O_2 can be reoxidized, as revealed by the three anodic peaks Epa, Ep_{a2}, and Ep_{a3} in Figure 7, according to the reactions

$$LiO_2^- = O_2 + Li^+ + e^- \qquad (E_{pa}) \qquad\qquad (9)$$

$$Li_2O_2 = O_2 + 2Li^+ + 2e^- \qquad (Ep_{a2}) \qquad\qquad (10)$$

$$2Li_2O = O_2 + 4Li^+ + 4e^- \qquad (Ep_{a3}) \qquad\qquad (11)$$

In DMSO-based electrolytes, the initial ORR process involving the formation of LiO_2 is reversible, whereas the two subsequent reactions leading to Li_2O_2 and Li_2O are

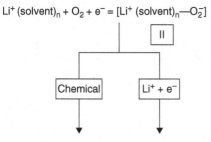

$Li^+ (solvent)_n + O_2 + e^- = [Li^+ (solvent)_n\!\!-\!\!O_2^-]$

½[Li_2O_2] + ½O_2 + n solvent Li_2O_2 + n solvent

SCHEME 2   ORR in Li^+-conducting organic electrolytes. (From [13].)

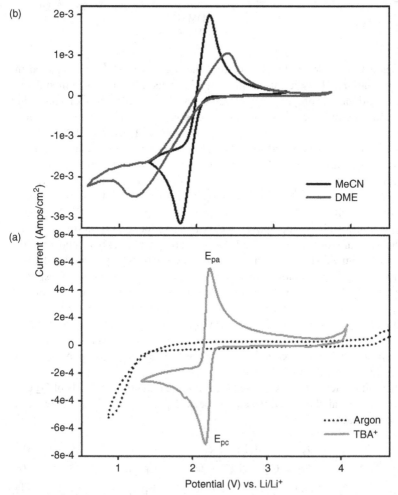

FIGURE 8   (a) Cyclic voltammograms for the reduction of oxygen in 0.1 M $TBAPF_6$ and the argon background (dotted) in DMSO. (b) Cyclic voltammograms for the reduction of oxygen in 0.1 M $TBAPF_6$/MeCN, and DME. Scan rate 100 mV/s. (From [13], with permission of the *Journal of Physical Chemistry.*)

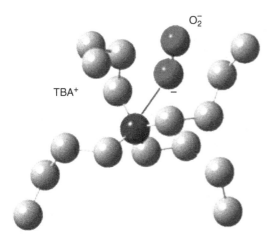

O_2^-

TBA$^+$

$-$

SCHEME 3   Ion pair formed between TBA$^+$ and O_2^-.

only quasireversible. This explains the high overpotentials required for charging of the Li–air battery and the need for catalysts to promote rechargeability.

The validity of the HSAB theory to explain ORR chemistry is exemplified further by O_2 electrochemistry in the presence of large cations such as tetrabutyl ammonium (TBA$^+$) ($r = 0.494$ nm), which has a relatively low positive charge density and hence it is a soft acid. Typical cyclic voltammetry results that we obtained [13] are presented in Figure 8. In all of the solvents we observed a reversible O_2/O_2^- reversible electrochemical couple, and the O_2^- formed resisted further reduction to peroxide and monoxide. This illustrates the exceptional stability of the TBA$^+$–O_2^- ion pair depicted in Seheme 3, formed between the soft acid TBA$^+$ and soft base O_2^-. Interestingly, TBA$^+$ is little solvated in organic electrolytes, and it exists as "naked" soft acids, free to form ion pairs with soft bases.

Raman spectroscopic analysis carried out in situ during O_2 electrochemistry has confirmed the aforementioned ORR mechanism that we first reported [23]. It was found that the LiO$_2$ formed has a finite lifetime in organic electrolytes and that it decomposes to Li$_2$O$_2$ as we described. Recent studies revealed that the LiO$_2$ formed reacts with organic solvents such as propylene carbonate, with the result that the ORR products in such electrolytes include organic carbonates [24]. The mechanism proposed for the reaction of LiO$_2$ with propylene carbonate is depicted in Scheme 4.

Our results as presented above suggest that one way to increase the stability of organic electrolytes for Li–air batteries is to use solvents with low donor numbers so that the lifetime of the initial product LiO$_2$ is decreased substantially, to the point of it not being present to react with the solvent.

3.3   Oxygen Reduction Reactions in Ionic Liquid Electrolytes for the Lithium–Air Battery

The use of room-temperature ionic liquids (RTILs) as electrolytes in Li–air batteries is attractive due to their low flammability, hydrophobic nature, low vapor

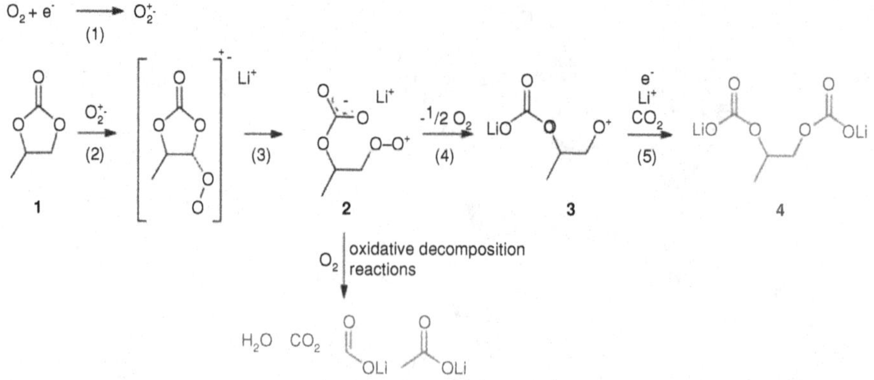

$$O_2 + e^- \longrightarrow O_2^{\cdot-} \quad (1)$$

$$2\,O_2^{\cdot-} + 2\,CO_2 \longrightarrow C_2O_6^{2-} + O_2 \quad (6)$$
$$C_2O_6^{2-} + O_2^{\cdot-} + 4\,Li^+ \longrightarrow 2\,Li_2CO_3 + 2\,O_2 \quad (7)$$

SCHEME 4 Mechanism of reaction between LiO_2 and propylene carbonate. (From [24].)

pressure, wide potential window, and high thermal stability. In particular, the extended anodic voltage window in RTILs is of interest since the presence of cathode cata-lysts, frequently employed in an Li–air system, can shorten the potential window of many organic electrolytes (e.g., carbonate solvents) and affect cell performance. Several electrochemistry studies of O_2 in ionic liquids have shown [25–28] that the one-electron reduction product superoxide O_2^- is the predominant ORR product without added Li salt. A couple of studies demonstrated limited charge–discharge cycles in Li–air cells utilizing RTILs [29,30]. However, all of those previous inves-tigations did not explore the detailed Li–O_2 cell chemistry, which persuaded us to elucidate the ORR and OER reactions and products in RTILs in the presence of

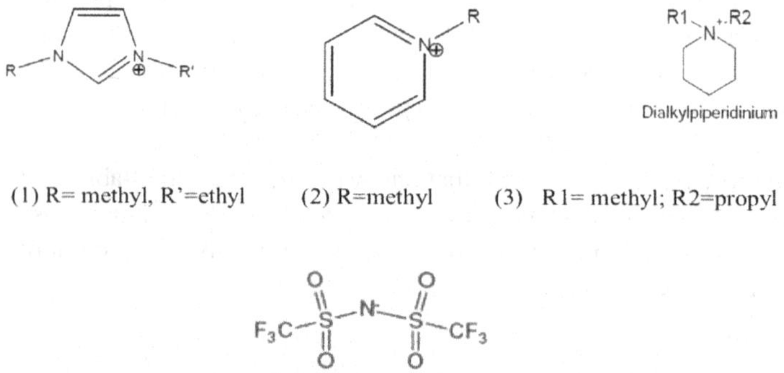

(1) R= methyl, R'=ethyl   (2) R=methyl   (3) R1= methyl; R2=propyl

(4) TFSI anion

SCHEME 5

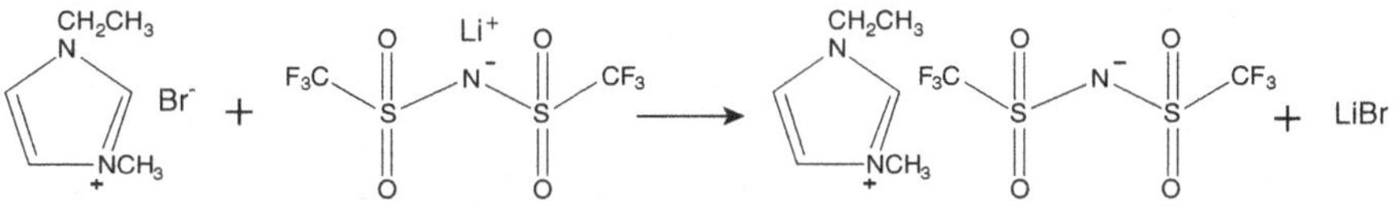

SCHEME 6   Synthesis of 1-ethyl-3-methylimidazolium bis(triflouromethanesulfonyl) imide.

added Li salt. RTILs investigated as electrolytes for Li batteries include 1-ethyl-3-methylimidazolium bis(trifluoromethylsulfonyl)imide (EMIImide) (1), alkyl (R = methyl or butyl) pyridinium imide (2), and methyl propyl piperidinium imide (3), depicted in Scheme 5 [31,32].

We performed ORR and OER investigations in 1-ethyl-3-methylimidazolium bis(triflouromethanesulfonyl)imide (EMITFSI), synthesized as shown by the reaction in Scheme 6. The material was isolated as a clear liquid identified by its characteristic ^{1}H NMR spectrum. Cyclic voltammetry revealed no electrochemical activity for the neat EMITFSI between 1 and 4 V vs. Li/Li$^+$. Cyclic voltammetry of an O$_2$-saturated solution revealed that ORR and OER on glassy carbon (GC) and gold electrodes in EMITFSI exhibited distinct differences in the presence of Li salts. Again, it is Li$^+$, not the anion, in the Li salt that influences the ORR mechanism. The one-electron O$_2$/O$_2^-$ reversible couple is observed on gold and GC electrodes in neat EMITFSI (Fig. 9). On the other hand, in the presence of added LiTFSI, the initially formed LiO$_2$ is not stable and decomposes to Li$_2$O$_2$. We also found that ORR and OER in Li$^+$-containing solution exhibit distinct differences between the Au and GC electrodes (Fig. 10). The voltammetric data on the Au electrode revealed a highly rechargeable multielectron ORR reaction yielding LiO$_2$ and Li$_2$O$_2$, which undergoes multiple cycles without electrode passivation [31].

The predominant O$_2$ electrochemistry in EMITFSI with and without added Li salt can be summarized according to the reactions depicted in Scheme 7. The ORR mechanism depicted in this scheme in RTIL is consistent with HSAB theory. The Like TBA$^+$, EMI$^+$ is a soft acid that stabilizes the one-electron reduction product O$_2^-$ as EMI$^+$–O$_2^-$. When Li$^+$ is present, due to stronger acidity it shows an affinity for O$_2^-$ to form LiO$_2$, but as predicted by HSAB theory, LiO$_2$ is not stable and decomposes to Li$_2$O$_2$ [32].

Our results show that O$_2$ reduction and evolution reactions in EMITFSI follow a markedly different reaction mechanism when Li salt is present. The nature of electrode affects ORR, with gold showing the ability for high-efficiency recharging of the oxygen electrode, as evidenced by multiple cycles without passivation.

Our data suggest that ionic liquids are promising electrolytes for rechargeable Li–air batteries. Further research and development should emphasize the selection of RTIL stable to both Li and oxygen electrodes, and the identification of appropriate electrode materials to facilitate mutlielectron ORR and OER. The relatively high viscosity of RTILs is a concern for the wide operating temperature range of the battery, but that could be overcome through the use of mixed electrolytes composed of RTIL and organic solvent.

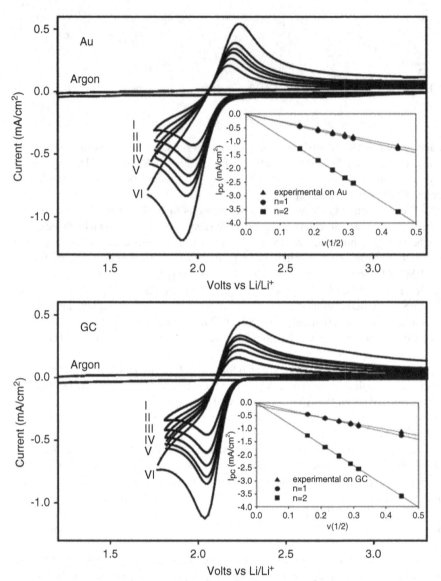

FIGURE 9 Cyclic voltammograms of O_2-saturated EMITFSI at Au and GC electrodes. Randles Sevcik plots (inset) for the cathodic peak currents include experimental data along with one- and two-electron theoretical fits [see equation (9)]. Scan rates: I = 25 mV/s, II = 45 mV/s, III = 65 mV/s, IV = 85 mV/s, V = 100 mV/s, VI = 200 mV/s. Argon-saturated solution scanned at 100 mV/s. (From [31], with permission of the *Journal of Physical Chemistry Letters*.)

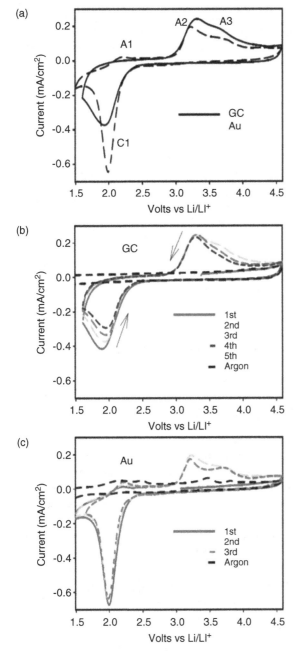

FIGURE 10   (a) Cyclic voltammograms in 0.025 LiTFSI doped O_2-saturated EMITFSI on Au and GC at 100 mV/s. Labeled peaks correspond to reactions described in Scheme 5. Initial cycles (b) on GC and (c) on Au. (From [31], with permission of the *Journal of Physical Chemistry Letters.*)

Neat EMITFSI

   Cathodic

$$EMI^+ + O_2 + e^- \rightarrow EMI^+ - O_2^{\cdot -} \qquad\qquad C1 \qquad\qquad [12]$$

   Anodic

$$EMI^+ - O_2^{\cdot -} \rightarrow EMI^+ + O_2 + e^- \qquad\qquad A1 \qquad\qquad [13]$$

0.025 M Li$^+$ in EMITFSI

   Cathodic

$$EMI^+ + O_2 + e^- \rightarrow EMI^+ - O_2^{\cdot -} \qquad\qquad C1 \qquad\qquad [14]$$

$$Li^+ + O_2 + e^- \rightarrow LiO_2 \qquad\qquad C1 \qquad\qquad [15]$$

$$2LiO_2 \rightarrow Li_2O_2 + O_2$$

   Anodic

$$EMI^+ - O_2^{\cdot -} \rightarrow EMI^+ + O_2 + e^- \qquad\qquad A1 \qquad\qquad [16]$$

$$LiO_2 \rightarrow Li^+ + O_2 + e^- \qquad\qquad A2 \qquad\qquad [17]$$

$$Li_2O_2 \rightarrow 2Li^+ + O_2 + 2e^- \qquad\qquad A3 \qquad\qquad [18]$$

SCHEME 7 ORR in EMITFSI (refer to Figures 9 and 10 for further explanations of equations).

4 LITHIUM–AIR CELLS

Investigations of Li–air fuel cells have been confined to laboratory testing with emphasis on identifying high-capacity carbon cathodes for optimum capacity utilization, stable electrolytes, and catalysts for both ORR and OER. Characterization of the discharge and charge reactions and products has also been given some emphasis. In all of these studies pure oxygen was used as the cathode active material, so the performances reported are for the most ideal conditions.

We recently performed a detailed study of the discharge reactions and rechargeability of Li/O$_2$ cells utilizing tetraethylene glycol dimethyl ether, $CH_3O(CH_2CH_2O)_4CH_3$ (TEGDME)/LiPF$_6$ electrolyte [33]. We also characterized the discharge product and the recharge reaction in the cell. All our studies were performed with uncatalyzed porous carbon cathodes prepared from Black Pearls 2000 (BP2000) carbon black, which allowed us to show that the Li–O$_2$ cell is rechargeable without a cathode catalyst.

The full discharge curves for two Li/O$_2$ cells obtained with BP2000 carbon electrodes exposed to a dry oxygen atmosphere are depicted in Figure 11. The carbon electrode had loadings between 7 and 8.5 mg/cm^2 of the BP 2000 carbon on Panex carbon cloth. The cell had an open-circuit voltage of about 3.1 V, which varied slightly from cell to cell, attributed to moisture contamination. The discharge capacity at 0.16 mA/cm^2 was 2760 mAh/g of the BP2000 carbon in the cathode. Increasing

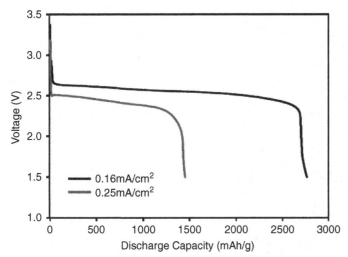

FIGURE 11   Li–air cell discharge curves at 0.25 (lower) and 0.16 upper mA/cm^2 in 1 M LiPF$_6$/TEGDME. Capacities are expressed per gram of carbon in the electrode. (From [33], with permission of the *Journal of the Electrochemical Society*.)

the current to 0.25 mA/cm^2 decreased the specific capacity to 1452 mAh/g of carbon. These values are very high considering that there was no catalyst in the cathode.

Following full discharge at 0.25 mA/cm^2, the cell was disassembled in the glove box and the carbon cathode was washed in acetonitrile to remove LiPF$_6$. Figure 12 shows the x-ray diffraction pattern of the fully discharged carbon electrode. The spectrum was assigned to that of Li$_2$O$_2$ based on the JCPDS data card (No. 7400115). We concluded that the discharge plateau observed at 2.5 to 2.6 V is due to Li$_2$O$_2$ formation.

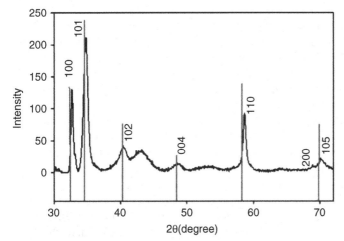

FIGURE 12   XRD pattern of fully discharged O$_2$ cathode to 2 V. (From [33], with permission of the *Journal of the Electrochemical Society*.)

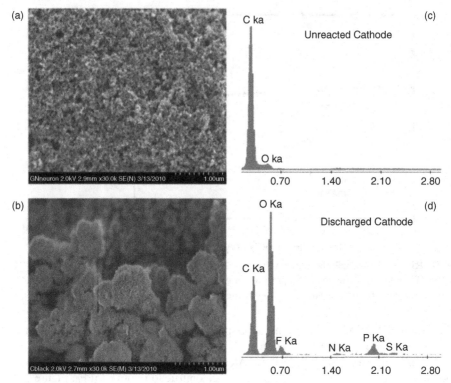

FIGURE 13 SEM micrographs of the air cathode (a and b) undischarged and discharged (c and d) at 0.13 mA/cm2 in oxygen. Scale bar of (a) and (c) is 1 μm and that of (b) and (d) is 10 and 20 μm, respectively. Energy-dispersive x-ray spectroscopy analysis of the discharged cathode is shown in (c) and (d). (From [33], with permission of the *Journal of the Electrochemical Society*.)

The scanning electron micrographs of surface morphology of the porous carbon electrode before and after discharge are shown in Figure 13a and b. Figure 13a shows individual particles of BP2000 carbon on the Panex substrate in the unused cathode. The average particle size of BP2000 carbon is 12 nm. Figure 13b reveals a much different cathode surface after discharge. The Li_2O_2 deposited in the carbon electrode pores appears as a very fine powder gathered into grains of larger particles. The product is evenly deposited on both BP2000 and the Panex substrate, resulting in high specific capacity. The deposit analyzed by energy-dispersive x-ray spectroscopy (EDAX) was found to be oxygen rich, which supports the presence of Li_2O_2 by x-ray diffraction (XRD) analysis.

Figure 14a depicts the discharge–charge cycle of another Li–air cell. After 20 h of discharge at 0.1 mA/cm^2, the capacity was 605 mAh/g under O_2, and a subsequent 20-h recharge to 4.6 V under argon yielded 595 mAh/g. The nearly identical discharge and charge capacities suggest a rechargeable Li–air cell. The charge curve shows a distinct endpoint and the oxidation of the electrolyte in our cell occurs at voltages higher than 5 V. The large hysteresis of about 2 V indicates a complex charging

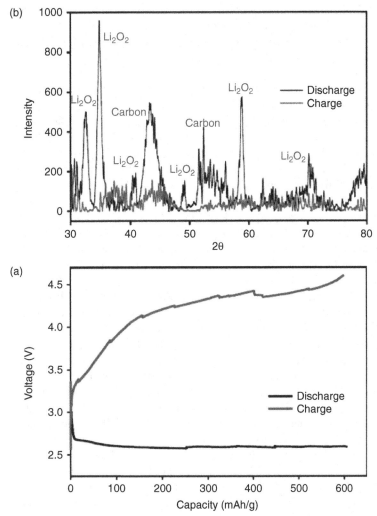

FIGURE 14   (a) Discharge and charge capacities of the Li/O_2 cell cycled at 0.1 mA/cm^2 at room temperature. Capacities are expressed per gram of carbon in the electrode. (b) XRD patterns of cathode discharged and charged. (From [33], with permission of the *Journal of the Electrochemical Society*.)

mechanism along with the relatively large impedance in the cell. The XRD pattern of the carbon electrode before and after the charge (Fig. 14b) demonstrated that Li_2O_2 was oxidized completely during charge. The XRD of the charged cathode showed no Li_2O_2 peaks. Thus, the principal discharge and charge reactions in Li–air cells are the reduction of O_2 to form Li_2O_2 and its reoxidation to release O_2, respectively. The mechanism of Li_2O_2 formation at the nearly constant voltage in the Li–air cell may occur through the electrochemical formation of LiO_2 and its subsequent decomposition, as described in equations (5) and (6).

It was found that, in general, a fraction of the discharge product in the cell could not be recharged and its amount increased with the depth of discharge of the cell [33]. Impedance of the cathode increased after discharge, indicating that the accumulation of the electronically insulating Li_2O_2 discharge product is probably the cause of the cell's recharge inefficiency.

When a Li–air cell is discharged, the Li_2O_2 formed on the carbon surface is pushed more and more into the electrode pores as the depth of discharge increases. When the cell is subsequently charged, a fraction of this Li_2O_2 located farther away from the carbon cathode surface loses electronic contact with the surface and becomes unable to be oxidized. The amount of the isolated Li_2O_2 increases with the depth of discharge so that limiting the thickness of the Li_2O_2 would be a way to increase Li–air cell rechargeability. This could be done by controlling the surface area of the carbon black and the porosity of the fabricated carbon electrode for optimum capacity utilization: for example, with very high surface carbon fabricated as thin electrodes having relatively high porosity and small pores sizes. In our Li_2–O_2 cell discussed above, we demonstrated more than 40 cycles, as shown in Figure 15. The Li–air cell cycle life would also be affected by the cycling efficiency of Li anode, as is well known for rechargeable Li anode batteries [34].

The thickness of the air cathode seems to play an important role in the discharge capacity of a Li–air cell. Thinner electrodes have been shown to provide higher capacity than thicker electrodes with the highest specific capacity obtained at an optimum thickness. This is understood by considering that the energy capacity of a Li–air cell depends on the ability of the electrolyte to access an adequate amount of O_2 at the reaction site on the carbon electrode immersed in it. Scanning electron microscopic images of discharged carbon electrodes have revealed that their pores on the air sides are usually filled, whereas those on the separator side are open. This implies that during cell discharge Li_2O_2 is deposited preferentially at the air side and blocks access of O_2 to the interior of the electrode, leading to lower specific capacity utilization in thicker electrodes. Indeed, Beattie et al. [35] found that the capacity increases as the carbon loading in the electrode is decreased, with the highest capacity (6000 mAh/g) obtained for an electrode with 1.9-mg/cm^2 carbon black loading. Clearly, although it is necessary to have an electrolyte with high O_2 solubility and diffusivity, it is also important to engineer the cell to prevent clogging of its pores by depositing the insoluble Li_2O_2 formed during discharge.

Sandhu et al. [36] and Yang et al. [37] found that diffusion of O_2 to the air cathode can be facilitated by increasing the oxygen partial pressure, in agreement with the requirement of adequate diffusion to the reaction sites. They increased the specific capacity of the carbon electrode by increasing the O_2 pressure.

Wan et al. discharged Li–air cells using air from the atmosphere as the cathode active reactant [38]. The cell was protected from water ingression by the use of a high-density polyethylene or poly(ethylene terephthalate)–based Melinex 302 film. The film was placed tight against the air cathode in contact with the electrolyte. The performance was superior for the cell with the Melinex film. The cell protected with this film was discharged in ambient air for 33 days and the specific energy of the fully sealed cells was calculated to be 362 Wh/kg. This specific energy for the packaged cell is modest compared to the theoretical specific energy, but the inactive

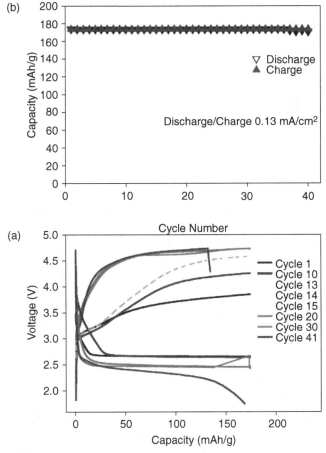

FIGURE 15   (a) The cycling data for a Li/O_2 cell at room temperature. The cell was discharged and charged for 2 h at 0.13 mA/cm2. Capacities are expressed as grams of BP2000 carbon + PVDF in the electrode. (b) Discharge–charge capacities as a function of cycle number for the same cell. (From [33], with permission of the *Journal of the Electrochemical Society.*)

materials in the cell were high compared to what a fully optimized battery might ultimately contain. This work, however, demonstrates that with an appropriate water-impermeable membrane to prevent moisture ingression, Li–air cells can be operated using atmospheric oxygen.

4.1   Catalysis of the Lithium–Air Battery

A number of authors have investigated the rechargeability of a Li–air battery utilizing a catalyst in the cathode. The catalysts studied include metals [39], metal oxides [14,39,40], metal complexes such as cobalt phthalocyanines [6], and PtAu nano-composites [41]. In general, catalysts increase the discharge voltage slightly, but the

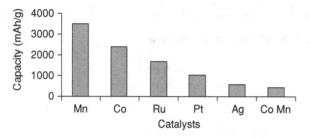

FIGURE 16   Specific capacities of air cathodes with various metal catalysts. (From [41].)

cell capacity is not improved significantly. This indicates that the catalysts investigated to date probably do not change the mechanism of the discharge reaction from those discussed earlier in the chapter, but they probably decrease the activation energy for oxygen reduction, which translates into a higher discharge voltage and a modest increase in capacity. Catalysts also enable rechargeability of a Li–air cell, as was demonstrated in the first paper on a nonaqueous Li–air battery [6], in which a cobalt phthalocyanine catalyst was used to recharge the Li–O_2 cell. Most studies on the rechargeability of a Li–air cell utilized a metal oxide catalyst, of which the mostly widely studied is MnO_2.

We first demonstrated [41] the catalytic activity of manganese and cobalt by discharging a series of Li–O_2 cells with carbon electrodes into which various metal catalysts were incorporated by impregnating the carbon with metal salts followed by thermal processing. Our results are presented in Figure 16. Bruce's group [14,39,40] studied the usefulness of Pt, $Li_{0.8}Sr_{0.2}MnO_3$, Fe_2O_3, NiO, Fe_3O_4, Co_3O_4, $CoFe_2O_4$, and various manganese oxides, including α-MnO_2, β-MnO_2 and λ-MnO_2. When Co_3O_4 was used as a catalyst, the redox potential of OER decreased significantly. Fe_3O_4, on the other hand, delivered a higher discharge capacity, as did the manganese oxides. Among the latter, α-MnO_2 nanowires showed the best electrochemical capacity and the lowest OER voltage. Using a bifunctional PtAu nanocomposite catalyst deposited on Vulcan carbon black, Lu et al. [42] found that Pt raised the discharge voltage with a modest increase in the cell capacity, while the Au in the composite electrode lowered the OER voltage.

A drawback of most of these studies of catalysis is that the Li–air cells contained electrolytes in propylene carbonate (PC), either alone or mixed with dimethoxy ethane. Since PC is now recognized to react with LiO_2, most of the catalysts studied to date are ambiguous as to what is being oxidized in Li–air cells. It is extremely important to identify electrolytes that are stable over the full voltage range of a rechargeable Li–air cell in order to determine unambiguously the catalytic activity for ORR and OER in a Li–air cell.

Giordani et al. found [43] a correlation between the oxidation of H_2O_2 by various metal oxide catalysts and Li_2O_2 oxidation in nonaqueous Li–air cells employing the same catalysts. They claimed that this provided a screening method for OER catalysts in nonaqueous electrolytes without the inference of solvent reactions. They concluded that α-MnO_2 is an efficient catalyst for a Li–air cell.

A recognized deficiency of the rechargeable Li–air battery is the large separation between the discharge and charge voltages, resulting in overall poor energy efficiency.

Data presently available suggest that a large part of this inefficiency can be reduced with the use of appropriate catalysts. The electronic insulating property of the Li_2O_2 discharge product also contributes to the poor rechargeability of a Li–air cell. A solution to this problem resides in advanced designs of porous electrode structures in which large amounts of Li_2O_2 formed during discharge can be stored without losing electronic contact with a carbon current collector to maintain high recharge efficiency.

4.2 Lithium Anode Protection

While it is important to protect the cell from water ingression from outside atmosphere, it may also be necessary to protect the Li anode inside the cell to mitigate its reactions with the electrolyte and the dissolved oxygen, thereby increasing cell cycling efficiency and life. To obtain an understanding of the reaction between the Li anode and the electrolyte in a Li–air battery, we examined [33] the Li anode from a cycled Li–air cell utilizing a $TEGDME/LiPF_6$ electrolyte. The morphology of the lithium electrode changed significantly after many charge–discharge cycles, with the anode surface covered by what appeared to be granular lithium particles, although there was bare Li foil underneath. The conversion of the Li foil into granules during cycling results in an increase in the overall surface area of the Li anode. The cycled Li electrode surface also becomes covered with a brown precipitate, probably the reaction product between the electrolyte and plated Li.

An approach to increasing the stability of a Li anode involves the use of an anode protection film sometimes called an artificial solid electrolyte interphase (SEI). It is recognized that an SEI formed from the reaction of Li with the electrolyte is necessary to protect the anode from its continued reaction with the electrolyte and afford stability to the battery. However, such an SEI does not seem to protect the Li from being converted into granular deposits, commonly known as Li dendrites, during cycling. Consequently, it has been proposed that an artificial SEI in the form of a Li-ion conducting membrane would better to protect the Li anode to provide high cycling efficiency and long cycle life. The use of Li anode protection films have been investigated by Visco et al. [44]. Li-ion conducting inorganic solid electrolytes such as lithium aluminum titanium phosphate, $Li_2O–Al_2O_3–TiO_2–P_2O_5$ (LATP), and lithium phosphorus oxynitride (LiPON) have been used. Since LATP reacts with Li metal, materials such as Li_3N, Li_3P, LiI, LiBr, LiCl, and LiPON were proposed for an intermediate layer between Li and LATP in a dual-layer arrangement in which LATP protects the Li from moisture and reactive electrolytes, while the intermediate layer protects the Li from reaction with LATP.

MaxPower [45] reported a cell design construct for a stable $Li–H_2O$ battery of which some components may be applicable to the nonaqueous Li–air battery. In this invention, the Li anode was made by packaging it in a multilayer flexible metal foil pouch, which was coated with a plastic layer and heat-sealed, and which hermetically encapsulated the Li electrode. The Li foil anode was protected on the active side by a flexible electrospun polyimide-based porous membrane, soaked by a nonaqueous Li-ion conducting electrolyte that was stable against lithium. There was a water-impermeable solid-state Li-ion conducting glass–ceramic layer facing the

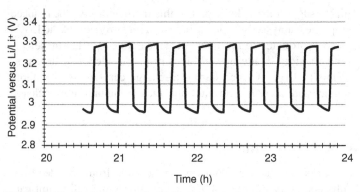

FIGURE 17 Performance of a Li–H_2O battery showing 10 cycles. (From [47], with permission of the *Journal of the Electrochemical Society*.)

air cathode. The resulting frame of the flexible pouch around the solid-state ceramic layer was also heat-sealed to the solid-state ceramic layer.

Yamamoto et al. reported [46] on a Li–H_2O battery with a protected anode comprised of a multilayer film of poly(ethylene oxide) (PEO)–LiTFSI/ LATP. The battery was shown to be rechargeable by demonstrating many low-depth charge–discharge cycles (Fig. 17).

To fabricate a practical rechargeable Li–air battery that operates with atmospheric air, the cell has to be protected not only from moisture but also from CO_2, as it will react with Li oxide discharge products to form Li_2CO_3, which may not be rechargeable. If the Li anode and the cell discharge products can be successfully protected from reactions with environmental contaminants, the same approach can be used to fabricate both Li–air and Li–H_2O batteries.

In this context, the design of a Li–air battery, in which the catalytic reduction of O_2 is done in an alkaline aqueous electrolyte while the metallic lithium anode is held in contact with a nonaqueous electrolyte and separated from the O_2 cathode compartment by means of a superionic conductor glass film (LISICON) for continuously reducing O_2, is intriguing in terms of developing a practical battery [47].

5 SOLID-STATE LITHIUM–AIR BATTERIES

Kumar et al. [48] reported a totally solid-state, rechargeable lithium–oxygen battery. The key component of this solid-state cell is a highly Li-ion conducting glass–ceramic (GC) electrolyte synthesized from lithium aluminum germanium phosphate (LAGP). The electrolyte has an ionic conductivity of 10^{-2} S/cm at 30°C. The cell consisted of a lithium anode and a carbon-based cathode separated by a composite solid electrolyte laminate comprised of the GC membrane and a polymer–ceramic (PC) membrane. The PC membrane of poly(ethylene oxide) (PEO)/LiBETI (8.5 : 1)–(1 wt% Li_2O) coupled the lithium anode and GC membrane electrochemically to minimize cell impedance, catalyze anodic reaction, and enhance cell rechargeability. The cathode and GC membrane were coupled by a PC membrane of the composition

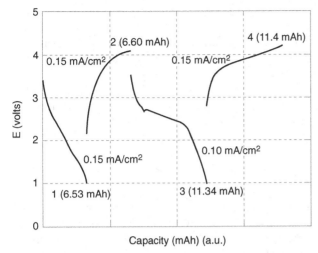

Capacity (mAh) (a.u.)

FIGURE 18   Discharge–charge profiles of a cell at current densities of 0.10 and 0.15 mA/cm². The charge current was maintained constant at 0.15 mA/cm² for both cycles. (From [48], with permission of the *Journal of the Electrochemical Society*.)

PEO/LiBETI ($8.5:1$)–(1 wt% BN). The GC membrane was synthesized from a superionic stoichiometry of the LAGP system. The open-circuit voltage (OCV) of the Li–O_2 cell was measured as a function of the state of charge and discharge, oxygen pressure on the cathode, and temperature. In an oxygen atmosphere, the OCV can vary from 2.6 to 3.6 V, depending on the state of charge. The OCV increased after a discharge and attained equilibrium OCV of about 3.1 V. The fully charged cell showed an OCV around 3.6 V, which slowly decreased to approach an equilibrium value of about 3.1 V. These OCV variations could be explained on the basis of slow diffusion effects. The equilibrium OCV of 3.1 V suggests that the electrochemical reaction in all probability proceeds with the reduction of O_2 to form Li_2O_2 as the reaction product, consistent with the OCV of the plastic Li–air cell reported by Abraham and Jiang [6].

Li–O_2 solid-state cells were discharged in the temperature range 30 to 105°C. Some cycles at 75°C are shown in Figure 18. The temperature-dependent discharge current density ranged from 0.025 to 0.15 mA/cm². In general, a lower discharge rate was used at temperatures below 75°C. Similarly, depending on the temperature, the charge rate varied from 0.0375 to 0.15 mA/cm². When the current density was lowered from 0.15 ma/cm² to 0.10 mA/cm², the cell capacity improved by over 70%. The coulombic efficiency for both the cycles was about 99%. This solid-state Li–air battery is an interesting concept that merits further development.

Recently, Hassoun et al. reported [49] a study of O_2 electrochemistry in a solid-state Li–air cell utilizing a PEO–$LiSO_3CF_3$–ZrO_2 composite electrolyte. The overall ORR process is consistent with the mechanism discussed earlier in the chapter. There is evidence for the formation of Li_2O_2 and then Li_2O as O_2 reduction products. Interestingly, the oxidation of these lithium oxides occurred at lower potentials with narrower separation between reduction and oxidation peaks than those found in organic liquid electrolytes. This makes further development of this solid-state Li–air battery worth pursuing.

Polymer electrolytes with high conductivity and optimum electrochemical and chemical stability are ultimately needed in building ambient-temperature Li–air batteries. In this context, a series of composite polymer electrolytes prepared from poly(vinylene difluoride) and LiTFSI solutions in low-molecular-weight poly(ethylene glycol) dimethyl ethers, and in which a Li–air battery was demonstrated, could serve as the starting point for further investigations [50].

6 PERSPECTIVE

Li–air batteries are the highest-energy-density electrochemical power sources that can practically be developed. They incorporate the positive attributes of both a fuel cell and a rechargeable battery [48]. Although Li–air and H_2–air cells function by the reduction of atmospheric oxygen, there are a few advantages associated with a Li–air cell. In a Li–air cell, the use of solid metallic lithium as the fuel minimizes the containment and transportation issues of hydrogen gas fuel for a H_2–air fuel cell. The solid lithium anode usually functions well over a wide temperature range without catalysts, whereas an expensive membrane electrode assembly that incorporates precious metal catalysts based on Pt or Pd, for example, must be used in H_2–air fuel cells to carry out the oxidation of hydrogen and reduction of oxygen at lower temperatures. The open-circuit voltage (OCV) of the Li–air cell is about 3.0 V, which is approximately three times the OCV of the H_2–air cell. The higher OCV translates into three times more power for a Li cell for an identical current density. The reactivity of lithium with an external contaminant such as moisture and CO_2, and with electrolyte inside the cell, is the most serious drawback of the Li–air battery. If these problems can be solved, a Li–air battery becomes an attractive alternative to the H_2–O_2 fuel cell while providing an advanced rechargeable power source.

We have only begun to scratch the surface of the fledging Li–Air battery research field. Realization of the Li–air dream will require long-term research. In particular, the development of a practical rechargeable Li–air battery will require active research and development of catalysts for O_2 reduction and lithium oxide rechargeability. Stable electrolytes are crucial to the success of a rechargeable Li–air battery with long cycle life. Ultimately, packaging of practical cells and batteries would require devoting considerable resources to engineering development, particularly in the areas of anode protection and prevention of moisture ingression into the cell.

Acknowledgments

The author acknowledges all his co-workers, past and present, who contributed to the research and development of the Li–air battery. Partial financial support for writing this chapter was provided by U.S. Army CERDEC through subcontract GTS-6-1-437. E-KEM Sciences, Needham, Massachusetts, also provided financial support.

REFERENCES

1. K. M. Abraham, Rechargeable Lithium Batteries for the 300-mile Electric Vehicles and Beyond, lecture at the 15th International Meeting on Lithium Batteries, Montreal, Quebec, Canada, June 2010.
2. Panasonic catalog: http://industrial.panasonic.com/www-ctlg/ctlg/qACA4000_WW.html.
3. Y. Wu and A. Manthiram, *Electrochem. Solid-State Lett.*, **9**(5), A221 (2006).
4. K.M. Abraham, Prospecting for a Counterpart of Moore's Law for Rechargeable Batteries, MRS Fall Meeting, Boston, Nov. 2006.
5. K. M. Abraham, Evolution of Lithium Batteries: Where Do We Go from Here ?, Workshop on Long Life Lithium-Ion Batteries, Jet Propulsion Lab and Caltech, Pasadena, CA, Feb. 2008.
6. K. M. Abraham and Z. Jiang, *J. Electrochem. Soc.*, **143**, 1 (1996).
7. K. M. Abraham and Z. Jiang, U.S. Patent 5,510,209 (1996).
8. K. M. Abraham, H. S. Choe, and D. M. Pasquariello, *Electrochim. Acta*, **43**, 2399 (1998).
9. K. M. Abraham, Z. Jiang, and B. Carroll, *Chem. Mater.*, **9**, 1978 (1997).
10. B. Girishkumar, A. C. McCloskey, S. Luntz, and W. Swanson, *J. Phys. Chem. Lett.*, **1**, 2193 (2010).
11. A. Kraytsberg and Y. Ein-Eli, *J. Power Sources*, **196**, 886 (2011).
12. J.-S. Lee, S. T. Kim, R. Cao, N.-S. Choi, M. Liu, K. T. Lee, and J. Cho, *Adv. Energy Mater.*, **1**, 34 (2011).
13. C. O. Laoire, S. Mukerjee, K. M. Abraham, E. J. Plichta, and M. A. Hendrickson, *J. Phys. Chem. C*, **114**, 9178 (2010).
14. A. Débart, A. J. Paterson, J. Bao, and P. G. Bruce, *Angew. Chem. Int. Ed.*, **47**, 4521 (2008).
15. C. Ó. Laoire, S. Mukerjee, E. J. Plichta, M. A. Hendrickson, and K. M. Abraham, *J. Phys. Chem. C*, **113**, 20127 (2009).
16. S. S. Zhang, X. Ren, and J. Read, *Electrochim. Acta*, **56**, 4544 (2011).
17. P. G. Bruce, L. J. Hardwick, and K. M. Abraham, *MRS Bull.*, **36**, 506 (2011).
18. Y.-C. Lu, Z. Xu, H. A. Gasteiger, S. Chen, K. Hamad-Schifferli, and Y. J. Shao-Horn, *J. Am. Chem. Soc.*, **132**, 12170 (2010).
19. J. J. Read, *J. Electrochem. Soc.*, **149**, A1190 (2002).
20. J. J. Read, *J. Electrochem. Soc.*, **153**, A96 (2006.)
21. J. Read, K. Mutolo, M. Ervin, W. Behl, J. Wolfenstine, A. Driedger, and D. Foster, *J. Electrochem. Soc.*, **150**, A1351 (2003).
22. D. Zhang, R. Li, T. Huang, and A. J. Yu, *J. Power Sources*, **195**, 1202 (2010).
23. Z. Peng, S. A. Freunberger, L. J. Hardwick, Y. Chen, V. Giordani, F. Bardé, P. Novák, D. Graham, J.-M. Tarascon, and P. Bruce, *Angew. Chem. Int. Ed.*, **50**, 6351 (2011).
24. S. A. Freunberger, Y. Chen, Z. Peng, J. M. Griffin, L. J. Hardwick, F. Barde, P. Novak, and P. G. Bruce, *J. Am. Chem. Soc.*, **133**, 8040 (2011).
25. I. M. AlNashef, M. L. Leonard, M. C. Kittle, M. A. Matthews, and J. W. Weidner, *Electrochem. Solid-State Lett.*, **4**, D16 (2001).
26. Y. Katayama, K. Sekiguchi, M. Yamagata, and T. Miura, *J. Electrochem. Soc.*, **152**, E247 (2005).
27. D. Zhang, T. Okajima, F. Matsumoto, and T. Ohsaka, *J. Electrochem. Soc.*, **151**, D31 (2004).
28. E. I. Rogers, X.-J. Huang, E. J. F. Dickinson, C. Hardacre, and R. G. Compton, *J. Phys. Chem. C*, **113**, 17811 (2009).
29. H. Ye and J. J. Xu, *ECS Trans.*, **3**, 73–81 (2008).
30. T. Kuboki, T. Ohsaki, and N. Takami, *J. Power Sources*, **146**, 766 (2005).
31. C. J. Allen, S. Mukerjee, E J. Plichta, M. A. Hendrickson, and K. M. Abraham, *J. Phys. Chem.Lett.* **2**, 2420 (2011).
32. C. J. Allen, S. Mukerjee, E J. Plichta, M. A. Hendrickson, and K. M. Abraham, *J. Phys. Chem.*, **116**, 20755 (2012)
33. C. O. Laoire, S. Mukerjee, E. J. Plichta, M. A. Hendrickson, and K. M. Abraham, *J. Electrochem. Soc.*, **158**, A302 (2011).
34. K. M. Abraham and S. B. Brummer, Secondary Li Cells, in *Lithium Batteries*, J. P Gabano, Ed., Academic Press, London, 1983, Chap. 14.
35. S. D. Beattie, D. M. Manolescu, and S. L. Blair, *J. Electrochem. Soc.*, **156**, A44 (2009).

36. S. S. Sandhu, J. P. Fellner, and G. W. Brutchen, *J. Power Sources*, **164**, 365 (2007).

37. X. H. Yang, P. He, and Y. Y. Xia, *Electrochem. Commun.*, **11**, 1127 (2009).

38. D. Wang, J. Xiao, W. Wu, and J.-G. Zhang, *J. Electrochem. Soc.*, **157**, A760 (2010).

39. A. Debart, J. Bao, G. Amstrong, and P. G. Bruce, *J. Power Sources*, **174**, 1177 (2007).

40. T. Ogazawa, A. Debart, M. Holzapfel, P. Novak, and P. G. Bruce, *J. Am. Chem. Soc.*, **128**, 1390 (2006).

41. A.Dobley, C. Morein, R. Roark, and K. M. Abraham, in *Proceedings of the 42nd Power Sources Conference*, Philadelphia, June 2006.

42. Y.-C. Lu, H. A. Gasteiger, E. Crumlin, R. McGuire, and Y. Shao-Horn, *J. Electrochem. Soc.*, **157**, A1025 (2010).

43. V. Giordani, S. A. Freunberger, P. G. Bruce, J.-M. Tarascon, and D. Larcher, *Electrochem. Solid State Lett.*, **13**, A180 (2010).

44. S. J. Visco, B. D. Katz, Y. S. Nimon, and L. D. DeJonghe, U.S. Patent 7,282,295, 2007.

45. U.S. patent application 2008/0102358.

46. Y. Wang and H. Zhou, *J. Power Sources*, **195**, 358 (2010).

47. T. Zhang, N. Imanishi, Y. Shimonishi, A. Hirano, J. Xie, Y. Takeda. O. Yamamoto, and N. Sammes, *J. Electrochem. Soc.*, **157**, A214 (2010).

48. B. Kumar, J. Kumar, R. Leese, J. P. Fellner, S. Rodrigues, and K. M. Abraham, *J. Electrochem. Soc.*, **157**, A50 (2010).

49. J. Hassoun, F. Croce, M. Armand, and B. Scrosati, *Angew. Chem. Int. Ed.*, **50**, 2999 (2011).

50. K. M. Abraham, Z. Jiang, and B. Carroll, *Chem. Mater.*, **9**, 1978 (1997).

CHAPTER 9

AQUEOUS LITHIUM–AIR SYSTEMS

Owen Crowther and Mark Salomon

1   INTRODUCTION

Metal–air batteries based on Zn, Mg, and Al anodes are a well-developed technology [1], but the use of Li anodes remained a long elusive goal until the breakthrough study by Abraham and Jiang [2]. The innovation reported by Abraham and Jiang is based on the use of nonaqueous electrolyte solutions, and advances in nonaqueous Li–air (O_2) systems are reviewed in Chapter 8. In the present chapter we deal specifically with aqueous systems, including the novel Li–seawater battery discussed below. Theoretical energies and capacities for selected metal–air systems were calculated from critically evaluated Gibbs energies of formation data, and the results are given in Table 1.

From the comparisons shown in Table 1, it is interesting to note that the aqueous Li–air cell will operate in seawater where sufficient O_2 exists to support a modest discharge rate of below about 0.1 mA/cm². For example, the O_2 solubility in ocean waters ranges from 250 to 200×10^{-6} mol/dm³ down to depths of around 500 m, after which there is a sharp decline to about 5×10^{-6} mol/dm³ (Pacific waters) and about 150×10^{-6} mol/dm³ (Atlantic waters) at depths of about 1000 m, followed by increasing O_2 concentration to 150 (Pacific) to 250×10^{-6} mol/dm³ (Atlantic) at

Lithium Batteries: Advanced Technologies and Applications, First Edition.
Edited by Bruno Scrosati, K. M. Abraham, Walter van Schalkwijk, and Jusef Hassoun.
© 2013 John Wiley & Sons, Inc. Published 2013 by John Wiley & Sons, Inc.

TABLE 1 Theoretical Specific Energy and Capacity Comparisons for Selected Metal–Air Systems

Aqueous metal–air systems (except where noted)	Open-circuit voltage (V)	Specific energy (Wh/kg)	Specific capacity (mAh/g)
$2Li + \frac{1}{2}O_2 \rightleftharpoons Li_2O$ (aprotic organic solution)	2.913	11,248[a]	3,862
$Li + \frac{1}{2}O_2 \rightleftharpoons \frac{1}{2}Li_2O_2$ (aprotic organic solution)	2.959	11,425[a]	3,862
$2Li + \frac{1}{2}O_2 + H_2SO_4 \rightleftharpoons Li_2SO_4 + H_2O$	4.274	2,046[a]	479
$2Li + \frac{1}{2}O_2 + 2HCl \rightleftharpoons 2LiCl + H_2O$	4.274	2,640[a]	616
$2Li + \frac{1}{2}O_2 + H_2O \rightleftharpoons 2LiOH$	3.446	5,789[a]	1,681
$2Li + H_2O$ (seawater) $\rightleftharpoons 2LiOH + \frac{1}{2}H_2$	2.512	9,701[b]	3,862
$Al + 0.75O_2 + 1.5H_2O \rightleftharpoons Al(OH)_3$	2.701	4,021[a]	1,489
$Mg + \frac{1}{2}O_2 + H_2O \rightleftharpoons Mg(OH)_2$	2.756	3,491[a]	1,267
$Zn + \frac{1}{2}O_2 \rightleftharpoons ZnO$	1.650	1,353[a]	820

Source: Data from [3].

[a]The molecular mass of O_2 is not included in these calculations because O_2 is freely available from the atmosphere and therefore does not have to be stored in the battery or cell.

[b]The molecular mass of H_2O is not included since it is freely available from seawater (pH 8.2) and does not have to be stored in the battery.

ocean depths of about 5000 m [4]. At high rates (≥ 0.1 mA/cm^2), the rate of diffusion of dissolved O_2 to the catalytic surface of the cathode becomes rate limiting, and the cathodic mechanism converts to the evolution of H_2. As shown in Table 1, the specific energy and capacity of this Li–seawater cell are higher than those for any other aqueous Li-, Zn-, Mg-, or Al-based cell, due to the fact that the molecular mass of water is not included in the calculations since water is freely available from seawater and does not have to be stored in the cell. A common problem with all the aqueous metal–O_2 cells shown in Table 1 is that of corrosion (i.e., reaction of the active metal anode with water to produce H_2). To prevent this corrosion, it is clear that water must be prevented from contacting the anode. For the Li anode, there are several approaches to the protection of the anode, the most promising of which is presently the single Li^+-conducting NASICON-type glass–ceramics $Li_2O + Al_2O_3 + TiO_2 + P_2O_5$ and $Li_2O + Al_2O_3 + Ge_2O + P_2O_5$ developed by Fu [5–7] and pioneered by Visco et al. for application to Li–air and Li–water systems [8]. Details on this and other "protected anode" approaches are discussed in Section 3.

Section 2 deals with mechanisms in acid and alkaline solutions, the nature of carbon materials and catalysts, and pressure and temperature. Both the carbon and catalyst will influence the rate and mechanism of the ORR (oxygen reduction reaction), and when a rechargeable system is of interest, the rate and mechanism of the OER (oxygen evolution reaction). The standard potentials for the reactions are summarized in Table 2 for both acid solutions [Equations (1) to (3)] and alkaline solutions (equations (4) to (6)). For all-carbon-based catalyst-free cathodes, two-electron reactions predominate, whereas for the preferred four-electron mechanism, a catalyst is essential. Due to the high positive potentials in an acid solution, most

TABLE 2   Half-Cell Reactions and Reduction Potentials for Oxygen

Electrochemical reaction	Potential (V) vs. the SHE	Eq.
$O_2 + 4H^+ + 4e^- \rightleftharpoons 2H_2O$	1.229	(1)
$O_2 + 2H^+ + 2e^- \rightleftharpoons H_2O_2$	0.695	(2)
$H_2O_2 + 2H^+ + 2e^- \rightleftharpoons 2H_2O$	1.763	(3)
$O_2 + 2H_2O + 4e^- \rightleftharpoons 4OH^-$	0.401	(4)
$O_2 + 2H_2O + 2e^- \rightleftharpoons H_2O_2 + 2OH^-$	−0.146	(5)
$\frac{1}{2}O_2 + H_2O + 2e^- \rightleftharpoons H_2O_2$	−0.534	(6)

transition metals will dissolve anodically, with the exception of the noble metals Pt, Pd, Au, Ru, and Rh. In alkaline solutions where the potentials for the ORR and OER are less positive (see Table 2), less expensive transition metals, such as Ni and Ag, and metal oxides, such as MnO_2, are stable and effective in catalyzing the four-electron mechanism. Another important issue in considering practical cathodes for the ORR and OER concerns the effect of surface area, catalyst, and stability of the carbon itself as a function of rate capability, O_2 partial pressure, and temperature. Due to the slow kinetics involved in oxygen electrochemistry, porous high-surface-area cathodes with appropriate catalysts are required to achieve practical current densities, which turns out to be a very difficult problem. For example, the operational temperature range required for important applications such as electric vehicles (EVs) is −40 to +56°C and up to 80°C (maximum), but to date there is limited literature demonstrating that the cathode for an aqueous-based fuel cell can provide practical current (re: power and rate) at or below 0°C. As of this writing, we found very few publications for Li–air where performance was studied at temperatures other than 25°C. Kowalczk et al. [9] reported performance at 0 °C in 3 mol/dm^3 KOH, He et al. [10] employed 0.1 to 2.0 mol/dm^3 LiOH solutions over the limited temperature range 25 to 55 °C and Zhang *et al.* [11] employed 1 mol/dm^3 LiCl at 60°C. Another issue addressed in Section 2 relates to noble metal catalyst degradation (e.g., Pt) due to dissolution and the electrochemical oxidation of the carbon matrix itself [12].

The third topic to be discussed in detail below concerns electrolyte solutions. Conductivities of aqueous acid and alkaline solutions are very high, and for H_2SO_4 and KOH solutions, the temperature range over which conductivities remain high is −60 to +80°C [1]. Corrosion and solubility of O_2, CO_2, and reaction products and corrosion in various acid and alkaline solutions are the major issues reviewed below. For example, it is well known that in nonaqueous solutions, the end of life of a Li–air cell is due to precipitation of reaction products such as Li_2O, Li_2O_2, and Li_2CO_3 within the pores of the porous cathode structure, as discussed in Chapter 8. In aqueous solutions, the question of precipitation of reaction products is yet to be fully resolved. For example, in pure water, the solubility LiOH is 5.2 mol/kg water, and for Li_2SO_4, the solubility is 3.1 mol/kg water [13]. Further complications arise due to the changing composition of the electrolyte solution upon discharge. For example, in alkaline solutions, water is consumed, thus decreasing the solubility of the product LiOH, and in acid solutions, water is a product of the discharge reaction, possibly increasing or at least maintaining the solubility of reactants such as Li_2SO_4 and CH_3COOLi [13,14]. For oxygen, literature on the solubility in various aqueous acid and alkaline solutions of interest to Li–air batteries is limited, but there are interesting publications dealing with O_2 solubility in seawater [4], [15], [16] and with aqueous H_2SO_4 and KOH as a function of concentration, temperature, and partial pressure of O_2 [15,16]. Note that the solubility decreases significantly as temperature increases, but also decreases significantly by a factor of 5 or 10 from pure water as H_2SO_4 and KOH concentrations increase to about 5 mol/dm^3 [15,16]. Important solubility data for LiOH or Li_2SO_4 in high concentrations of KOH and H_2SO_4 are not yet available in the literature.

2   LITHIUM–AIR POSITIVE ELECTRODES FOR AQUEOUS CATHOLYTE SYSTEMS

Oxygen is reduced by lithium ions in an air electrode during discharge of a Li–air battery via the ORR. However, the catholyte solution also is a reactant in an aqueous Li–air system, unlike a cell using a nonaqueous electrolyte. The OER occurs during charging of secondary cells, producing Li^+, O_2, and electrons (e^-) in the air electrode. The air electrode usually consists of carbon, a catalyst, and possibly a binder. The carbon is used as a support for the catalyst and to provide electronic conduction. Catalysts are used to promote the $4e^-$ ORR with increased activities and also to enhance OER activity in secondary cells. Binders are used to introduce hydrophobicity and to increase the mechanical strength of the electrode, although it is not necessary for all materials used as electrodes (e.g., "buckypapers"). The active electrode may be attached to a gas diffusion layer (GDL) to allow rapid uniform diffusion of O_2 to active reaction sites. For a secondary Li–air battery, this layer must have both hydrophobic and hydrophilic pores. The hydrophobic pores supply O_2 during ORR, and the hydrophilic pores transport water from the active electrode to stop flooding.

   In this section we review air electrodes that have been used in aqueous Li–air cells as reported in recent literature. Studies of aqueous Li–air cells are scarce, since this is a new battery system, particularly when using "dual" electrolytes. However, the cathode of the primary Li–air cell is analogous to the air–O_2 electrode of a polymer electrolyte fuel cell (PEMFC). Furthermore, the positive electrode of a secondary Li–air cell operates in a fashion similar to that of a unitized regenerative fuel cell (URFC) in that both ORR and OER occur at a single electrode. Pettersson et al. provided a good review of a URFC system, with a focus on bifunctional catalysts [17]. However, the presence of Li^+ in the catholyte solution reduces the ORR performance of the air electrode of a Li–air cell compared to a PEMFC. Liu and Xing showed that the intrinsic kinetic activity decreased by 40% for a commercial Pt/C electrode at 0.8 V vs. NHE when Li^+ ions were added to 1 mol/dm^3 H_2SO_4 as shown in Figure 1 [18]. They found that this was caused by Li^+ reducing the rate of O_2 diffusion through the electrolyte and not by interactions between Li^+ and Pt.

   One advantage of the aqueous compared to the nonaqueous system is that gas diffusion electrode technology developed for PEMFCs can be used in the air electrode. The gas diffusion electrode contains hydrophilic pores filled with electrolyte solution to transport the Li^+ (in the case of Li–air) and hydrophobic pores filled with gas to

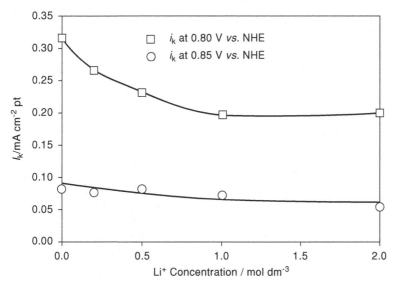

FIGURE 1   Influence of Li^+ kinetic current at different concentrations on the intrinsic activity of Pt for ORR at different polarization potentials.

transport O_2 to active reaction sites. The model accepted used to define the GDL for fuel cells is that the electrolyte solution only partially fills the pores of the electrode. Within each pore, a meniscus is formed, resulting in a three-zone interface (solid + liquid + gas) where a thin layer of electrolyte solution exists on the cathode surface above the meniscus. Since O_2 is transported through the electrolyte solution to the catalytic sites on the cathode, it is accepted that it is this thin film above the meniscus where the majority of the ORR current is occurring. The difficulty in defining the three-phase interface is complicated by uncertainties in the effect of atmospheric pressure and partial pressure of O_2, uneven electrode porosities and surfaces, and the possibility of precipitation of reaction products: for example, Li_2SO_4 in aqueous H_2SO_4 solutions and $LiOH + Li_2CO_3$ in neutral and alkaline solutions. Clearly, any precipitation of reaction products will negate the advantage of a three-phase interface, as the cathode will then operate in a flooded state. However, nonaqueous organic electrolyte solutions used in Li–air batteries tend to wet even the hydrophobic pores, and thus the cathode always operates in a flooded state [19]. To our knowledge, no true gas diffusion electrode exists for the nonaqueous Li–air system, so the air electrode is always "flooded" with electrolyte.

The initial disclosure of the dual electrolyte aqueous Li–air system by PolyPlus used a commercial Zn–air cathode as the positive electrode in 1 mol/dm^3 LiOH [8]. Similarly, early work was reported by MaxPower, Inc. using a commercial Alupower air electrode [9]. Primary Li–air cells using these air electrodes were able to discharge for several days in either 3.0 mol/dm^3 KOH at 0°C or 5.25 mol/dm^3 H_2SO_4 at 0.1 mA/cm^2 and 25°C [9]. The discharge of both of these cells is shown in Figure 2.

Zhang et al. have demonstrated Li–air cells that use a Pt-black-catalyzed air electrode similar to that used in a fuel cell [11,20]. Their initial cell using 1 mol/dm^3

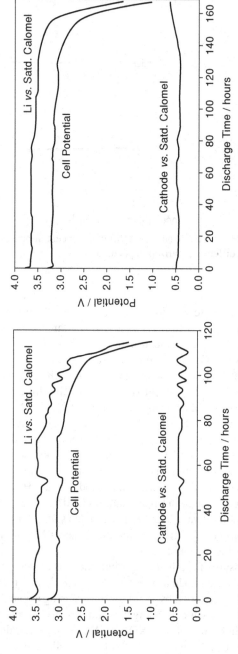

FIGURE 2   a. Li–air discharge performance (a) in 3.0 mol/dm³ KOH at 0°C and (b) in 5.25 mol/dm³ H₂SO₄ at 25°C.

196

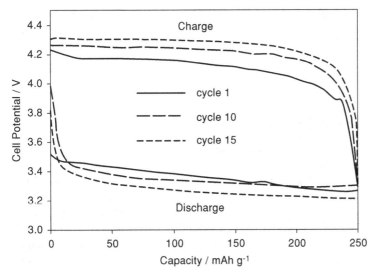

FIGURE 3 Charge–discharge performance of Li‖PEO/LTAP/HOAc–H$_2$O–LiOAc (sat)‖carbon air cell at 0.5 mA/cm^2 and 60°C. The cell was sealed in a high-pressure vessel with 3 atm of air to suppress evaporation of the electrolyte.

LiCl catholyte solution was able to discharge and charge at 60°C and current densities of 0.05, 0.10, and 0.25 mA/cm^2 [11]. At the two lower current densities, the discharge and charge potentials were fairly flat, but at the higher current density of 0.25 mA/cm^2, the charge voltage was not flat but continued to increase until the cell was shut down. This continual increase in cell potential during charge is probably due to Cl$_2$ coevolution (with O$_2$), as discussed in Sections 3 and 4. Later they studied the cycling behavior of a similar cell using an acetic acid–based catholyte solution [20]. The cell discharged at a rate of 0.5 mA/cm^2 at a potential of 3.46 V and demonstrated an energy density of 779 Wh/kg based on the mass of Li anode and acetic acid catholyte solution [20]. The cell demonstrated good secondary performance for 15 cycles with a Li–air cell using a positive electrode similar to a fuel cell in a buffered acetic acid solution, as shown in Figure 3 [20]. However, since the air electrode contained a Pt-black catalyst, it is probably not economically feasible for many applications.

Wang and Zhou [21] have demonstrated a Li–air system in 1 mol/dm^3 KOH (presumably at room temperature) using a Mn$_3$O$_4$-catalyzed air electrode. The electrode was made by mixing a prepared Mn$_3$O$_4$–carbon composite with poly(tetrafluoroethylene) and then roll-pressing to obtain a continuous electrode. They then laminated a gas diffusion layer consisting of 60% acetylene black and 40% PTFE onto this catalyst layer. This cell was discharged at room temperature and a rate of 0.5 mA/cm^2. The cell discharged over 500 h at a constant voltage of approximately 2.8 V [21]. They state that the delivered capacity for an electrolyte solution in what appears to be a flow-type cell is 50,000 mAh/g, based on the total mass of the catalytic layer [21]. However, the true capacity is actually much lower since the mass of the catholyte solution should also be included in this system when comparing it to nonaqueous Li–air cells, since the catholyte must be treated as a reactant. A second titanium electrode to support the OER process was later added to the system to demonstrate rechargeability [22]. Recently, they have begun to look at different catalysts, such as Cu, TiN, and graphene, which are discussed in greater detail below [23–25].

2.1 Temperature and Pressure Effects

In this section we discuss the effect of temperature and pressure on Li–air positive electrode performance. Increasing the temperature generally should improve the kinetics of the reaction and thus result in smaller overvoltages for the ORR and OER. An increase in temperature will also increase the electrolyte conductivity. However, increasing the temperature will lead to quicker evaporation of the catholyte solution and will increase the rate of negative interactions between the catholyte solution and LiC–GC. It may also lead to quicker mechanical degradation of the cell. Operating PEMFCs at temperatures of $0°C$ and slightly below ($\sim -5°C$) is possible provided that the cell is purged at room temperature with dry air or N_2 before being subjected to low temperatures (below freezing) prior to cold startup [26]. However, it should be noted that the catholyte solution should preferably be chosen so that it does not freeze, as discussed in Section 4, as the cell discharge voltage will be lower since the ORR kinetics decreases significantly with decreasing temperature. He et al. studied the effect of temperature on the performance of a Li–air cell using the Mn_3O_4-catalyzed air electrode described above. They found that the maximum power density increased nearly linearly as the temperature was raised from $25°C$ to $55°C$, as seen on the left-hand axis of Figure 4 [27]. The maximum power density was defined as the open-circuit voltage divided by the resistance multiplied by 2 [27]. However, many different factors can account for this increase in performance, as described above.

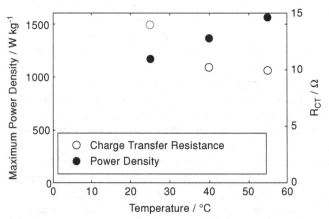

FIGURE 4 Effect of temperature on maximum power density and charge transfer for resistance for a primary Li–air cell with a Mn_3O_4/C composite air electrode and KOH catholyte solution.

Therefore, the authors used electrochemical impedance spectroscopy to separate cell resistances. The charge transfer resistance decreases significantly between 25 and 40°C and then reaches a nearly constant value as the temperature is raised to 55°C, as shown on the right-hand axis of Figure 4 [27]. Zhang et al. operated the cells described above at 60°C [11,20]. This temperature was used to increase the conductivity of the poly(ethylene oxide) (PEO)-based solid electrolyte used as a Li/LTAP interlayer to enable it to support a moderate rate discharge with decreased ohmic losses. The melting point of the PEO-based electrolyte is ~60°C.

The simplest Li–air battery would use a passive, air-breathing electrode that consumes O_2 from ambient air. It is possible to use bipolar plates, compressors, and so on, found in fuel cells to increase the partial pressure of O_2 in the air electrode, leading to better cell performance [28]. However, this is not straightforward, as the mass and volume of the additional equipment must be accounted for in the final power and energy densities [28]. Zhang et al. operated the cells described above at 3 atm of O_2 to minimize electrolyte evaporation [11,20]. There are no quantitative studies on the effect of pressure on aqueous Li–air performance in the literature at this time, although Yang and Xia have reported that the air electrode performance at higher rates was found to improve by increasing the O_2 partial pressure for a nonaqueous Li–air system [29]. From a theoretical viewpoint, the effect of pressure can be demonstrated using the Nernst equation. For the 4e⁻ ORR mechanism in acid media (see Table 2), the Nernst equation is

$$E = E^0 - \frac{2.303\,RT}{nF} \log \frac{a_{H_2O}^2}{a_{H^+}^4 p_{O_2}} \tag{7}$$

where p_{O_2} is the partial pressure of O_2 and all other terms have their usual significance. For constant pH and water activity and a temperature of 298.2 K, the difference in potential for the 4e⁻ ORR mechanism, ΔV, is

$$\Delta V = E_2 - E_1 = \frac{0.05916}{n} \log \frac{p_2}{p_1} \tag{8}$$

For $p_1 = 1$ atm of O_2, $p_2 = 0.21$ atm and $n = 4$, equation (8) yields $\Delta V = -0.0100$ V. Similarly, for the 2e⁻ ORR mechanism (see Table 2) where $n = 2$, the potential difference $\Delta V = -0.0200$ V.

2.2 Carbon in Air Electrodes

There are abundant publications on the selection of carbons for the air electrode of a Li–air battery for a nonaqueous system [30–34]. However, the reaction products for such a system have very low solubility in the electrolyte solution and thus will precipitate in the pores of the carbon electrode [30]. This can lead to premature end of life, as the transport of Li⁺ and O_2 to active carbon reaction sites is blocked by the precipitate [30]. Therefore, a large pore volume available to accommodate the reaction product is necessary for good performance of a nonaqueous Li–air system. In contrast, the reduction products of an aqueous Li–air system usually have moderate to high solubilities, depending on the nature of the electrolyte solution used. Therefore, the problem of pore blocking, described above, does not appear to be as severe in an aqueous acidic system as in an aqueous alkaline system, as discussed in Section 4.

As discussed above, the limited available research publications for aqueous Li–air positive electrodes have either used a commercial air electrode designed for a different system or a catalyzed carbon black air electrode [8,20–25]. However, we can look to technology for the PEMFC and URFC to determine an ideal air electrode for primary and secondary Li–air systems. Song et al. did this in a recent review of nanostructured materials for lithium and lithium–air batteries [35]. Carbon cathodes are known to reduce O_2 via the $2e^-$ mechanism during the ORR reaction to form peroxide [36]. The desired $4e^-$ mechanism for the ORR mechanism is greatly enhanced through the use of a catalyst added to the carbon-based cathode. Carbon is commonly used as a catalyst support in PEMFCs. The ideal catalyst support has a high surface area to disperse catalyst, high electronic conductivity, and is stable under ORR and OER (for secondary batteries) conditions. The instability of catalyst supports is well known to cause PEMFC performance degradation [37]. Carbon blacks are known to degrade over time under standard PEMFC operating conditions [1,35–37]. This problem occurs even more quickly at the higher potentials needed for the OER [17,36]. Highly ordered carbons such as carbon nanotubes and carbon aerogels have demonstrated higher stability than carbon blacks [37,38]. Graphene has also recently been demonstrated as a catalyst support [39]. In addition to carbon-based materials, some oxides can also be used as catalyst supports [17,36]. For example, $Ti_{0.9}Nb_{0.1}O_2$ has demonstrated good stability as a catalyst support in a URFC [40].

Zhu et al. have demonstrated the buckypaper air electrode for use in both non-aqueous Li–air batteries [41] and PEMFC [42]. Buckypaper electrodes decorated with catalysts are also intriguing for use in aqueous Li–air systems. Buckypapers (Fig. 5), are fabricated by filtering mixtures of single-walled carbon nanotubes

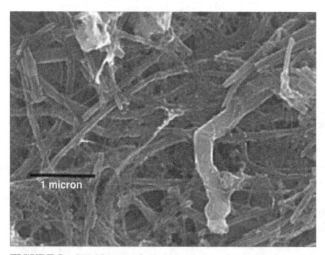

FIGURE 5 SEM image of a buckypaper surface. (From [41], with permission of the Electrochemical Society.)

(SWNTs) and carbon nanofibers (CNFs) [41]. The porosity, surface area, and conductivity can all be tailored by adjusting the ratio of SWNTs to CNFs [41]. They have better stability than carbon black in H_2SO_4 [42]. They also require no binder or current collector [41]. The lack of the need for a metal current collector is a special advantage, since many of the common current collectors (e.g., aluminum, stainless steel, or titanium grids) corrode in acidic aqueous electrolytes.

2.3 Catalysts

As discussed briefly above, catalysts are used to increase the activity of ORR and OER. Monofunctional catalysts increase the activity of just one of the reactions; for example, IrO_2 is known to improve the rate of OER when added to a carbon but does not improve ORR. A bifunctional catalyst, often some type of alloy, such as Pt/IrO_2 in URFCs, improves both the ORR and OER. Zhou and others have studied different catalysts in aqueous Li–air systems [21–25]. Their initial publication used a Mn_3O_4 and carbon black electrode, described in greater detail above [21]. However, they did not run the cell with carbon black alone, so it is difficult to determine the impact of the catalyst for this cell. They next showed that Cu could catalyze ORR via the copper corrosion mechanism by discharging a Li–air cell with a Cu foil cathode in 1 mol/dm³ $LiNO_3$ purged with either O_2 or N_2 [22,23]. Next, they studied a TiN/C composite air electrode and found that it has ORR activity in an acetic acid–based catholyte approaching a Pt foil, as shown in Figure 6 a [24].

Figure 6b shows that the cell was able to discharge over 250 h at a rate of 0.5 mA/cm² [24]. They also studied the use of graphene nanosheets (GNSs) as a catalyst in a secondary Li–air cell [25]. The electrode consisted of 90% GNS, 7% PTFE, and 3% acetylene black (AB). They compared the primary and secondary discharges of this in 1 M $LiNO_3$ + 0.5 M LiOH aqueous electrolyte with a 93% AB air electrode and a commercial 20 wt% Pt-black. Figure 7 shows that the discharge voltage at a rate of 0.5 mA/cm² of the GNS-based electrode is only 50 mV less than that of the Pt-black electrode and 250 mV higher than that of the AB-based electrode [25]. They also found that the GNS-based cathode was able to cycle, and that heat-treating the GNSs improved the reversibility [25]. Both TiN and graphene are promising air electrode materials for an aqueous Li–air system.

Jörissen has provided a good review on bifunctional catalysts for metal–air batteries [36]. This review focuses on bifunctional catalysts for use in acidic electrolytes, since this electrolyte does not have the inherent problem of carbonate precipitation. Readers interested in catalysts for alkaline systems are referred to Jörissen or to the literature concerning the zinc–air system. Neburchilov et al. recently reviewed the air electrode for zinc–air cells [43].

Platinum or a platinum metal alloy such as Pt–Au is considered the best-performing catalyst for ORR in acidic electrolytes [44]. Since Pt and Au are relatively rare and costly, care must be taken to minimize Pt loading while maintaining high electroactivity [44]. Chen et al. recently reviewed nonprecious metal ORR catalysts for PEMFCs [45]. The transition metal oxides studied by Debart et al. [46] for nonaqueous Li–air systems are not typically stable in aqueous acidic electrolytes [36]. While Pt exhibits high catalytic activity for the ORR, it is not a particularly

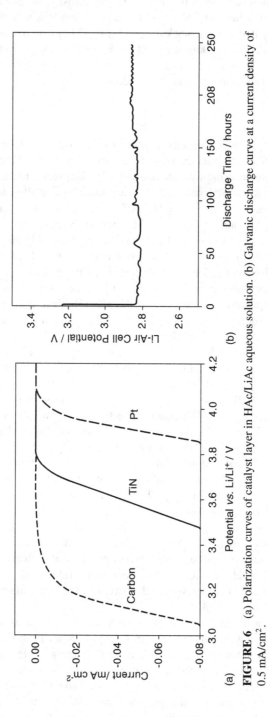

FIGURE 6 (a) Polarization curves of catalyst layer in HAc/LiAc aqueous solution. (b) Galvanic discharge curve at a current density of 0.5 mA/cm^2.

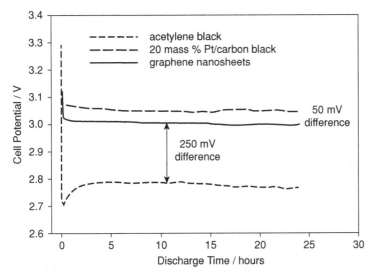

FIGURE 7   Discharge curves of 20 wt% Pt/CB, GNSs, and AB at a current density of 0.5 mA/cm^2 for 24 h (probably at room temperature).

good catalyst for the OER. Much improved OER activity can be achieved by alloying Pt with other precious metals, such as Au, Ir, and Ru. For example, at 75°C in 0.5 mol/dm^3 H$_2$SO$_4$, the onset potential of the OER on Pt is 1.6 V vs. NHE and is around 1.45 V for a Pt$_{0.7}$Ir$_{0.3}$ alloy, as shown in Figure 8 [47].

Chen et al. [48] studied 715 unique combinations of Pt, Ru, Os, Ir, and Rh and found that the ternary catalyst Pt$_{4.5}$Ru$_4$Ir$_{0.5}$ offered the best combination for improved electroactivity and stability in an acidic electrolyte. Other potential OER catalysts

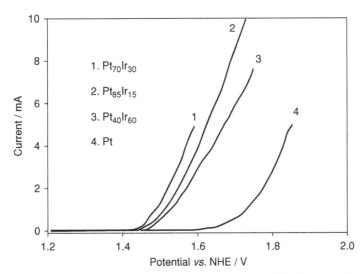

FIGURE 8   Effect of Ir content in Pt–Ir catalysts on OER (in 0.5 mol/dm^3 H$_2$SO$_4$ at 75°C).

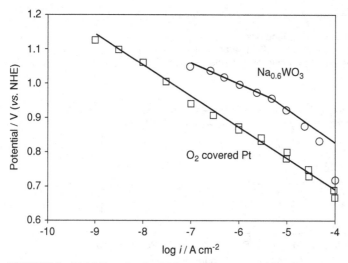

FIGURE 9 Tafel lines for the ORR at oxide-covered Pt and $Na_{0.6}WO_3$ cathodes in 0.05 m H_2SO_4 solution. (The plots for the two catalysts are based on data digitized directly from Figure 6 of [51] and from [52].)

are RuO_x, RhO_x, and $Na_xPt_3O_4$ [17,36]. These can be combined with Pt to obtain a bifunctional catalyst.

The "bronzes" are low-cost materials that we find interesting as both ORR and OER catalysts. The bronzes are nonstoichiometric mixed oxides with the general formula M_xTO_3, where M is a group IA or IIA metal with x values between 0.5 and 0.9, and T is one of the transition metals W, Nb, or Ta. The bronzes have electronic conductivities approaching those of metals. For example, the resistivity of the bronze $Na_{0.46}WO_3$ is 81 $\mu\Omega \cdot cm$ at room temperature [49]. The use of bronzes to catalyze the $4e^-$ reaction was proposed some 40 to 45 years ago [50–52], but this approach has essentially been neglected since that time. In acids, the Na_xWO_3 bronzes are highly stable at the ORR and OER potentials and exhibit electrocatalytic activity comparable to that of Pt. For example, the exchange current density for the ORR on Na_xWO_3 in a solution of pH 0.3 under a 1 atm pressure of O_2 ranges from 0.5 $\times$ 10^{-9} to 5 $\times$ 10^{-9} A/cm^2, which is higher than 10^{-10} to 10^{-11} A/cm^2 for Pt in 0.1 N H_2SO_4. The bronzes of niobium and tantalum are highly stable in acid solutions and also exhibit exchange current densities superior to those on Pt [50–52]. Not only are the exchange current densities for the ORR and OER on bronzes at least equal to but generally superior to Pt, indicating improved reversibility, but the overpotentials are also lowered significantly, as shown in Figure 9.

3 THE LITHIUM ANODE

The key issue in commercializing an aqueous Li–air battery is the prevention of corrosion of lithium by water. This is a major and continuing focus for present-day research and development. This is accomplished by using a barrier layer that allows the transport of Li ions (Li^+) to the positive electrode while preventing water from reaching the Li-metal electrode. The initial attempt [53] to commercialize the Li–air battery, which relied on a strong alkaline solution to form a thick passive KOH film on the Li, preventing further corrosion, was wisely abandoned, for obvious reasons. Other more recent researchers attempted to eliminate corrosion using a hydrophobic polymer membrane between the Li metal and catholyte solution [54], but again abandoned the effort since water transport through the polymer could not be prevented. In 2004, the PolyPlus Battery Company developed breakthrough technology that allows for safe operation of aqueous Li–air batteries [8,55]. This approach uses a Li ion–conducting glass ceramic (LiC-GC) to separate the Li metal from the positive electrode, which contains an aqueous electrolyte solution. The LiC-GC allows the transport of Li^+ to the positive electrode during cell discharge but blocks water in the catholyte solution from reaching the Li metal and reacting. The most successful approach to date utilizes single Li^+–conducting NASICON-type glass ceramics $Li_2O + Al_2O_3 + TiO_2 + P_2O_5$ and $Li_2O + Al_2O_3 + Ge_2O + P_2O_5$ developed by Fu and patented by Ohara Corporation [5–7]. The cell utilizing this approach is

$$
\left\| \begin{array}{c} \text{Metallic Li} \\ \text{anode} \end{array} \right\| \begin{array}{c} Li^+\text{-conducting} \\ \text{interlayer} \end{array} \left\| \begin{array}{c} \text{LiC-GC} \\ \text{ceramic} \end{array} \right\| \begin{array}{c} \text{Aqueous} \\ \text{solution} \end{array} \left\| \begin{array}{c} \text{Air } (O_2) \\ \text{cathode} \end{array} \right\|
\qquad (9)
$$

Due to the reaction of Li with LiC-GC, an interlayer is dispersed between the Li and the LiC-GC. A detailed review is presented below concerning the anode compartment itself (including the interlayer between the Li and the LiC-GC).

Ideally, the LiC-GC would be compatible with Li and not require an interlayer. However, this review focuses on the Ohara Corporation material since it is the only material presently commercially available. In addition to its resistance to water penetration, for a solid-state material, this material exhibits reasonably high conductivity at ambient temperatures. At 25°C its conductivity is about 1.75×10^{-4} S/cm and over the temperature range 25 to 513°C, the conductivity of the LiC-GC of reference 7 can be calculated from the following smoothing equation (Kowalczk et al. [4]:

$$
\sigma = 29.38 \exp\left(\frac{-7046}{1.987T}\right) \qquad (10)
$$

Although equation [10] is strictly valid over this temperature range, conductivities at lower temperatures calculated from this equation are extrapolated estimates but informative; we find that at 15°C, $\sigma \approx 1.3 \times 10^{-4}$ S/cm and at 0°C, $\sigma \approx 6.8 \times 10^{-5}$ S/cm. For application to EVs for which battery operation at $-40°C$ is a requirement, the conductivities of the LiC-GC ceramics will be problematic, which is a major motivation for continued research on new ceramic materials. However, in view of the success and commercial availability of Ohara's LiC-GC materials, here we discuss only the properties of this ceramic.

Both types of LiC-GC available [5,6] will react with Li and therefore require an interlayer that is composed of a material that is compatible with both Li metal and the

LiC-GC. This can be as simple as a Li battery separator (e.g., Celgard 2400) soaked in a common battery electrolyte such as 1.2 mol/dm^3 lithium hexafluorophosphate (LiPF$_6$) in ethylene carbonate/ethylmethyl carbonate (3 : 7 by mass), or it can be a Li-stable ceramic such as LiPON. In addition, it is necessary to obtain a highly stable hermetic seal on the anode compartment to prevent water ingress around the seals that will corrode Li metal. The seal must be stable over a large temperature range and not dissolve in the electrolyte solution when a long cycle life is required. This seal design, referred to as a "protected anode" can be realized by using either a solid fixture with an O-ring or a flexible laminated pouch material that is heat sealed to the LiC-GC, such as that recently patented by MaxPower [56]. In a Li–air or Li–seawater battery, upon cell discharge Li is consumed, creating a vacuum, and a flexible laminate pouch that is filled with a liquid electrolyte is necessary so that the pouch will collapse, minimizing the pressure differences between the anode compartment and the surrounding environment and thus preventing the LiC-GC from rupturing, which would result in a catastrophic ingress of water.

Additional concerns regarding the protected anode involve the stability of LiC-GC in acid and alkaline solutions as a function of temperature, and how stability considerations will serve as a guide for specific applications either as a primary or a rechargeable Li–air battery. For secondary systems, Li dendrites will become a serious problem after repeated cycling [57]. Therefore, the interlayer should be chosen to stop or at least slow dendrite initiation and propagation. The interlayer can be composed of any material that is stable with both the negative electrode and LiC-GC, including organics, polymers, ionic liquids, and solid ionic conductors. PolyPlus Battery Company's initial patent demonstrated air cells using Ohara LiC-GC with 1 mol/dm^3 LiPF$_6$ in propylene carbonate (PC) with a microporous polymer separator such as Celgard serving as the interlayer [8]. They also demonstrated charging of a carbon negative electrode with 1 mol/dm^3 LiPF$_6$ in an ethylene carbonate/dimethyl carbonate anolyte solution that was lithiated by a 1 mol/dm^3 HCl + 2 mol/dm^3 LiCl catholyte prior to first discharge [8]. They note that a carbon or metal alloy negative electrode can be lithiated chemically or electrochemically prior to cell construction [8]. In a subsequent patent they disclosed various solid interlayers and methods for fabricating them [55]. They demonstrated stable interlayers with relative low resistances consisting of very thin LiPON, Cu$_3$N, and red phosphorus [55]. However, these layers were all formed via sputtering or evaporation processes which are not practical for large-scale production.

Zhang et al. fabricated a Li anode consisting of lithium, a solid polymer electrolyte, and Ohara's LiC–GC [11,20]. They used an interlayer consisting of PEO and LiTFSI in the ratio 1 Li/18 O [11]. The anode demonstrated no increase in resistance over a period of 30 days. However, this cell needed to be operated at 60°C where a PEO-based solid polymer electrolyte (SPE) melts; this minimum high temperature is required to provide high conductivity, to allow for practical current densities. They later used electrochemical impedance spectroscopy to determine that the cell resistance was dominated by the Li/SPE and SPE/LiC-GC interfaces [20]. Therefore, the approach taken by Zhou et al. using a liquid organic electrolyte interlayer may be more promising, especially for the primary system [22]. They used 1 mol/dm^3 LiClO$_4$ in an ethylene carbonate (EC) and dimethylcarbonate (DMC) mixture with a microporous polymer separator [22]. However, they did not try to recharge the cell electrochemically. For a secondary cell, dendrite formation will occur quicker in a liquid organic electrolyte (compared to a SPE or a material such as LiPON).

In a number of studies by Yamamoto and others, the stability of the LiC-GC was studied by changes in conductivity and structure (by x-ray diffraction and cyclic voltammetry) after immersion in various aqueous solutions at 50°C for periods of up to 3 weeks [58–60]. The results indicated that the ceramic degraded in solutions of strong acids (e.g., H_2SO_4) and bases (e.g., KOH), whereas solutions of acetic acid, acetic acid saturated with lithium acetate, binary water + LiCl, and LiOH saturated with LiCl were stable. However discharge and charge performance were reported in only two papers for the cell given in equation (9). Cycle performance for the Li‖PEO/LiC-GC/HOAc–H_2O–LiOAc (sat)/‖carbon air cell at 0.5 mA/cm² and 60°C in reference 20, was discussed above, and the Li‖PEO–LiTFSI‖LiC-GC‖1 mol/dm³ cell in reference 11 was cycled three times at 60°C, as shown in Figure 10. Cycle performance at the lower current densities was satisfactory, but not at the higher current density of 0.25 mA/cm², where the charging potential did not reach a stable voltage but continued to increase until charging was terminated after 1 h.

As mentioned above and discussed in more detail in the Section 4, it is most probable that co-Cl_2 evolution is occurring along with O_2 evolution. It is therefore apparent that a chloride-containing aqueous electrolyte solution cannot be used in a long-life rechargeable Li–air cell or battery.

The use of strong acids and bases in a rechargeable battery is still a very workable medium for use in a rechargeable Li–air system. First, electrolyte solutions

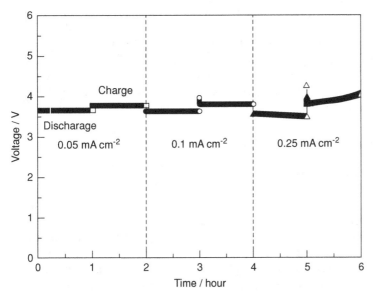

FIGURE 10 Charge–discharge performance of the Li/PEO–LiTFSI/LiC–GC/1 mol/dm³ LiCl cell at various current densities. (From [11], with permission of the Electrochemical Society.)

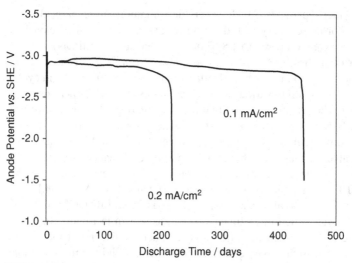

FIGURE 11   Long-term discharge of a protected Li anode in seawater. (Data from [61].)

based on H_2SO_4 and KOH are odorless compared to acetic acid solutions, and existing data show that cell performance in these electrolyte solutions is excellent, at least for a primary system. For example, Figure 2a and b demonstrate that cells could be discharged continuously for 5 to 7 days at 0.1 mA/cm² in KOH at 0°C and H_2SO_4 at 25°C, and Figure 11 shows that the LiC-GC ceramic cell can operate successfully in seawater continuously for over 400 days at room temperature [61]. Note that seawater is slightly alkaline (pH 8.2) [4] and the amount of concentration of NaCl is, on average, 0.420 mol/dm³. At 0.1 mA/cm² the seawater cell discharged continuously for 414 days until all Li was depleted (100% efficiency), and at 0.2 mA/cm² the seawater cell discharged continuously for 218 days until all Li was depleted. There are no literature reports indicating that the cell represented by equation (9) will not cycle satisfactorily in strong acid or alkali solutions. The bottom line is that KOH and H_2SO_4 electrolyte solutions can operate effectively in both primary and rechargeable LiC-GC ceramic-based Li–air cells.

4   ELECTROLYTE SOLUTIONS

The selection of an electrolyte solution is generally governed by several factors, such as electrolyte solution conductivity, oxygen solubility in the electrolyte solution, stability of the LiC-GC membrane in the various electrolyte solutions, and possible precipitation of reaction products upon cell discharge. Another important factor is whether the the Li–air battery is intended for a primary or rechargeable application. For a primary system, almost any electrolyte solution will suffice, such as solutions containing acetic acid saturated with lithium acetate [59], LiCl + HCl [59], LiCl [11], or H_2SO_4 and KOH [8,9]. For a rechargeable aqueous Li–air cell, the most practical electrolyte solution is probably H_2SO_4 solutions and, where the problem

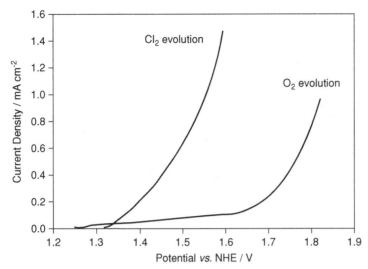

FIGURE 12 Current vs. potential curves for Cl_2 and O_2 evolution at a graphite anode in 5.3 mol/dm^3 NaCl solution. (From [63].)

of atmospheric CO_2 can be tolerated. Any electrolyte solution containing Cl^- ions is bound to evolve Cl_2 on recharge. Even though the standard electrode potential for the reaction $Cl_2(g) + 2e^- \rightleftharpoons 2Cl^-$ (1.358 V) is more positive than that for the $4e^-$ oxygen reaction [1.229 V; equation (1) in Table 2) significant Cl_2 evolution will occur in addition to O_2 evolution upon attempting to recharge the Li–air cell. This will occur since the OER overpotential is high, and its exchange current density is 10^{-10} to 10^{-12} A/cm^2 at Pt electrodes depending on pH [52,62]. Thus, whereas O_2 evolution may initiate first, because of the high exchange current density for Cl_2 evolution (10^{-3} A/cm^2), Cl_2 evolution becomes predominate at higher current densities, as shown in Figure 12 [63].

At the present time, aqueous Li–air cells which have exhibited promise for commercial applications are based on the two electrolyte solutions of H_2SO_4 and KOH. Both solutions are highly conductive over a very large temperature range, exhibit reasonable capability for dissolution of oxygen, and are odorless compared to aqueous solutions containing high concentrations of organic acids needed to ensure low-temperature operation (e.g., acetic acid). Figures 13 and 14 show the high conductivities of these binary solutions over a wide temperature range, and Figure 15 shows the solubility of O_2 in H_2SO_4 and KOH solutions.

Figure 15 clearly shows that the solubility of O_2 decreases significantly from that of pure water as the concentration of acid or base increases, more dramatically in KOH solutions. The solubility of O_2 at 1 atm and 25°C in pure water is given as the Ostwald coefficient $L = 0.03104$, and in 2.65 mol/dm^3 H_2SO_4, $L = 0.0192$ [15,16]. Based on the O_2 solubilities shown in Figure 15, it appears that H_2SO_4 solutions are preferred over KOH for use in an aqueous Li–air battery. As noted by Read et al. [67], the transport properties of O_2 in solution are dependent upon solubility, viscosity,

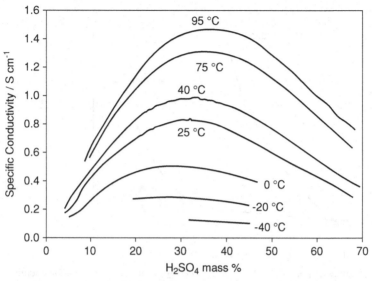

FIGURE 13   Specific conductivity of binary $H_2SO_4 \cdot H_2O$ solutions. (Data from [1, 64].)

and electrolyte concentration, and all affect the discharge capacity. These are key considerations in optimizing electrolyte solutions for future development of aqueous Li–air cells.

In calculating theoretical specific energies and specific capacities in Table 1, the reaction products were taken to be completely ionized under standard-state conditions. However, for real systems, these quantities will depend on a number of factors,

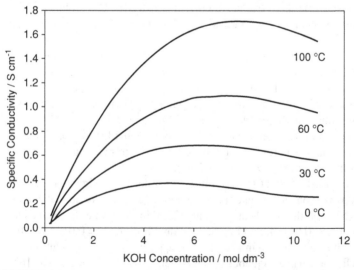

FIGURE 14   Specific conductivity of binary $KOH \cdot H_2O$ solutions. (Data from [65].)

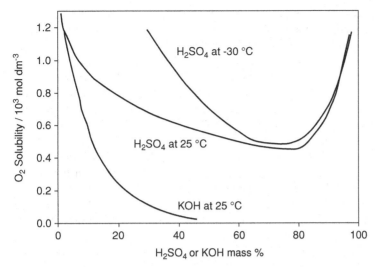

FIGURE 15   Oxygen solubility in binary $H_2SO_4 \cdot H_2O$ and binary $KOH \cdot H_2O$ solutions. (Data from [66].)

such as activities, ion association, and solubility as a function of temperature, where the solid phase is often a hydrate depending on temperature. For example, the solubility of LiCl at 25°C is 19.1 mol/kg and the solid phase is the monohydrate, and at −40 °C the solubility is 11.52 mol/kg and the solid phase is $LiCl \cdot 3H_2O$ [68]. Both Li_2SO_4 with a solubility of 3.13 mol/kg in pure water at 25°C [69], and LiOH with a solubility of 5.2 mol/kg H_2O in pure water [13], will both precipitate as monohydrates at room temperature, nominally 25°C. Important data needed for optimization of the electrolyte solution concentrations, such as the solubilities of LiOH in KOH and Li_2SO_4 in H_2SO_4 over a wide temperature range (−40 to +80 °C), are not yet available in the literature. Other complex equilibria exist, particularly for the Li_2SO_4 + H_2SO_4 + H_2O system, where the solid phases reported include species such as $LiHSO_4$ and $LiH_3(SO_4)_2$ in addition to Li_2SO_4 and the monohydrate $Li_2SO_4 \cdot H_2O$ over the temperature range 12.5 to 30°C [70–72].

Another important issue of concern for alkaline-based aqueous Li–air cells is the effect of water consumption or production upon cell discharge [13,14]. Given that the discharge product in alkaline solution is the monohydrate, we can write the overall cell reaction as

$$2Li + \tfrac{1}{2}O_2 + 3H_2O \rightarrow 2LiOH \cdot H_2O \qquad (11)$$

Since water is consumed in this reaction, it follows that the concentration of LiOH increases as discharge proceeds and will therefore precipitate in a much shorter time than expected based on its solubility in pure water (5.2 mol/kg H_2O at 25°C). The net result is that precipitation of $LiOH \cdot H_2O$ will deposit in the pores of the cathode, leading to a decrease in specific capacity similar to the effect of lithium oxides on specific capacity in nonaqueous Li–air systems. Thus, as shown by Zheng et al., large porosities of the cathode in aqueous alkaline Li–air cells are also important to maximize specific capacity [14]. Stevens et al. did report precipitation of LiOH and Li_2CO_3 in a rechargeable alkaline Li–air cell [73]. These authors used several types of Li–air cells given in equation (9) with the Ohara LiC–GC solid electrolyte. In one cell an organic electrolyte solution (not specified) was used for the interlayer and in another cell the interlayer was LiPON deposited onto the LiC–GC membrane. Using the organic interlayer solution and 2 and 5 mol/dm^3 LiOH as the catholyte and electric fuel cathodes, cells discharged in the atmosphere did not perform well, the major factor being the precipitation of Li_2CO_3 in the pores of the cathode. The lifetime of the cell discharged at 0.10 mA/cm^2 was extended by a factor of 100 by placing an anionic membrane to the air side of the cathode. Cells using the LiPON-covered LiC-GC membrane showed improved cycling at low rates, but limitations persisted, due to Li_2CO_3 and LiOH precipitation. It is interesting to note that the majority of LiOH precipitation occurred on the surface of the LiC-GC membrane and that dendrite formation was a contributing factor, limiting the number of cycles to about 100 cycles at low rates (0.1 mAh/cm^2).

In acid solutions where Li_2CO_3 formation does not occur, the situation is more complex than that for alkaline solutions. If the solid discharge product is the anhydrous electrolyte, water is a discharge product, resulting in a decrease in concentration of the electrolyte product, as shown for Li_2SO_4 in Table 1. Also, there will be a decrease in the H_2SO_4 composition of the electrolyte solution. On the other hand, if the discharge product in H_2SO_4 solutions is the monohydrate, water is neither consumed nor produced, as indicated by the following reaction:

$$2Li + \tfrac{1}{2}O_2 + H_2SO_4 \rightarrow Li_2SO_4 \cdot H_2O \qquad (12)$$

However, determining the true nature of the effect of water production is complicated by the probable production of $LiHSO_4$ and possible production of $LiH_3(SO_4)_2$ and Li_2SO_4 in addition to the monohydrate, but details on the phase diagram for the $Li_2SO_4 + H_2SO_4 + H_2O$ system over the temperature and concentration range of interest for aqueous Li–air systems are not presently available. Further complications exist in H_2SO_4 solutions and probably also in other aqueous acid solutions. As discussed above, Liu and Xing showed that the intrinsic kinetic activity decreased by 40% for a commercial Pt/C electrode when Li^+ were added to 1 mol/dm^3 H_2SO_4 (see Fig. 1) [18]. They attributed this to the decrease in the O_2 limiting diffusion current as the Li^+ concentration increases.

5 CONCLUSIONS

In this chapter we have reviewed the highlights of present-day advances in the development of aqueous Li–air (Li–O_2) cells and batteries, to identify areas for future research and development. Based on the review, summaries of the technical areas that require advanced R&D for a commercially viable Li–air battery are outlined below.

The Anode The key to any aqueous Li–air system is the prevention of direct water contact with the anode, and the most important innovation for this purpose was the introduction of Li^+-ion conducting glass ceramics (LiC-GC). These LiC-GC materials are generally fragile, highly resistive at low temperatures, and may possibly be somewhat reactive in strong alkali and acid solutions; present studies do not indicate stability problems for a rechargeable system, but future R&D will resolve this question since the high resistance of the present technology for LiC-GC materials is a major factor limiting the rate capability of aqueous Li–air cells. New approaches are focusing on stable new LiC-GC materials that address these problems. In view of the fragility and limited size of present-day LiC-GC materials, future R&D on larger and more flexible LiC-GC membranes will be very beneficial. Based on existing literature, it is doubtful that Li^+-ion conducting organic-based polymers can meet this challenge. Li dendrite formation is a continuing problem for which no satisfactory solution has yet been demonstrated.

The Cathode   For a primary Li–air battery, present and future cathodes are highly dependent on the well-advanced technologies for aqueous fuel cell and Zn–air systems. For primary systems, the selection of suitable electrolyte solutions is less problematic than for rechargeable systems. Catalysts, particularly for the OER, are a key issue in developing a commercially viable rechargeable system. There is a critical need to replace expensive precious metal catalysts with low-cost materials such as oxides which are stable in aqueous solution at the highly positive potentials of the OER. While aqueous electrolyte solutions are known to be highly conductive at temperatures from -60 to $+95°C$, a question remains as to whether the ORR and OER will operate efficiently (or at all) at the minimum required temperature of $-40°C$ for EVs.

Electrolyte Solutions   The nature of the composition of electrolyte solutions remains a complex issue, mainly due to uncertainties in the diffusion of O_2 and the solubility of O_2 along with discharge reaction products in composite electrolyte solutions such as $KOH + LiOH + H_2O$ and $H_2SO_4 + Li_2SO_4 + H_2O$. The solubilities of reaction products of LiOH and Li_2SO_4 are not particularly high in pure water, and in alkaline or acidic solutions, the solubilities will decrease as temperatures are lowered. It is expected that these reaction products, particularly in alkaline-based cells, will be a major factor in limiting specific capacities due to precipitation within the pores of carbon-based cathodes, analogous to the limitations of lithium oxide precipitates in nonaqueous electrolyte solutions. The limited number of studies for aqueous Li–air cells from 25 to 60°C in acid solutions have not reported a loss in specific capacity due to the precipitation of discharge reaction products, but it is expected to have an effect.

REFERENCES

1.   T. B. Reddy, Ed., *Handbook of Batteries*, 4th ed., McGraw-Hill, New York, 2011.
2.   K. M. Abraham and Z. Jiang, *J. Electrochem. Soc.*, **143**, 1 (1996).
3.   A. J. Bard, R. Parsons, and J. Jordan, *Standard Potentials in Aqueous Solution*, Marcel Dekker, New York, 1985.

4. F. J. Millero, *Chemical Oceanography*, 2nd ed., CRC Press, Boca Raton, FL, 2002.
5. J. Fu, *Solid State Ionics*, **96**, 195 (1997).
6. J. Fu, *Solid State Ionics*, **104**, 191 (1997).
7. J. Fu, U.S. Patent 6,485,622, Nov. 26, 2002 and earlier patents cited therein.
8. S. J. Visco, B. D. Katz, Y. S. Nimon, and L. C. De Jonghe, U.S. Patent 7,282,295, Oct. 16, 2007.
9. I. Kowalczk, J. Read, and M. Salomon, *Pure Appl. Chem.*, **79**, 851 (2007).
10. P. He, Y. Wang, and H. Zhou, *J. Power Sources*, **196**, 5611 (2011).
11. T. Zhang, N. Imanishi, S. Hasegawa, A. Hirano, J. Xie, Y. Takeda, O. Yamamoto, and N. Sammes, *J. Electrochem. Soc.*, **155**, A965 (2008).
12. L. Li and Y. Xing, *J. Electrochem. Soc.*, **153**, A1823 (2006).
13. J. P. Zheng, P. Andrei, M. Hendrickson, and E. J. Plichta, *J. Electrochem. Soc.*, **158**, A43 (2011).
14. J. P. Zheng, R. Y. Liang, M. Hendrickson, and E. J. Plichta, *J. Electrochem. Soc.*, **155**, A432 (2008).
15. R. Battino, T. R. Rettich, and T. Tominago, *J. Phys. Chem. Ref. Data*, **12**, 163 (1983).
16. R. Battino, Ed., *Oxygen and Ozone, IUPAC Solubility Data Series*, Vol. 7, Pergamon Press, Oxford, UK, 1984.
17. J. Pettersson, B. Ramsey, and D. Harrison, *J. Power Sources*, **157**, 29 (2006).
18. H. Liu and Y. Xing, *Electrochem. Commun.*, **13**, 646 (2011).
19. A. Kraytsberg and Y. Ein-Eli, *J. Power Sources*, **196**, 886 (2011).
20. T. Zhang, N. Imanishi, S. Hasegawa, A. Hirano, J. Xie, Y. Takeda, O. Yamamato, and N. Sammes, *Chem. Commun.*, **46**, 1661 (2010).
21. Y. Wang and H. Zhou, *J. Power Sources*, **195**, 358 (2010).
22. H. Zhou, Y. Wang, H. Li, and P. He, *ChemSusChem*, **3**, 1009 (2010).
23. Y. Wang and H. Zhou, *Chem. Commun.*, **46**, 6305 (2010).
24. P. He, Y. Wang, and H. Zhou, *Chem. Commun.*, **47**, 10701 (2011).
25. E. Yoo and H. Zhou, *ACS Nano*, **5**, 3020 (2011).
26. H. Meng and B. Ruan, *Int. J. Energy Res.*, **35**, 2 (2011).
27. P. He, Y. Wang, and H. Zhou, *J. Power Sources*, **196**, 5611 (2011).
28. J. Larminie and A. Dicks, *Fuel Cell Systems Explained, 2nd ed.*, Wiley, Hoboken, NJ, 2003.
29. X. Yang and Y. Xia, *J. Solid State Electrochem.*, **14**, 109 (2009).
30. J. Read, *J. Electrochem. Soc.*, **149**, A1190 (2002).
31. J. Xiao, D. Wang, W. Xu, D. Wang, R. E. Williford, J. Liu, and J.-G. Zhang, *J. Electrochem. Soc.*, **157**, A487 (2010).
32. O. Crowther et al., *J. Power Sources*, **196**, 1498 (2011).
33. O. Crowther et al., "Lithium Metal-Air Battery Technology, *Proceedings of the 44th Power Sources Conference*, June 14–17, 2010, p. 151.
34. C. Tran, J. Kafle, X.-Q. Yang, and D. Qu, *Carbon*, **49**, 1266 (2010).
35. M.-K. Song, S. Park, F. M. Alagir, J. Cho, and M. Liu, *Mater. Sci. Eng. R* (2011), doi:10.106/j.mser.2011.06.001.
36. L. Jörissen, *J. Power Sources*, **155**, 23 (2006).
37. E. Antolini, *Appl. Catal. B*, **88**, 1 (2009).
38. X. Wang, W. Li, Z. Chen, M. Waje, and Y. Yan, *J. Power Sources*, **158**, 154 (2006).
39. R. Kou, Y. Shao, D. Wang, M. H. Engelhard, J. H. Kwak, J. Wang, V. V. Viswanathan, C. Wang, Y. Lin, Y. Wang, I. A. Aksay, and J. Liu, *Electrochem. Commun.* **11**, 954 (2009).
40. G. Chen, C. C. Waraksa, H. Cho, D. D. Macdonald, and T. E. Mallouka, *J. Electrochem. Soc.*, **150**, E423 (2003).
41. G. Q. Zhang, J. P. Zheng, R. Liang, C. Zhang, B. Wang, M. Hendrickson, and E. J. Plichta, *J. Electrochem. Soc.*, **157**, A953 (2010).
42. W. Zhu, J. P. Zheng, R. Loang, B. Wang, C, Zhang, G. Au, and E. J. Plichta, *J. Electrochem. Soc.* **156**, B1099 (2009).
43. V. Neburchilov, H. Wang, J. J. Martin, and W. Qu, *J. Power Sources*, **195**, 1271 (2010).
44. C. Song and J. Zhang, Electrocatalytic Oxygen Reduction Reaction, in *PEM Fuel Cell Electrocatalysts and Catalyst Layers: Fundamentals and Applications*, J. Zhang, Ed., Springer-Verleg, New York, 2008.
45. Z. Chen, D. Higgins, A. Yu, L. Zhang, and J. Zhang, *Energy Environ. Sci.*, **4**. 3167 (2011).
46. A. Débart, J. Bao, G. Armstrong, and P. G. Bruce, *J. Power Sources*, **174**, 1177 (2007).

47. H.-Y. Jung, S. Park, and B. N. Popov, *J. Power Sources*, **191**, 357 (2009).
48. G. Chen, D. A. Delafuente, S. Sarangapan, and T. E. Mallouk, *Catal. Today*, **67**, 341 (2001).
49. I. Chaitanya Lekshmi and M. S. Hegde, *Mater. Res. Bull.*, **40**, 1443 (2005).
50. D. B. Sepa, A. Damjanovic, and J. O'M. Bockris, *Electrochim. Acta*, **12**, 746 (1967).
51. A. Damjanovic, D. Sepa, and J. O'M. Bockris, *J. Res. Inst. Catal. Hokkaido Univ.*, **16**, 1 (1968).
52. A. Damjanovic, Mechanistic Analysis of Oxygen Electrode Reactions, in *Modern Aspects of Electrochemistry*, Vol. 5, J. O'M. Bockris and B. E. Conway, Eds., Plenum Press, New York, 1969.
53. E. L. Littauer and K. C. Tsai, *J. Electrochem. Soc.*, **123**, 771 (1976).
54. D. T. Welna, D. A. Stone, and H. R. Allcock, *Chem. Mater.*, **18**, 4486 (2006).
55. S. J. Visco, Y. S. Nimon, B. D. Katz, and L.C. De Jonghe, U.S. Patent 7,491,458, Feb. 17, 2009 and other patents cited therein.
56. I. Kowalczyk, W. Eppley, M. Salomon, D. Chua, and B. Meyer, U.S. Patent 7,842,423, Nov. 30, 2010.
57. D. H. Doughty, in *Linden's Handbook of Batteries, 4th ed.*, editor T. B. Reddy, Ed., McGraw-Hill, New York, 2010, p. 27.1.
58. S. Hasegawa, N. Imanishi, T. Zhang, J. Xie, A. Hirano, T. Takeda, and O. Yamamoto, *J. Power Sources*, **189**, 371 (2009).
59. Y. Shimonishi, T. Zhang, P. Johnson, N. Imanishi, A. Hirano, Y. Takeda, O. Yamamoto, and N. Sammes, *J. Power Sources*, **195**, 6187 (2010).
60. Y. Shimonishi, T. Zhang, N. Imanishi, D. Im, D. J. Lee, A. Hirano, Y. Takeda, O. Yamamoto, and N. Sammes, *J. Power Sources*, **196**, 5128 (2011).
61. S. J. Visco, E. Nimon, B. Katz, M.-Y. Chu, and L. De Jonghe, Protected Lithium Electrodes as Universal Anodes for Ultra-High Energy Density Batteries and Drug Delivery Systems, presented at the symposium *Lithium Mobile Power 2008*, Las Vegas, NV, Dec. 8–9, 2008.
62. J. O'M. Bockris, Electrode Kinetics, in *Modern Aspects of Electrochemistry*, Vol. 1. J. O'M. Bockris and B. E. Conway, Eds., Plenum Press, New York, 1954.
63. V. S. Bagotsky, *Fundamentals of Electrochemistry*, Wiley, Hoboken, 2006, NJ, p. 278.
64. A. N. Campbell, E. M. Kartzmark, D. Bisset, and M. E. Bednas, *Can. J. Chem.*, **31**, 303 (1953).
65. R. J. Gilliam, J. W. Graydon, D. W. Kirk, and S. J. Thorpe, *J. Hydrogen Energy*, **32**, 359 (2007).
66. K. E. Gubbins and R. D. Walker, Jr., *J. Electrochem. Soc.*, **112**, 469 (1965).
67. J. Read, K. Mutolo, M. Ervin, W. Behl, J. Wolfenstine, A. Driedger, and D. Foster, *J. Electrochem. Soc*, **150**, A1351 (2003).
68. R. Cohen-Adad and J. W. Lorimer, Eds., *Alkali Metal and Ammonium Chlorides in Water and Heavy Water (Binary Systems)*, IUPAC Solubility Data Series, Vol. 47, Pergamon Press, Oxford, UK, 1991.
69. J. A. Rard, S. L. Clegg, and D. A. Palmer, Lawrence Livermore National Laboratory report UCRL-JRNL-227057, *Isopiestic Determination of the Osmotic and Activity Coefficients of $Li_2SO_4(aq)$ at $T = 298.15$ and 323.15 K*, Jan. 4, 2007.
70. G. C. A. van Dorp, *Z. Phys. Chem.*, **73**, 289 (1910).
71. C. Montemartini and L. Losana, *Ind. Chim., Roma*, **4**, 107, 199, 291 (1928).
72. V. G. Shevchuk and V. A. Storozhenka, *Zh. Neorg. Khim.*, **6**, 1654 (1970).
73. P. Stevens, G. Toussaint, G. Caillon, P. Viaud, P. Vinatier, C. Cantau, O. Fichet, C. Sarrazin, and M. Mallouki, *ECS Trans.*, **28**, 1 (2010).

CHAPTER 10

POLYMER ELECTROLYTES FOR LITHIUM–AIR BATTERIES

Nobuyuki Imanishi and Osamu Yamamoto

1 INTRODUCTION

Lithium–air batteries are attractive energy storage systems in view of their high energy density for application in electric vehicles, the calculated energy density of which is comparable with that of gasoline. Now two types of lithium–air batteries have been developed: a nonaqueous system and an aqueous system. Two general possible cell reactions for the nonaqueous system are

$$2Li + O_2 = Li_2O_2 \tag{1}$$

$$4Li + O_2 = 2Li_2O \tag{2}$$

The reversible cell voltage is 2.959 V for reaction (1) and 2.913 V for reaction (2) [1]. It was demonstrated that reaction (1) is reversible, whereas reaction (2) is irreversible [2]. The energy density of 3457 Wh/kg is calculated from reaction (1) at the discharged state. At the charged state, the energy density is 11,425 Wh/kg, because oxygen is not included in the cell. The average calculated energy density in the charged state is 7441 Wh/kg. In an aqueous lithium–air system, water molecules are involved in the cell reaction according to the reaction

$$4Li + 6H_2O + O_2 = 4(LiOH \cdot H_2O) \tag{3}$$

Lithium Batteries: Advanced Technologies and Applications, First Edition.
Edited by Bruno Scrosati, K. M. Abraham, Walter van Schalkwijk, and Jusef Hassoun.
© 2013 John Wiley & Sons, Inc. Published 2013 by John Wiley & Sons, Inc.

The reversible cell voltage for reaction (3) is calculated to be 3.856 V from the thermodynamic data for a neutral solution, but the experimental open-circuit voltage (OCV) with a Pt-black oxygen electrode was 3.43 V in 1 M LiCl + 0.004 M LiOH and approximately 3.0 V in saturated LiCl and LiOH aqueous solutions [3]. The energy density for reaction (3), calculated using the cell voltage of 3.0 V, is 1910 Wh/kg with oxygen and 2371 Wh/kg without oxygen, and the average energy density is 2141 Wh/kg. The energy density of the aqueous system is less than half that of the nonaqueous system, but several times higher than that of lithium-ion batteries with a carbon anode and a lithium-metal oxide anode such as $LiCoO_3$. The calculated energy density of the aqueous lithium–air system suggests the possibility of developing a battery with an energy density of 700 Wh/kg, which is comparable to that of internal conversion engines.

Lithium reacts vigorously with water to produce LiOH and hydrogen gas; therefore, to avoid this reaction, most research on lithium–air batteries has focused on using aprotic organic electrolytes. However, there are some severe problems that must still be addressed, such as lithium corrosion by water and CO_2 ingression when operated in air (which shortens lifetime), precipitation of high resistance lithium oxide reaction products (low capacity), and high polarization during charge and discharge processes (low energy conversion efficiency) [4]. These problems could be removed by employing a water-stable lithium-metal electrode and an aqueous electrolyte.

The water-stable lithium-metal electrode was addressed preliminarily by the concept of a composite lithium anode with a three-layer construction proposed by Visco et al. in 2004 [5]. This electrode concept adopts a water-stable lithium-ion conducting solid electrolyte as a protective layer that covers and isolates the lithium metal from direct contact with the aqueous electrolyte. At present, a NASICON type of solid electrolyte of $Li_{1+x}Ti_{2-x}Al_x(PO_4)_3$ (LTAP) is used as the water-stable lithium-ion conducting solid electrolyte. The typical conductivity of a dense LTAP plate provided by the Ohara Corporation in Japan is 3.5×10^{-4} S/cm at 25°C and 1.4×10^{-3} S/cm at 60°C [6]. However, LTAP is unstable in direct contact with lithium metal and produces a high-resistance layer [7], so that a buffer layer must be used between lithium metal and the LTAP plate. The requirements of the buffer layer are as follows: (1) stable with lithium metal, (2) high lithium-ion conductivity, (3) ease of preparation, and (4) free from dendrite formation between the lithium metal and the buffer layer. Visco et al. [5] employed lithium nitride and lithium phosphorus oxynitride (Lipon) prepared by a sputtering technique as buffer layers. Zhang et al. [8] proposed a poly(ethylene oxide) (PEO)-based polymer electrolyte as the buffer layer. The polymer electrolyte is one of the best candidates for the buffer layer, due to the good compatibility with lithium metal, high lithium-ion conductivity above the melting point, flexibility, and ease of preparation. In this chapter, the properties of the water-stable lithium electrode with a polymer electrolyte buffer layer, dendrite formation between the lithium metal and the polymer electrolyte, and the electrode performance of the water-stable lithium electrode with the polymer electrolyte buffer layer for lithium–air batteries are introduced.

2   INTERFACE RESISTANCE BETWEEN THE POLYMER ELECTROLYTE AND LITHIUM METAL

Figure 1 shows an queous lithium–air system proposed by Zhang et al. [8]. The air electrode has been studied extensively for rechargeable zinc–air batteries and alkaline fuel cells, and a high-performance electrode has been developed in a concentrated alkaline solution. The key issue of the aqueous lithium–air system is on the lithium electrode side. The lithium metal is covered with a PEO-based polymer electrolyte, $PEO_{18}Li(CF_3SO_2)_2N$ ($PEO_{18}LiTFSI$), and an LTAP water-stable lithium conducting solid electrolyte. A lithium ion–containing aqueous electrolyte was used as the electrolyte because LTAP is unstable in pure water but stable in an aqueous electrolyte containing lithium-ions [7]. The cell voltage of $Li/PEO_{18}LiTFSI/LTAP/1$ M LiCl aqueous solution/Pt, air was approximately 3.7 V at 60°C. The lithium metal contacts the polymer electrolyte; therefore, the potential between the lithium metal and the polymer electrolyte is 3.040 V. The decomposition potential of PEO–LiTFSI was reported to be approximately 3.9 V vs. Li/Li^+. The potential between lithium metal and PEO–LiTFSI is less than 3.7 V, and the polymer electrolyte could be considered to be stable in this battery system, although the detailed potential distribution is still under study.

One of the important requirements for batteries in electrical vehicles is a low energy loss at a high current drain. Therefore, the cell resistance should be as low as possible. Lithium-ion conducting polymer electrolytes generally exhibit a high interface resistance with lithium metal, due to the formation of a passivation layer. Typical examples of impedance profiles of the $Li/PEO_{18}LiTFSI/Li$ and $Li/PEO_{18}LiTFSI$–10 wt% $BaTiO_3/Li$ symmetrical cells at 60°C are shown in Figure 2, where the thickness of the polymer electrolyte was about 170 μm [8]. The electrical conductivity of $PEO_{18}LiTFSI$ is 5×10^{-4} S/cm at 60°C. These impedance spectra show a small semicircle in the high-frequency range and a large semicircle in the low-frequency

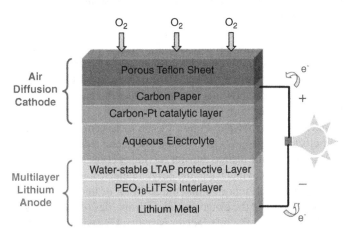

FIGURE 1   Aqueous lithium–air battery system.

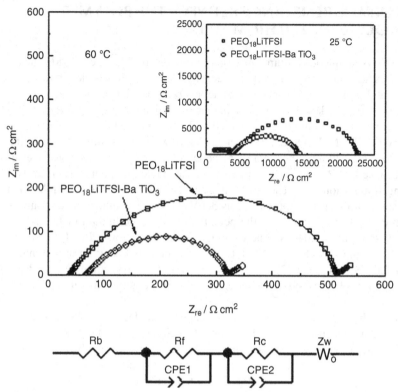

FIGURE 2 Impedance spectra of Li/PEO$_{18}$LiTFSI/Li and Li/PEO$_{18}$LiTFSI–BaTiO$_3$/Li at 60°C.

range. The small semicircle in the high-frequency range could be assigned as the grain boundary resistance of the polymer electrolyte because a similar semicircle was observed in the same frequency range for the Au/PEO$_{18}$LiTFSI/Au cell.

The high-frequency intercept of the large semicircle with the real axis can be attributed to the total resistance of the composite electrolyte. The diameter of the large semicircle is associated with the overall interface resistance (R_i), which consists of two parts: the resistance of the passivation film (R_f), formed on the lithium-metal electrode surface by reaction with the polymer electrolyte and the charge transfer resistance (R_c) of the Li$^+$ + e$^-$ = Li reaction. The resistances of each component were analyzed using the equivalent circuit shown in Figure 2 to be 41 $\Omega \cdot$ cm^2 for the bulk resistance of the polymer electrolyte, 439 $\Omega \cdot$ cm^2 for the interfacial resistance of the passivation layer, and 35 $\Omega \cdot$ cm^2 for the charge transfer resistance; thus, the dominant interface resistance between the lithium metal and the polymer electrolyte is from the passivation film. Compaano et al. [10] reported that the addition of finely dispersed ceramic powder significantly improved the electrochemical properties of the polymer electrolyte. Zhang et al. [8] found that the interfacial resistance between the lithium metal and PEO$_{18}$LiTFSI was improved by the addition of nanosized BaTiO$_3$ into the polymer electrolyte, as shown in Figure 2; Li/PEO$_{18}$LiTFSI–10 wt% BaTiO$_3$/Li has

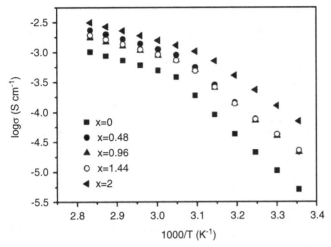

FIGURE 3 Temperature dependence of the electrical conductivity for PEO$_{13}$LiTFSI–xPP13TFSI as a function of x.

a low cell impedance. The addition of nanosized BaTiO$_3$ results in a reduction of the interfacial resistance to 196 $\Omega \cdot$ cm^2. The reduction in interfacial resistance between the lithium metal and the polymer electrolyte by the addition of BaTiO$_3$ into the polymer electrolyte was proposed by Sun et al. [11].

The polymer electrolyte composite and a room-temperature ionic liquid (RTIL) enhance the electrical conductivity of the polymer electrolyte at room temperature [12]. Many types of RTILs have been reported, some of which have poor stability with lithium metal. Kim et al. [13] reported that the addition of an ionic liquid of N-alkyl-N-methylpyrrolidinum bis(trifluoromethylsulfonylimide) into PEO$_{10}$LiTFSI enhanced the electrical conductivity of the polymer electrolyte ($>10^{-4}$ S/cm at 20°C) and decreased the interfacial resistance with lithium metal (400 $\Omega \cdot$ cm^2 at 40°C). Liu et al. [14] also reported enhancement of the conductivity and a large decrease in the interfacial resistance between the polymer electrolyte and lithium metal in a composite of PEO$_{18}$LiTFSI and an ionic liquid. Figure 3 shows the temperature dependence of the conductivity for a composite polymer electrolyte of PEO$_{18}$LiTFSI and N-methyl-N-propylpiperidinum TFSI (PP13TFSI) as a function of the PP13TFSI content. The electrical conductivity of PEO$_{18}$LiTFSI–xPP13TFSI increases with increasing x. The conductivity of PEO$_{18}$LiTFSI–2.0PP13TFSI at room temperature is 7.1×10^{-4} S/cm, which is approximately one order of magnitude lower than that of PP13TFSI and approximately one order of magnitude higher than that of PEO$_{18}$LiTFSI. A composite electrolyte with a PP13TFSI content higher than 2.0 was viscous, and self-supported films with thicknesses of approximately 200 μm could not be prepared. Arrhenius plots of PEO$_{18}$LiTFSI have an inflection at 55°C, which reflects the well-known crystalline-to-liquid phase transition. The phase transition temperature decreases slightly by addition of the RTIL into PEO$_{18}$LiTFSI. The lack of any significant change in the activation energy for the composite polymer electrolyte suggests that lithium cations are still controlled by the local relaxation and segmental

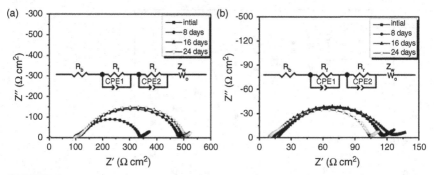

FIGURE 4 Impedance spectra for (a) Li/PEO$_{18}$LiTFSI/Li and (b) Li/PEO$_{18}$LiTFSI–1.44PP13TFS/Li at 60°C as a function of the storage time.

motion of the PEO chains. In the composite electrolyte of PEO$_{18}$LiTFSI–xPP13TFSI, TFSI$^-$ and PP13$^+$ are mobile in addition to Li$^+$. The lithium-ion transport number for PEO$_{18}$LiTFSI–xPP13TFSI obtained from polarization measurements was decreased with increasing x, from 0.29 for $x = 0$ to 0.21 for $x = 2.0$. The decrease may be due to the increase of the TFSI$^-$ content in the composite electrolyte.

Figure 4 shows typical impedance profiles of Li/PEO$_{18}$LiTFSI/Li and Li/PEO$_{18}$LiTFSI–1.44PP13TFS/Li cells at 60°C as a function of the storage time at 60°C. The cell resistance of the Li/PEO$_{18}$LiTFSI–1.44PP13TFSI/Li cell (Fig. 4b) is considerably lower than that of the Li/PEO$_{18}$LiTFSI/Li cell (Fig. 4a) and shows no change with the storage time. Sakaebe et al. [15] reported that the initial Li/PP13TFSI/Li cell resistance of 30 Ω · cm^2 at 60°C increased to 60 Ω · cm^2 after storage for 3 days. Kim et al. [13] also found that the Li/PEO$_{100}$0.96PYRTFSI/Li cell showed an initial resistance of 400 Ω · cm^2, which increased to 800 Ω · cm^2 after storage for 24 days at 40°C. Compared to these results, the Li/PEO$_{18}$LiTFSI–1.44PP13TFSI/Li cell resistance is stable with storage time. The impedance spectra were analyzed using the equivalent circuit shown in Figure 4.

Analysis of the impedance spectrum for Li/PEO$_{18}$LiTFSI–1.44PP13TFSI/Li at the initial stage yielded a passivation film resistance of $R_f = 68.4$ Ω · cm^2 and a charge transfer resistance of $R_c = 30.1$ at 60°C, which are comparably less that $R_f = 199$ Ω · cm^2 and $R_c = 49$ Ω · cm^2 obtained for Li/PEO$_{18}$LiTFSI/Li at 60°C. The polymer electrolyte bulk resistance of PEO$_{18}$LiTFSI, $R_b = 110$ Ω · cm^2, is reduced to 18 Ω · cm^2 by the addition of PP13TFSI. The passivation layer resistance for Li/PEO$_{18}$LiTFSI/Li increased significantly with the storage time, but that for Li/PEO$_{18}$LiTFSI–1.44PP13TFSI/Li decreased slightly with the storage time. The interface resistance of Li/PEO$_{18}$LiTFSI–xPP13TFSI/Li is dependent on x; R_f decreased with increasing x up to $x = 1.44$ and that for $x = 2.0$ was almost the same as that for $x = 1.44$. The temperature dependence of the interface resistance exhibited no inflection for the phase transition temperature as with the electrical conductivity shown in Figure 3, which indicates that the interface resistance is not affected by the nature of the polymer electrolyte matrix but by the characteristics of the passivation layer. The activation energy estimated from the inverse of the passivation layer resistance vs. temperature curves decreased slightly from 78 kJ/mol for Li/PEO$_{18}$LiTFSI/Li to 69 kJ/mol for Li/PEO$_{18}$LiTFSI–1.44PP13TFSI/Li, which is larger than that for the bulk conductivity of PEO$_{18}$LiTFSI–1.44PP13TFSI in the high-temperature region and lower than that in the low-temperature region. The activation energy for the charge transfer between lithium and PEO$_{18}$LiTFSI–xPP13TFSI ($\sim$79 kJ/mol) showed no significant dependence on x.

The low and steady overpotential on the lithium metal/polymer electrolyte interface is an important factor for the development of a high-power-density lithium–air battery. Cell potential vs. time curves at various current densities for Li/PEO$_{18}$LiTFSI/Li and Li/PEO$_{18}$LiTFSI–1.44PP13TFSI/Li indicated that the overpotentials were lower for the latter than for the former and did not increase with polarization time. The Li/PEO$_{18}$LiTFSI/Li cell exhibited a high overpotential that increased with time at high current density.

3 DENDRITE FORMATION AT THE LITHIUM METAL/POLYMER ELECTROLYTE INTERFACE

Lithium metal is a promising negative anode for rechargeable batteries, because of its high specific capacity of 3861 mAh/g and high negative potential (-3.05 V vs. NHE). However, the performance of this anode material is severely affected by capacity loss and safety hazards, both due to dendrite growth during the recharging process. Short circuiting by dendrite formation in a Li/ethylene carbonate–ethylmethyl carbonate–dimethyl carbonate ($1:1:1$)-LiPF$_6$/Li cell was observed at 0.2 h during polarization under 1.0 mA/cm^2 at 15°C with a 1.0-mm-thick separator [16]. The period until short circuit is too short and the specific weight capacity of the lithium electrode calculated from the total weight of a 10-μm-thick copper foil current collector and the lithium metal deposited on the copper foil is only 22 mAh/g. A gel-type polymer electrolyte comprised of 30 wt% PEO and poly(propylene oxide) copolymer with 70 wt% ethylene carbonate/propylene carbonate ($1:1$ weight ratio) provided improved suppression of dendrite formation with dendrites appearing at 1 h under polarization at 0.5 mA/cm^2 with a 1.0-mm-thick separator [17]. This phenomenon occurs even in polymer electrolytes, although to a lesser extent than in liquid electrolytes [18]. The Li/PEO–LiTFSI/Li cell showed dendrite growth at 0.6 h with a polarization of 0.7 mA/cm^2 where the separator was approximately 1 mm thick [19]. Lithium dendrite growth is dependent on the current density; therefore, during practical applications the dendrite formation onset time should be more prolonged at higher current density. The water-stable lithium electrode for the lithium–air battery proposed by Zhang et al. [8] exhibited excellent stability in an aqueous electrolyte. This electrode consisted of a lithium sheet as the active material, a PEO$_{18}$LiTFSI polymer electrolyte buffer layer, and a water-stable NASICON-type LTAP solid electrolyte. The polymer electrolyte protects the electrode from the direct reaction of lithium metal and LTAP, which is unstable in contact with lithium metal. This three-layer lithium electrode has been confirmed to have stability in water and to exhibit reversible lithium stripping and deposition with low polarization [3,8]. However, the problem of dendrite formation on the lithium electrode remains when charging at high current density and for long-term cycling.

The mechanism for dendrite formation has been investigated by Chazalviel [20]. and Leger et al. [21]. According to Chazalviel, a positive space charge appears in the vicinity of the negative electrode when the ionic concentration falls to zero, which indicates a local space charge near the lithium electrode, and instability of the interface, such as dendrite growth. This happens after a time τ_S (Sand time [22]), which varies as the square of the current density J according to the equation

$$\tau_S = \frac{\pi e^2 D [(\mu_a + \mu_c)/\mu_a]^2 C_0^2}{4J^2} \tag{4}$$

where e is the electronic charge, C_0 the initial concentration, μ_a and μ_c the anionic and cationic mobility, respectively, and D the ambipolar diffusion coefficient:

$$D = \frac{\mu_a D_c + \mu_c D_a}{\mu_a + \mu_c} \tag{5}$$

where D_c and D_a are the anionic and cationic diffusion constants, respectively. The Sand time, τ_S, corresponds to the dendrite formation onset time, t_0. Brissot et al. [23]. reported that the dendrite growth in Li/PEO$_{20}$LiTFSI corresponded to the model proposed by Chazalviel in the current density range 0.02 to 0.3 mA/cm^2 [20]. Imanishi's group [14,24,25]. examined the dendrite formation for Li/PEO$_{18}$LiTFSI/Li, Li/PEO$_{18}$LiTFSI-nano-SiO$_2$/Li, Li/PEO$_{18}$LiTFSI-PP13TFSI/Li, and Li/PEO$_{18}$LiTFSI–PP13TFSI-nano-SiO$_2$/Li cells using direct in situ observation under galvanic conditions at 60°C and compared the effect of both filler and RTIL addition. The dendrite formation onset time was dependent on the current density. The Li/PEO$_{18}$LiTFSI/Li cell exhibited dendrite formation after 125 h at 0.1 mA/cm^2 and 15 h at 0.5 mA/cm^2. Dendrite formation was suppressed by the addition of nanosized SiO$_2$ and RTIL of PP13TFSI into PEO$_{18}$LiTFSI. Figure 5 shows typical dendrite growth in a Li/PEO$_{18}$LiTFSI-nano-SiO$_2$/Li visualization cell

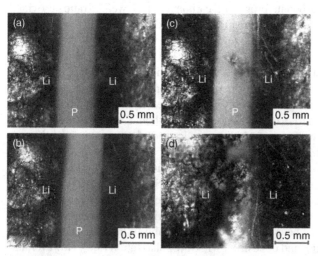

FIGURE 5 Dendrite growth in a visualization Li/PEO$_{18}$LiTFSI–nano-SiO$_2$/Li cell at 60°C under 0.5 mA/cm^2 at (a) 0, (b) 25, (c) 32, and (d) 42 h.

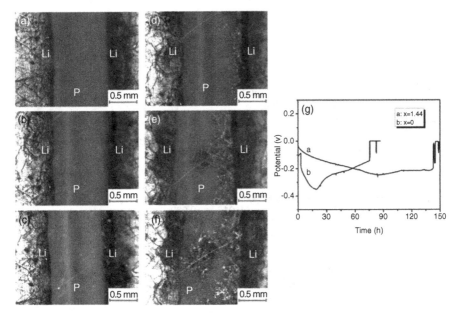

FIGURE 6   Dendrite growth in the Li/PEO$_{18}$LiTFSI–1.44PP13TFSI/Li cell at 60°C under 0.3 mA/cm^2 at (a) 0, (b) 78, (c) 85, (d) 100, (e) 130, and (f) 145 h. (g) Cell potential vs. polarization time curve.

at 60°C and 0.5 mA/cm^2 at various polarization periods, where the thickness of the polymer electrolyte was about 1.0 mm. After polarization for 32 h at 0.5 mA, dendrite formation was observed, and short circuiting occurred by dendrite formation after 42 h. The addition of PP13TFSI into PEO$_{18}$LiTFSI was effective in prolonging the dendrite formation onset time.

Figure 6 shows dendrite growth in a Li/PEO$_{18}$LiTFSI–1.44PP13TFSI/Li cell under 0.3 mA/cm^2 at (a) 0, (b) 78, (c) 85, (d) 100, (e) 130, and (f) 145 h, and the cell potential vs. polarization time curve at 60°C [14]. The dendrite formation onset time and short-circuit time were observed after 85 and 145 h, respectively. The small signal in the polarization curves after 85 h in Figure 6g is due to dendrite formation, as determined by microscopic observation. The Li/PEO$_{18}$LiTFSI/Li cells with and without PP13TFSI showed an abrupt voltage change after 145 and 75 h of polarization, respectively, due to short circuiting of the lithium dendrites, as confirmed by microscopic observation.

According to equation (4), the dendrite formation onset time (t_0) follows a power law as a function of the current density (J). The log t_0 vs. log J curve for the Li/PEO$_{18}$LiTFSI–1.44PP13TFSI/Li cell at 60°C showed good linearity in the range 0.1 to 0.5 mA/cm^2 with a slope of −1.25. At these current densities, the dendrite growth could not be explained by the Chazalviel model [20], but local inhomogeneities of the ion concentration, which indicates that local fluctuations of current density may be an important factor in dendrite formation, as indicated by Brissot et al. [23]. Equation (4) suggests that the dendrite formation onset time t_0 is

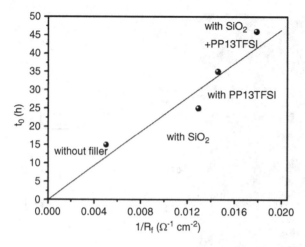

FIGURE 7   Relationship between the dendrite formation onset time (t_0) and the inverse of the specific interface layer resistance between Li and a PEO-based electrolyte (R_f) at 60°C.

proportional to the diffusion coefficient, which is related to the interface resistance from the Nernst–Einstein equation.

Figure 7 shows the relationship between the dendrite formation onset time at 0.5 mA/cm^2 and the inverse of the specific interface layer resistance R_f between Li and PEO$_{18}$LiTFSI with and without additives. A good linear relationship is observed in the t_0 vs. R_f^{-1} curve, which suggests that the high-lithium-ion-conductivity polymer electrolyte may offer a dendrite-free interface between the lithium metal and the polymer electrolyte.

The primary advantage of a lithium-metal anode is to provide a high theoretical capacity of 3861 mAh/g, which is 10 times higher than that of the carbon anode in conventional lithium-ion batteries. However, the capacity is limited by dendrite formation. Table 1 summarizes the onset times for dendrite formation between the lithium metal and various electrolytes. Lithium-metal electrodes have been used in practical applications with a copper foil current collector. The specific weight capacity of the lithium electrode calculated from the total weight of a 10-μm-thick copper foil and lithium metal deposited until the dendrite formation is shown in the table and compared to that of liquid and gel-type electrolytes. The specific capacities of a lithium-metal electrode with a copper substrate in contact with PEO$_{18}$LiTFSI–1.44PP13TFSI–10 wt% nano-SiO$_2$ were calculated to be 1459 mAh/g at 1.0 mA/cm^2 and 2204 mAh/g at 0.1 mA/cm^2, which is significantly higher than 109 mAh/g for a gel-type polymer electrolyte and 22 mAh/g for a liquid electrolyte at 1.0 mA/cm^2 and at room temperature. The specific capacity of 1459 mAh/g calculated is about 30% that of lithium metal and approximately four times higher than that of the typical carbon anode in conventional lithium-ion batteries. After 30 cycles at 0.3 mA/cm^2 for 30 h, the capacity was calculated as 798 mAh/g, and the interface between the lithium metal and PEO$_{18}$LiTFSI–1.44PP13TFSI–10 wt% nano-SiO$_2$ showed no dendrite formation [25]. The composite polymer electrolyte is quite attractive as a protective layer between the lithium metal and the LTAP water-stable lithium-ion-conducting solid electrolyte as a lithium electrode for lithium–air batteries. However, a higher rate capability is required for use in a battery for electric vehicles.

TABLE 1   Summary of the Dendrite Formation Onset Time for Li/Electrolyte/Li Cells

Electrolyte	Temperature ($°C$)	Distance of electrodes (mm)	Current density (mA/cm^2)	Onset time (h)	Capacity[a] (mAh/g)	Ref.
$PEO_{18}LiTFSI$	60	1.0	1.0.	10	866	24
			0.5	15	688	
			0.1	125	1025	
$PEO_{20}LiTFSI$	80	1.0	0.7	0.5	39	19
				0.05	28	
$PEO_{18}LiTFSI/$ nano-SiO_2	60	1.0	1.0	10–15	866–1171	24
			0.5	25	1025	
			0.1	205	1444	
$PEO_{18}LiTFSI/$ PP13TFSI	60	1.0	1.0	17	1270	14
			0.5	35	1300	
			0.1	434	2148	
$PEO_{18}LiTFSI/$ PP13TFSI/ nano-SiO_2	60	1.0	1.0	21	1459	25
			0.5	46	1543	
			0.1	460	2204	
EC/DMC/EMC/ $LiPF_6$[b]	15	1.0	1.0	0.2	22	16
PAN/PC/EC/ $LiClO_4$[c]	Room temperature		1.0	1	109	17

[a]Calculated by the weight of the lithium metal deposited at the onset time, with a 10-μm-thick copper substrate as a current collector.
[b]Ethylene carbonate–dimethyl carbonate–ethylmethyl carbonate–$LiPF_6$.
[c]Poly(acrylonitrile)–propylene carbonate–ethylene carbonate–$LiClO_4$.

4   WATER-STABLE LITHIUM ELECTRODES WITH A POLYMER ELECTROLYTE BUFFER LAYER FOR LITHIUM–AIR BATTERIES

The water-stable multilayer lithium anode proposed by Zhang el al. [8] consists of lithium metal, a polymer electrolyte, and a water-stable lithium conducting solid electrolyte as shown in Figure 1. At present, water-stable lithium conducting solid electrolytes with high lithium-ion conductivity are limited to NASICON-type LTAP [26] and garnet-type $Li_7La_3Zr_2O_{12}$ [27] (LLZ). The conductivity of the $Li_{1+x-y}Ti_{2-x}Al_xSi_yP_{1-y}O_{12}$ ($x \sim 0.25$, $y \sim 0.1$) glass ceramic and $Li_7La_3Zr_2O_{12}$ was reported to be 3.3×10^{-4} S/cm and 2.44×10^{-4} S/cm at room temperature, respectively. Imanishi and colleagues reported that LTAP [28] and LLZ [29] showed no change of x-ray diffraction patterns and significant conductivity degradation by immersion in water at 50°C for 1 week. The degradation in conductivity may be due to the exchange reaction of Li ions and protons. However, LTAP and LLZ were stable in an aqueous solution containing LiCl. The time dependence of the impedance spectra of a Li/$PEO_{18}LiTFSI$–10 wt% $BaTiO_3$/LTAP/aqueous 1 M LiCl/Pt cell at 60°C was examined, where a Pt-black counterelectrode was used and the thickness of the LTAP layer was about 250 μm. The cell resistance was increased by only

10% over one month, which suggests excellent stability of the multilayer Li anode construction in an aqueous electrolyte. The impedance parameters were analyzed using an equivalent circuit, and the interface resistance between the polymer electrolyte and LTAP was estimated to be about 42 $\Omega \cdot cm^2$, which is comparable to the sum of the bulk resistances of the polymer electrolyte and LTAP. The polarization resistance calculated from the lithium electrode potential vs. current density curves for the Li/PEO$_{18}$LiTFSI–10 wt% BaTiO$_3$/LTAP/aqueous 1 M LiCl/Pt, air cell at 60°C was 170 $\Omega \cdot cm^2$ for the discharge curve in the current density range 0.1 to 1.0 mA/cm, which was comparable to the total cell resistance estimated by the impedance method. In the high-current-density region (>1.3 mA/cm^2), appreciable polarization was observed during the discharge process (lithium stripping from lithium metal). This polarization could be postulated as being due to the lithium-ion diffusion process in the electrode.

A proto-type lithium–air rechargeable cell with a water-stable lithium-metal anode of Li/PEO$_{18}$LiTFSI/LTAP/CH$_3$COOH–LiCH$_3$COO–H$_2$O/carbon, air was demonstrated by Zhang et al. [30] in which acetic acid and lithium acetate aqueous solutions were used, because the water-stable LTAP solid electrolyte is unstable in strong basic and acidic solutions. Figure 8 shows typical cycling performance of the cell at 0.5 mA/cm^2 and 60°C. Stable charge and discharge curves were observed for this cell. A discharge capacity of 225 mAh/g was obtained for 56% CH$_3$COOH utilization.

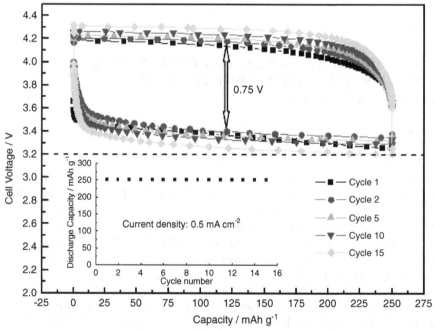

FIGURE 8 Charge–discharge performance of Li/PEO$_{18}$LiTFSI/LTAP/CH$_3$COOH– CH$_3$COOLi–H$_2$O/carbon. The amount of CH$_3$COOH was defined as 1 mg.

The estimated energy density from the weight of the lithium anode and CH_3COOH is as high as 779 Wh/kg and should be compared to that estimated for a typical lithium-ion battery with a $LiCoO_2$ cathode and graphite anode (380 Wh/kg). In this aqueous type of lithium–air system, the capacity is dependent on the amount of CH_3COOH, because CH_3COOH is the active material for the cell reaction according to

$$2Li + \tfrac{1}{2} O_2 + 2CH_3COOH = 2CH_3COOLi + H_2O. \qquad (6)$$

The energy density calculated from this cell reaction and the open-circuit voltage of 3.69 V observed is 1320 Wh/kg, which is 30% lower than that for reaction (3) with water as the active material.

Recently, Shimonishi et al. [31] found that the water-stable lithium-ion conducting LTAP solid electrolyte is stable in saturated LiCl and LiOH aqueous solutions, which suggests that these solutions could be used as the electrolyte in aqueous lithium–air batteries. Figure 9 shows the storage time dependence of the $Li/PEO_{18}LiTFSI$–10 wt% $BaTiO_3/LTAP$/saturated LiCl and LiOH aqueous solution/Pt,air cell at 60°C.

The cell resistance showed no significant change after storage for 2 weeks at 60°C; the three-layer lithium metal electrode of $Li/PEO_{18}LiTFSI$–10 wt% $BaTiO_3/LTAP$ is stable in saturated LiCl and LiOH aqueous solutions. The polarization behavior of the lithium composite electrode at 60°C is shown in Figure 10. Polarization for lithium insertion and stripping up to 4 mA/cm^2 is due primarily to the charge transfer resistance between the lithium metal and the passivation layer, and at a current density as high as 5 mA/cm^2, lithium diffusion through the passivation layer and polymer electrolyte may be the rate-determining step. The interface resistances

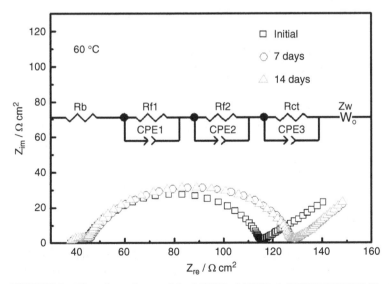

FIGURE 9   Time dependence of the $Li/PEO_{18}LiTFSI$–1.44PP13TFSI/LTAP/saturated LiCl and LiOH aqueous solution/Pt,air cell resistance at 60°C.

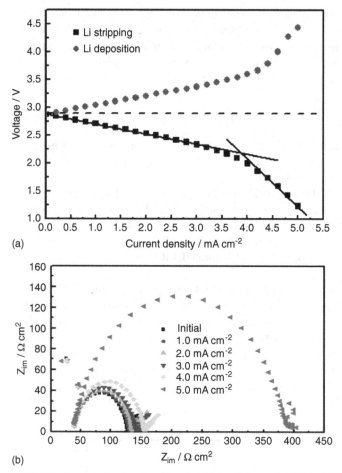

FIGURE 10 Polarization curves and cell impedance of Li/PEO$_{18}$LiTFSI–1.44PP13TFSI/LTAP/saturated LiCl and LiOH aqueous solution/Pt,air at 60°C.

increase significantly at 5 mA/cm^2, as shown in Figure 10. To improve the high rate polarization, the lithium diffusion coefficient in the passivation layer and polymer electrolyte, especially in the polymer electrolyte, should be increased. Therefore, a high lithium-ion conducting polymer electrolyte without lithium dendrite formation should be developed.

5 CONCLUSIONS

An aqueous lithium–air rechargeable battery that consists of a lithium-metal anode, a polymer electrolyte interface, a water-stable lithium-ion conducting solid electrolyte, a saturated LiCl aqueous solution, and an air cathode is a promising power storage system for electric vehicles, because it has the potential to be developed into a high-specific-energy-density battery with a capacity as high as 700 Wh/kg; however, there are many problems to be resolved to realize an acceptable battery system. The most important point is to improve the low power density by high polarization at the lithium electrode. The high polarization is caused mainly by the high interface resistance between the lithium metal and the polymer electrolyte. At present, a composite lithium-metal electrode can pass a current up to 4 mA/cm^2 without high polarization; however, batteries for electric vehicle applications require higher rate capability, as much as several tens of milliamperes per square centimeter. Thus, there is a challenge to develop a polymer electrolyte with high lithium-ion conductivity and low passivation film resistance between the lithium metal and the polymer electrolyte.

REFERENCES

1. Y. Cu, H. A. Gasteiger, M. C. Parent, V. Chlloyan, and Y. Shao-Horn, *Electrochem. Solid-State Lett.*, **13**, A69 (2010).
2. J. Read, *J. Electrochem. Soc.*, **149**, A1190 (2002).
3. T. Zhang, N. Imansihi, A. Hirano, Y. Takeda, and O. Yamamoto, *Electrochem. Solid-State Lett.*, **14**, A15 (2010).
4. T. Ogasawara, A. Debart, M. Holzapfel, P. Novak, and P. G. Bruce, *J. Am. Chem. Soc.*, **128**, 1390 (2006).
5. S. J. Visco, E. Nimon, B. Katz, L. C. D. Jonghe, and M. Y. Chu, 12th International Meeting on Lithium Batteries, Nara, Japan, 2004, Abstr. 53.
6. T. Zhang, N. Imanishi, A. Hirano, Y. Takeda, and O. Yamamoto, *Electrochem. Solid-State Lett.*, **14**, A45 (2011).
7. N. Imanishi, S. Hasegawa, T. Zhang, A. Hirano, Y. Takeda, and O. Yamamoto, *J. Power Sources*, **185**, 1392 (2008).
8. T. Zhang, N. Imanishi, S. Hasegawa, A. Hirano, J. Xie, Y. Takeda, O. Yamamoto, and N. Sammes, *Electrochem. Solid-State Lett.*, **12**, A132 (2009).
9. Q. Li, N. Imanishi, A. Hirano, Y. Takeda, and O. Yamamoto, *J. Power Sources*, **110**, 38 (2002).
10. F. Compaano, F. Croce, and B. Scrosati, *J. Electrochem. Soc.*, **138**, 1918 (1991).
11. H. Y. Sun, Y. Takeda, N. Imanishi, O. Yamamoto, and H.-J. Sohn, *J. Electrochem. Soc.*, **147**, 2462 (2000).
12. J.-H. Shin, W. A. Henderson, and S. Passeini, *Electrochem. Commun.*, **5**, 1016 (2003).
13. G.-T. Kim, G. B. Appetecchi, F. Alessandrini, and S. Passerini, *J. Power Sources*, **171**, 861 (2007).
14. S. Liu, N. Imanishi, T. Zhang, A. Hirano, Y. Takeda, O. Yamamoto, and J. Yang, *J. Electrochem. Soc.*, **157**, A1092 (2010).
15. H. Sakaebe, H. Matsumoto, and K. Tasumi, *J. Power Sources*, **146**, 693 (2005).
16. H.E. Park, C.H. Hong, and W. Y. Yoon, *J. Power Sources*, **178**, 765 (2008).
17. T. Matusi and K. Takeyama, *Electrochim. Acta*, **40**, 2165 (1995).
18. S. Megahed and B. Scrosati, *Interface*, **4**, 34 (1995).
19. C. Brissot, M. Rosso, J.-N. Chazalviel, and S. Lascaud, *J. Power Sources*, **81–82**, 925 (1999).
20. J.-N. Chazalviel, *Phys. Rev.*, **A42**, 7355 (1990).
21. C. Leger, J. Elezgaray, and F. Argoul, *Phys. Rev.*, **E58**, 7700 (1998).
22. H. J. S. Sand, *Philos. Mag.*, **1**, 45 (1901).
23. C. Brissot, M. Rosso, J.-N. Chazalviel, P. Bandry, and S. Lascaud, *Electrochim. Acta*, **43**, 1569 (1998).
24. S. Liu, N. Imanishi, T. Zhang, A. Hirano, Y. Takeda, O. Yamamoto, and J. Yang, *J. Power Sources*, **195**, 6847 (2010).
25. S. Liu, N. Imanishi, T. Zhang, A. Hirano, Y. Takeda, O. Yamamoto, and J. Yang, *J. Power Sources*, **196**, 7681 (2011).

26. H. Aono, E. Sugomoto, Y. Sadaoka, N. Imanaks, and S. Adachi, *J. Electrochem. Soc.*, **136**, 500 (1989).

27. A. Murugan, V. Tangadural, and W. Weppner, *Angew. Chem. Int. Ed.*, **46**, 7778 (2007).

28. S. Hasegawa, N. Imanishi, T. Zhang, J. Xie, A. Hirano, Y. Takeda, and O. Yamamoto, *J. Power Sources*, **189**, 371 (2009).

29. Y. Shimonishi, A. Toda, T. Zhang, A. Hirano, N. Imanishi, O. Takeda, and O. Yamamoto, *Solid State Ionics*, **183**, 48 (2011).

30. T. Zhang, N. Imansihi, Y. Shimonihsi, A. Hirano, Y. Takeda, O. Yamamoto, and N. Sammes, *Chem. Commun.*, **46**, 1661 (2010).

31. Y. Shimonishi, T. Zhang, T. Imanishi, D.M. Im, D.O. Lee, A. Hirano, Y. Takeda, O. Yamamoto, and N. Sammes, *J. Power Sources*, **196**, 5128 (2011).

CHAPTER **11**

KINETICS OF THE OXYGEN ELECTRODE IN LITHIUM–AIR CELLS

Michele Piana, Nikolaos Tsiouvaras, and Juan Herranz

1 INTRODUCTION

The oxygen electrode is the heart and peculiarity of a Li–air cell. This does not mean that the other components of the battery are less important: The entire system must be balanced and developed harmoniously. This only means that the reason for the strong appeal of Li–air batteries resides in their positive electrode, which consumes oxygen, potentially from the atmospheric air. This could be the breakthrough that

Lithium Batteries: Advanced Technologies and Applications, First Edition.
Edited by Bruno Scrosati, K. M. Abraham, Walter van Schalkwijk, and Jusef Hassoun.
© 2013 John Wiley & Sons, Inc. Published 2013 by John Wiley & Sons, Inc.

gives Li batteries more chances for wide applications, for example, in long-range fully electric vehicles.

In current Li-ion batteries, improvement in the cathode capacity could lead to a remarkable increase in energy density in the entire battery, even more than that achieved by improving its working voltage to about 5 V vs. Li$^+$/Li (V_{Li}) pursued in high-voltage cathode concepts. Unfortunately, the topotactic insertion–deinsertion chemistry of current positive electrodes limits their specific energy to the current 120 Wh$_{total}$/kg$_{pack}$ and in long-term projection to 200 Wh$_{total}$/kg$_{pack}$ [1].

Use of the oxygen reduction reaction (ORR) at the positive electrode could be virtually limited by the anode theoretical specific capacity (in the case of Li metal it is 3860 Ah$_{theor}$/kg$_{Li}$) if the positive electrode weight is not taken into account. This is, of course, purely an ideal number, not meaningful in reality: One should consider not only the weight of the positive electrode and of the products formed during discharge, but also the space available in the electrode for them [2,3]. A better estimate for a Li–air cathode gives, in comparison to the currently available positive electrodes ($\approx$ 160 Ah/kg$_{cathode}$), an improvement between five- and eightfold in terms of specific capacity and between four- and sixfold in terms of gravimetric energy density [3]. For fully packaged batteries this could result in a three- to fourfold increase in specific energy compared to state-of-the-art Li-ion batteries [4].

To fully exploit the capacities of an oxygen electrode, the mechanisms that rule and limit its functioning must be fully understood. First, the fundamental reactions for all types of Li–air batteries are taken into account (Section 2), dealing with thermodynamics and, most important, kinetic paths for them. Starting from an understanding of the discharge behavior, the kinetic effects on the ORR must be clarified (Section 3), not only to choose the catalysts with the lowest overpotential (Section 3.1) but also to increase the current density per gram or specific surface of carbon/catalyst (Section 3.2), still maintaining high specific capacity. Such understanding should help in improving the poor rate capability that is really a weakness of the Li–air battery, since the increase in current density, typically from 0.04 to 0.10 mA/cm^2 [2,5–7] up to a maximum of 1.0 mA/cm^2, results in a significant loss in capacity. The structural origin of the catalytic effect should also be understood (Section 3.3) in order to design and produce the best catalyst for the ORR. The stabilization or reactivity of intermediate and discharge products in the electrolyte chosen is another important point to discuss, possibly to optimize the performance of Li–air batteries (Section 3.4 and 3.5). All the aspects cited above are important pieces of a puzzle that will show the limitations during discharge (Section 3.6) and could allow improving specific capacity and rate capability.

On the other hand, if charge and cyclability of the system are not acceptable, the system itself will lose the most important advantage of the secondary Li batteries and its chance to be used as power source in commercial fully-electric vehicles.Improving this aspect is important in evaluating the catalytic effect on the oxygen evolution reaction (OER) during charge (Section 4).The most important aspect is the decreasing of charge potential to improve the round-trip efficiency, usually very low also at low current density: $\approx$ 70% for MnO$_2$ [8] and $\approx$ 75% for PtAu [9]. In this chapter we give an overview of all the aspects mentioned above.

2 THERMODYNAMICS AND KINETIC PATHS IN THE VARIOUS LITHIUM–AIR CELL TYPES

Four different types of Li–air batteries have been shown and classified in the literature [10], according to the electrolyte used (Fig. 1): (a) aprotic (nonaqueous), (b) solid-state, (c) aqueous, and (d) mixed aqueous/aprotic. The same lithium metal used as a negative electrode is common to the various typologies. Despite its problems with dendrite formation and consequent safety limitation, lithium metal is still very interesting in a high-power-density device, due to its high specific capacity, high electronegativity, and low working potential. The kinetics of lithium dissolution–deposition is very fast, so that it is practically limited by the migration of Li^+ ions through the solid electrolyte interphase (SEI) formed in contact with the electrolyte [11–13]. Thus, if the decomposition of electrolyte could be inhibited by a passivating layer, lithium metal can be a counterelectrode neither kinetic nor capacity limited, suitable to studying the behavior of the positive electrode. It has been, indeed, the most used negative electrode in the literature on Li–air and has been a good starting point to study the oxygen electrode, also a common component in the various types of Li–air cells. Each type is characterized by the different electrolyte and thus different electrode reactions, thermodynamics, and kinetic mechanism.

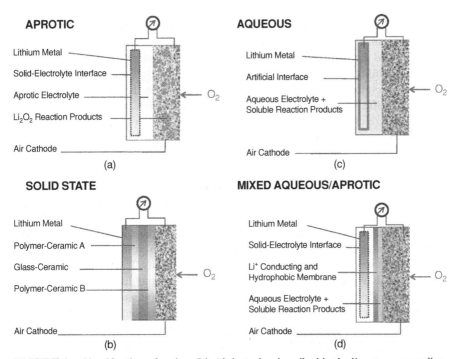

FIGURE 1 Classification of various Li–Air batteries described in the literature, according to the types of electrolytes used. (From [10] with permission.)

2.1   Aprotic (Nonaqueous) and Full Solid-State Electrolyte

The Li–air cell consisting of only one compartment full of aprotic electrolyte is the type most studied so far [10,14,15]. Its architecture has a lot in common with Li-ion technology, the only difference being the presence of oxygen in the system. The discovery of the interesting peculiarities of Li–air batteries happened when a battery technician accidentally introduced a little oxygen into a Li/graphite half-cell [16]. The interesting increase in potential intrigued him and he reported it to his group leader (K. M. Abraham). A more accurate analysis was carried out, with the consequent first publication [2]. The use of liquid or gel electrolyte characterizes this type, for which the reported possible overall reactions and their standard potentials vs. Li are [4,17]

$$2Li(s) + O_2(g) \rightleftharpoons Li_2O_2(s) \qquad E_{rev} = 2.96 \; V_{Li} \qquad (1)$$

$$4Li(s) + O_2(g) \rightleftharpoons 2Li_2O(s) \qquad E_{rev} = 2.91 \; V_{Li} \qquad (2)$$

As shown in Figure 1a, the formation of a solid oxide at the oxygen electrode, insoluble in aprotic solvent, is the result of the discharge reaction (the ORR), while the charge consists of the reversed oxygen evolution reaction (OER) from the oxides. A third possibility for the O_2 is to form Li superoxide, but this is considered more an intermediate than a final discharge product [18,19], due to its high instability.

Another configuration for Li–air cells shown in the literature [20] uses all solid-state components, the electrolyte included. The two electrodes are separated by a solid glass–ceramic Li^+ conductor, sandwiched between two polymer–ceramic composites to protect it from reduction in contact with lithium and to improve contact at the interfaces. Even in the positive electrode the glass–ceramic Li^+ conductor is used to replace the liquid electrolyte. The cell exhibited excellent thermal stability and rechargeability in the range from 30 to 105°C. This cell is not affected by water contamination since it has been tested with pure oxygen, but the authors claim that it would be possible, after some modifications in the cell design, to use it also with air. This probably refers to the use of a barrier to water already on the positive electrode, to keep the battery purely aprotic, avoiding the water, to react with discharge products and lithium. The working mechanism of such a cell is similar to the mechanism using liquid aprotic electrolyte.

At this point it is noteworthy to point out the differences between a real Li–air battery, using dry air, and a Li–O_2 battery, using pure oxygen. The use of air has the theoretical tremendous advantage of avoiding carrying the oxidant together with the battery [1]. On the other hand, this advantage is only on paper, taking into account that the discharge products will remain inside the system at the end of discharge, increasing its total weight and decreasing its energy density. Unfortunately, the use of air has the drawback of decreasing the reversible potential due to the low oxygen concentration. Another consequence of the lower partial pressure of oxygen in air, in combination with the use of a liquid aprotic electrolyte, could be the poorer rate capability, due to the lower solubility of oxygen in the electrolyte. A possible solution to this could be the use of a solid electrolyte or ionomer in the positive electrode, as in the full solid-state Li–air battery described earlier. Most studies carried out until now dealt with the Li–O_2 battery more than pure Li–air; this is because the Li–air battery is a brand new subject and all its working mechanisms are still to be understood. For this purpose, a Li–O_2 system is a simpler and better starting point aiming at an initial understanding and system modeling. In fact, exactly for these reasons, in this chapter we deal mainly with Li–O_2 cells.

Understanding the kinetic mechanism for ORR is one of the first, very important steps. In the absence of proton sources, the very first kinetic step is considered the reversible reduction of oxygen to a superoxide ion radical [3,19,21–23]:

$$O_2 + e^- \rightarrow O_2^{\cdot-}(ads) \qquad E_{rev} = 2.71 \ V_{Li} \tag{3a}$$

followed by the reaction with Li^+ to form the intermediate LiO_2:

$$O_2^{\cdot-}(ads) + Li^+(sol) \rightarrow LiO_2(ads) \ or \ LiO_2(sol) \tag{3b}$$

which can be adsorbed on the electrode or dissolved in the electrolyte. In the case of dissolution in a small amount of electrolyte, as in a working cell, it can disproportionate to

$$2LiO_2(sol) \rightarrow Li_2O_2 + O_2 \tag{4}$$

In the case of the reactions of LiO_2 adsorbed at the electrode, there are two possible paths [3]. If the electrode surface is weakly chemisorbing oxygen, as in the case of pure carbon or gold surfaces, the lithium superoxide can easily be electrochemically oxidized to peroxide:

$$LiO_2(ads) + Li^+(sol) + e^- \rightarrow Li_2O_2(s) \tag{5a}$$

This step can be followed by a further reduction to lithium oxide, possible in the range of the discharge potential of a Li–air cell (2.8 to 2.0 V_{Li}) [3]:

$$Li_2O_2(s) + 2Li^+(sol) + 2e^- \rightarrow 2Li_2O(s) \qquad E_{rev} = 2.86 \ V_{Li} \tag{5b}$$

Very recently, Peng et al. [24] collected in situ surface-enhanced Raman spectroscopic data on gold surfaces, to gather evidence of the proposed mechanisms and discriminate among disproportionation and electrochemical oxidation of the lithium superoxide formed. They provided evidence for LiO_2 detection at 2.2 V_{Li} and showed that it disproportionates to lithium peroxide in the time scale of minutes [equation (4)]; furthermore, they discarded the steps involving the electrochemical reduction of superoxide to peroxide, claiming that it would happen at much lower potentials than 2.2 V_{Li}. In conclusion, they provided evidence for an ORR path at first through a reversible electrochemical process, followed by two chemical processes, the third completely irreversible. Conversely, with strong oxygen chemisorption on the electrode surface, as in the case of Pt, Pd, or Ru, the ORR mechanism could proceed as follows [3]:

$$LiO_2(ads) + Li^+(sol) + e^- \rightarrow Li_2O(s) + O(ads) \tag{6a}$$
$$O(ads) + Li^+(sol) + e^- \rightarrow LiO(ads) \tag{6b}$$
$$LiO(ads) + Li^+(sol) + e^- \rightarrow Li_2O(ads) \tag{6c}$$

The formation of Li_2O at the cathode has been reported, but under certain experimental conditions. Read [25], performing a study varying the discharge rate and using mass balance calculations while monitoring the value of mAh/mL of O_2, has reported that Li_2O is formed during the discharge while working with low oxygen contents. However, although Li_2O formation is thermodynamically possible, so far there has not been conclusive spectroscopic data of its formation by either Raman or XANES studies [24,26].

The OER mechanism during charge depends on the starting discharge product. The oxidation of Li_2O is obviously more difficult than that of Li_2O_2, since in the first case an oxygen–oxygen bond must be formed again, while in the second it is already formed. In other words, the OER overpotential could be inversely related to the oxygen chemisorption strength of the electrode surface used in the discharge, and the formation of Li_2O will make rechargeability more difficult. The mechanism of lithium peroxide oxidation directly to oxygen has been compared to H_2O_2 decomposition, within reason [27]. Peng et al. [24] also evaluated a possible path through superoxide ion, using propylene carbonate as a superoxide scavenger, that would yield CO_2 as a final product; since no carbon dioxide was detected, they concluded that the mechanism for OER is simply the electrochemical oxidation of Li_2O_2 to oxygen [equation (2)]. In that case, two paths are considered for the OER [23]: a fast formation of singlet oxygen [equation (7a)], the excited state at $\approx 4\ V_{Li}$, or, more slowly, to triplet oxygen, the ground state at 2.96 V [equation (7b)]:

$$Li_2O_2 \rightarrow O_2^* + 2Li^+ + 2e^- \qquad (7a)$$

$$Li_2O_2 \rightarrow O_2 + 2Li^+ + 2e^- \qquad (7b)$$

After discussion of the OER mechanism, a fundamental thermodynamic quantity characterizing electrochemical reversible system, called the *round-trip efficiency*, must be introduced. It is defined as the ratio between the operating voltages during discharge and charge at a known current density:

$$\varepsilon(i)_{RT} = \frac{E(i)_{charge}}{E(i)_{discharge}}$$

It compares the energy obtained by discharging with the energy needed to charge the electrochemical system. Compared to Li-ion cells, based on topotactic insertion–deinsertion of Li, Li–O_2 cells have a much lower round-trip efficiency; the energy efficiency could be improved by decreasing the overpotential of ORR and OER, through the employment of a catalyst.

Apart from the low energy efficiency, the biggest challenges to face in this aprotic electrolyte configuration are the diffusion of oxygen to lithium metal, potentially changing its SEI and the reactivity toward the electrolyte, and the very strong reactivity of a superoxide ion radical that can attack the electrolyte on the cathode side.

2.2   Aqueous and Mixed Aprotic/Aqueous Electrolyte

An aqueous Li–air battery can only be possible using a solid barrier toward water for the negative electrode, protecting it from the parasitic corrosion reaction with water [28]:

$$Li(s) + H_2O(l) \rightarrow LiOH(sol) + \tfrac{1}{2}H_2(g) \qquad (5)$$

In fact, the corrosion of lithium metal is on the order of several A/cm^2 in neutral aqueous electrolyte and decreases to tens of mA/cm^2 in a strong alkaline medium.

The SEI on lithium and/or a gel–polymer electrolyte is not sufficient to stop the parasitic reaction of water over time. To establish an artificial, protective interface (shown schematically in Fig. 1c), the use of a lithium metal electrode protected by a solid electrolyte has been introduced by the PolyPlus Battery Company [29,30]; it is characterized by a water-stable and impermeable solid electrolyte consisting of a glass–ceramic Li^+ conductor having a Nasicon-type structure. It has the drawback of showing a conductivity of 10^{-3} to 10^{-4} S/cm and of being unstable in contact with lithium metal. The first limit is intrinsic to the material, but the second has been solved using another conductive interlayer between a solid electrolyte and lithium metal. A similar type of cell is the hybrid aprotic/nonaqueous cell [31]; the only difference from the previous configuration is the use of a separator soaked with aprotic liquid electrolyte as an interlayer between the lithium and the solid ionic conductor.

Use of two different compartments for the anode and cathode allows choosing the medium in which the ORR can occur. This results in more possibilities for the overall reactions and relative theoretical cell voltage vs. Li:

Alkaline: $2Li(s) + H_2O(l) + \frac{1}{2}O_2(g) \rightleftharpoons 2LiOH(s)$ $E_{rev} = 3.35\ V_{Li}$ (9a)

Acidic: $2Li(s) + 2H^+(sol) + \frac{1}{2}O_2(g) \rightleftharpoons H_2O(l) + 2Li^+(sol)$ $E_{rev} = 4.28\ V_{Li}$ (9b)

Seawater $2Li(s) + H_2O(l) + \frac{1}{2}O_2(g) \rightleftharpoons 2LiOH(s)$ $E_{rev} = 3.79\ V_{Li}$ (9c)
(pH 8.2):

These various types of aqueous Li-air batteries can take the electrolyte from ambient, as in a seawater Li–air battery, or carry it in a reservoir [32]. In the first case the batteries don't need to carry the electrolyte with them but lose their rechargeability; in the second case the specific capacity is sacrificed and limited by the aqueous electrolyte carried inside the cell, but rechargeability is still possible.

The ORR path is strongly dependent on the proton activity and is described by a four-electron multistep process medium; the single possible steps are as follows [33,34]:

Acid medium:

   Direct reduction:

$$O_2(g) + 4H^+(sol) + 4e^- \rightarrow 2H_2O(l)\qquad E_{rev} = 4.27\ V_{Li}\quad (10a)$$

   Indirect reduction:

$$O_2(g) + 2H^+(sol) + 2e^- \rightarrow H_2O_2(l)\qquad E_{rev} = 3.74\ V_{Li}\quad (10b)$$
$$H_2O_2(l) + 2H^+(sol) + 2e^- \rightarrow 2H_2O(l)\qquad E_{rev} = 4.80\ V_{Li}\quad (10c)$$

Alkaline medium:

   Direct reduction:

$$O_2(g) + 2H_2O(l) + 4e^- \rightarrow 4OH^-(sol)\qquad E_{rev} = 3.44\ V_{Li}\quad (11a)$$

   Indirect reduction:

$$O_2(g) + H_2O(l) + 2e^- \rightarrow HO_2^-(sol) + OH^-(sol)\quad E_{rev} = 2.98\ V_{Li}\quad (11b)$$
$$HO_2^-(sol) + H_2O(l) + 2e^- \rightarrow 3O\ H^-(sol)\qquad E_{rev} = 3.91\ V_{Li}\quad (11c)$$

However, the initial transfer of one electron to form a superoxide ion radical [equation (3a)] is a very important step, even the rate-determining step (RDS), which is also true for ORR in the aqueous system [3]. The dependence of ORR kinetics with pH has been claimed as evidence of the fact that equation (3a) is the RDS [3,34,35]. The reason for this resides in the fact that it can be possible only with an RDS not dependent on the pH: equation (3a) is the only step clearly independent of the pH, which is converse to all other possible reactions in the presence of water. This can easily be shown in a Pourbaix diagram of the ORR to hydrogen peroxide [34,36] (Fig. 2a). As the pH is increasing (Fig. 2b) the distance between the constant potential of superoxide formation (line 6) and the ORR to peroxide (lines 4 and 5) is decreasing, improving the kinetics of the process. The effect of the catalyst on RDS and the kinetics of both ORR and OER, due to its interaction with oxygen or lithium oxides, is obviously very important and is the main subject discussed in the following sections. We focus there on aprotic Li–O_2 batteries since this is the newest and recently most studied configuration and since the aqueous case has been discussed broadly, from a mechanistic point of view, in all the fuel cell literature.

3 OXYGEN REDUCTION REACTION KINETICS

The reduction of O_2 is probably one of the most studied reactions in electrochemistry in both aqueous and aprotic media. Recent developments in the field of Li–air batteries have brought a renewed and increasing interest in the ORR in aprotic electrolytes, so far monopolized almost exclusively by ORR in aqueous media, which is of interest for fuel cell technology. In the following section we discuss in more detail how the use of catalysts, their loading, and different current rates influence the ORR in aprotic-electrolyte-based Li–O_2 batteries, highlighting the most important experimental results in this field as well as the challenges that remain.

3.1 Catalyst Effect

The use of a catalyst for the ORR is extensive in literature, and in this section the most important results for materials used so far in Li–O_2 cathodes are explored and presented, according to the composition of these catalysts: (1) nonnoble metal, (2) metal-oxide, (3) noble metal, and (4) pure carbon.

Nonnoble Metal Catalysts The advantages of using catalysts in a Li–O_2 system with aprotic electrolytes were clear from the introduction of this technology, as presented in a publication by Abraham and Jiang in 1996 [2]. A battery built with the positive electrode open to either dry air or pure oxygen demonstrated three cycles with good coulombic efficiency and relatively high discharge potentials. This system employed a composite positive electrode of cobalt phthalocyanine on Chevron carbon (5:95 wt%) and a gel–polymer electrolyte consisting of poly(acrylonitrile) (PAN)–EC/propylene carbonate (PC)–$LiPF_6$. Metal phthalocyanines have been studied widely as an ORR electrocatalyst in fuel cells [37] and as an ORR catalyst in a Li–O_2 battery; they presented one of the highest discharge potentials reported so

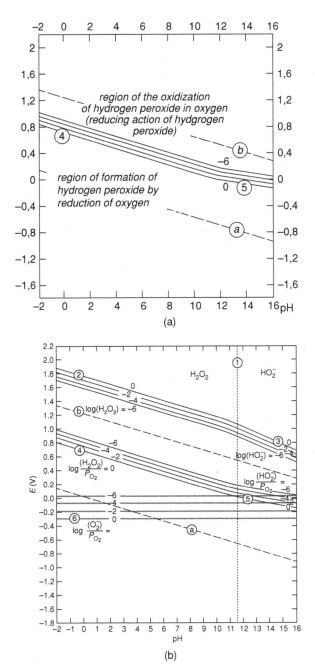

FIGURE 2 Pourbaix (a) (from [36], with permission) and modified Pourbaix (b) (from [34], with permission) diagrams for hydrogen peroxide with equilibria for the superoxide–oxygen reaction (labeled 6 in the diagram) added; other notations for equilibria as in reference [36].

far: 2.85 V at 0.05 mA/cm^2. These experiments demonstrated the beneficial effect of the use of a catalyst, although to our knowledge there are no other results using cobalt–phthalocyanine as an ORR catalyst in a Li–O$_2$ system.

Further studies on nonnoble metal catalysts were performed in a recent work by Ren et al. [38], who reported the best ORR activity so far, using a CuFe catalyst supplied by Acta SpA, Italy [39–41]. In this work they are comparing three positive electrodes, two with different types of carbon and one with the CuFe catalyst ($\approx$ 3 wt%) supported on Ketjen black carbon. The cells with CuFe-catalyzed Ketjen electrodes demonstrated a higher cell discharge voltage at all the rates tested, reaching a potential plateau of almost 2.9 V at the slowest rate (0.05 mA/cm^2). The authors claim the independency of the catalyst activity from the cell's discharge capacity. They report that the capacity of the battery is related to the cathode pore volume and not to the use of a catalyst, since they showed that at low current densities both catalyzed and noncatalyzed Ketjen electrodes presented the same capacity, assuming that the low catalyst loading and the heat treatments do not change the carbon porosity and surface area in the case of catalyzed carbon. The authors explained this finding through an ORR mechanistic model in which the limiting factor is the deactivation of catalytic sites due to product precipitation.

Metal-Oxide Catalysts A systematic screening of possibly interesting materials has been reported in several papers published by Bruce's group [8,42,43], although their main criteria for catalyst evaluation are battery capacity and cyclability. All materials studied represented a small weight percentage in the positive electrode, typically 2.5 mol%, and they were all materials with high ORR activities in alkaline media. In a 2006 paper by Ogasawara et al. [43], the study focused on manganese-oxide materials such as electrolytic manganese dioxide (EMD), β-MnO$_2$, Mn$_2$O$_3$, and Mn$_3$O$_4$. In this work a positive effect on battery performance was observed, although it was related more to cyclability than to ORR activity, as the discharge potential was not affected by the use of a catalyst (2.6 V at 70 and 100 mA/g$_{carbon}$, 1 M LiPF$_6$ in PC). In a 2007 study [42], transition metal oxides were studied and it was found that Fe$_3$O$_4$ and CuO present good capacity retention through 10 cycles, whereas Fe$_3$O$_4$ also presented the highest initial capacity. In addition, catalysts currently used in fuel cell cathodes, such as platinum and lanthanum manganite, La$_{0.8}$Sr$_{0.2}$MnO$_3$, were also tested and presented the lowest initial capacity and the highest capacity loss through cycling. Conflicting results have, however, been reported regarding these two materials [4,44]. Cathodes, all with different catalyst loading, were cycled in cells with 1 M LiPF$_6$ in PC electrolyte and presented similar discharge voltages at about 2.6 V at 70 mA/g$_{carbon}$. In a publication by Debart et al. [8] the study was centered on the promising MnO$_2$ materials (α, β, γ, λ), showing that the α-MnO$_2$ has a higher capacity ($\approx$ 3000 mAh/g$_{carbon}$ and an average discharge potential $\approx$ 2.6 V at 70 mA/g$_{electrode}$ in 1 M LiPF$_6$ PC electrolyte) and better capacity retention than that of the other manganese structures. The performance of a bulk and nanowire α-MnO$_2$ was also compared. The α-MnO$_2$ nanowires were most effective in the formation and decomposition of Li$_2$O$_2$, although once again without a considerable decrease of ORR overpotential. The reason that in terms of capacity the α-MnO$_2$ performs better than the remaining manganese dioxides in this work is considered to be related to its structure. This manganese polymorph has a hollandite structure which consists of 2×2 tunnels formed by edge- and corner-sharing MnO_6 octahedra [45]. The ability of manganese dioxides to intercalate lithium has been reported in the past [43]; nevertheless, the advantage of this material is considered to be not only Li intercalation but also its ability to accommodate both Li^+ and O^{2-} within its tunnel structure. In this way, lithium oxides can be incorporated into the tunnels, leading to better reversibility of the process.

The highest capacity achived using manganese oxide catalysts, 4750 mAh/g_{carbon}, has been reported by Cheng and Scott [46]. In this work they compare the catalytic effect of three different manganese materials, EMD, MnO_x, and MnO_x, heat-treated at 300°C and supported on three different carbon materials, Norit (800 m^2/g), acetylene black (75 m^2/g), and Super P (62 m^2/g). In all cases the Mn content was in the range of 28 wt%. The highest capacity was observed for the heat-treated MnO_x material supported on the Norit carbon. The authors claim that the increased capacity of the heat-treated catalyst is more related to the removal of impurities during heating, resulting in increased porosity of the material, than to the catalyst effect and the reaction kinetics. However, the structural changes in the MnO_x during heat treatment are not mentioned by the authors. In addition, the discharge plateaus for all three manganese catalysts (2.6 V at a rate of 70 mA/g_{carbon} and 1 M $LiPF_6$ in PC electrolyte) present only minor differences compared to the results shown by Debart et al.

Lower overpotentials using α-MnO_2 catalyst were reported by Thapa and Ishihara [47]. Using a mixture of α-MnO_2 and Teflonized acetylene black at the molar ratio 9:1, they observed a very high discharge potential between 2.9 and 2.7 V; however, at low discharge rates of 0.025 mA/cm^2 in a 1 M LiTFSI EC/diethyl carbonate (DEC) (3 : 7 v/v) electrolyte. In this work, the mesoporous α-MnO_2 used performs the dual function of being both the catalyst for the reaction and the storage medium for the reaction products. As a storage medium, it performs poorly compared with materials where the α-MnO_2 is supported on high-surface-area carbons, showing 350 mAh/$g_{catalyst}$ at the low discharge rate of 0.025 mA/cm^2.

Noble Metal Catalysts   Noble metal nanoparticles used as catalysts for Li–O_2 batteries have been studied extensively by several groups. The research group of Shao Horn, using 40 wt% Au/C and 40 wt% Pt/C, found that gold showed very high activity for ORR and low activity for OER, whereas for platinum it was the opposite. Gold and platinum catalysts presented discharge plateaus at 2.7 and 2.5 V, respectively, at 100 mA/g_{carbon} and 1 M $LiClO_4$ in PC–dimethoxyethane (DME) electrolytes [9,26]. To explain the different discharge potentials observed for each catalyst, it is mentioned that different catalysts may yield different reaction products (Li_2O, Li_2O_2, LiO_2) with different reduction plateaus. Combining the catalytic effects of these noble metals, they prepared a bifunctional 40 wt% AuPt(1 : 1)/C electrocatalyst, which demonstrated a high discharge plateau at 2.7 V and an improved round-trip efficiency, although no results on cyclability were reported [9]. Interestingly, in a later publication [48] by the same group, the importance of the electrolyte in the ORR kinetics is highlighted. Using the same Au and Pt catalyst as in previous work, the ORR activity was evaluated in a DME-based electrolyte. In this work the Pt electrocatalyst is considerably more active than gold, presenting lower overpotentials, while Pd seems to be the most active ORR catalyst. The authors explain that this inconsistency derives from the reactivity of the carbonate-based electrolytes used in the earlier studies. The parasitic reactions taking place during discharge between the carbonate electrolyte and the ORR intermediates can greatly influence the discharge voltages and distort the evaluation of a catalyst.

The combination of noble metals and oxide catalysts seems promising as reported in recent work by Thapa et al. [47,49]. Employing mesoporous β-MnO_2 and α-MnO_2 with Pd as a composite positive electrode, they observed a considerable improvement on round-trip efficiency for both manganese structures. It is also shown that even small amounts of Pd can change the battery capacity and charge–discharge voltages, demonstrating the advantages of both nanostructures and noble metal catalysts. The Pd/α-MnO_2/TAB (Teflonized acetylene black) (molar ratio 15 : 75 : 10) presents a discharge plateau close to 2.8 V at 0.025 mA/cm^2 in an electrolyte of 1 M $LiPF_6$ in PC.

In a more detailed study on metal and metal oxides, Cheng and Scott [50] compared their activities as ORR catalysts. The work is carried out with Pd, PdO, Ru, RuO_2, Pt, and MnO_2, all supported on carbon and prepared in house. The catalyst contents were, in all cases, around 40 wt% and the carbon support used was Norit, which provided the best results reported in a previous publication by the same group [46]. The authors compare the results provided by Pd and PdO catalysts in Li–O_2 cells, using as the electrolyte $LiPF_6$ in PC. The oxide presents a higher discharge potential and, at the same time, a lower capacity; according to the authors this is due to its lower level of dispersion compared to the Pd catalyst. The capacity retention is better for the oxide catalyst because of its increased stability compared to Pd, although both catalysts are progressively more agglomerated during cycling. At a discharge rate of 70 mA/g_{carbon}; the catalyst with the higher discharge potential is RuO_2 at 2.9 V, followed by Ru (2.7 V), Pt (2.7 V), PdO (2.7 V), and Pd (2.6 V). The lowest discharge potential is reported for MnO_2 (2.5 V) which, nevertheless, seems to be the most stable, since it retains the capacity better than the rest of the catalysts and presents the highest capacity.

Pure Carbon as a Catalyst Considering the similar discharge potentials observed in most manganese and transition-metal oxide catalysts reported in the literature [42], it is argued that this does not mean that the ORR kinetics are not catalytically sensitive but, on the contrary, that the kinetics are dominated by the high catalytic activity of carbon, which in most of these studies represents the largest part of the electrode loading [4]. This argument is enhanced by the results by Thapa et al. presented above [47]. Considering that the same carbon can also act as an ORR catalyst in the presence of Li^+, there is ample literature regarding different types of materials [5,7,23,25,38,44,51–58]. Nevertheless, carbon electrodes are usually studied as storing materials for the solid discharge products, and their catalytic activity toward oxygen reduction is not evaluated. That is due primarily to the fact that the discharge potentials seem not to be affected by the various carbons and are usually in the range 2.5 to 2.7 V. In addition, Rotating Disk Electrode (RDE) experiments carried out on glassy carbon electrodes in aprotic media do not seem to suggest a high level of catalytic activity [19,48]. However, working with high-surface-area carbons, the overall activity is expected to be enhanced; this is why the carbon itself is already a good ORR catalyst.

Advanced carbon materials have also been reported in the literature, but without great improvements. Zhang et al. [59], using single-walled carbon nanotubes supported on carbon-nanofiber buckypaper, show rather high discharge plateaus at low rates (0.1 mA/cm^2), close to 2.7 V in 1 M LiPF$_6$ PC/THF 1 : 1. Nevertheless, they argue that the discharge process is dominated by the oxygen concentration gradient inside the electrode, without considering a catalytic effect of the material; similar results have also been presented by other authors [60]. Li et al. [58] studied the catalytic activity of a variety of carbonaceous materials, comparing nitrogen-doped carbon nanotubes (NCNTs) and normal nanotubes. The NCNTs present slightly higher ORR catalytic activities than those of the nondoped carbon nanotubes. Nevertheless, the discharge plateau for NCNTs is 2.51 V at 75 mA/g$_{carbon}$ in 1 M LiPF$_6$ PC/EC (1 : 1 wt%) which is considerably lower than that reported in other literature results using carbon cathodes.

General Remarks   As shown above, it has been claimed in the literature that catalysts are able to improve Li–O$_2$ battery performance in three distinct areas: reducing battery overpotentials [2–4,9,26,38,48,61], increasing battery capacity [8,24,42,43, and improving battery cyclability [8,43,58]. The catalytic effect on battery overpotentials is the most straightforward of these criteria. A catalyst facilitates the oxygen reduction, which results in production at lower overpotentials of O$_2$$^{\cdot-}$ species, which react with Li$^+$ during battery discharge. Therefore, the catalyst that presents the highest discharge potentials is the best ORR catalyst. However, during the catalyst overpotential evaluation, the effect of electrolyte degradation is a factor that has to be taken into account, as shown by Lu et al. [48].

The increase in capacity related to a catalyst is something that many authors are claiming, although usually without reporting the mechanism by which this is achieved. Adding a catalyst, the number of catalytic sites would be increased, but the storage capacity of discharge products would usually not be affected. An exception to this is the electrode prepared by Thapa and Ishihara [47], which consisted primarily of α-MnO$_2$. When the catalyst represents only a small volume of the electrode, it can affect the capacity through two mechanisms: (1) higher selectivity toward products that provide higher capacity (e.g., Li$_2$O instead of Li$_2$O$_2$), and (2) formation of products that do not block active sites (see Section 3.5). To date, the only mechanism through which a catalyst influences the positive electrode capacity is due to the coordination of Li$^+$ and O^{2-} inside the porous structure of the α-MnO$_2$ reported by Debart et al. [8]. This mechanism would result in a higher selectivity of the catalyst toward certain Li–O$_2$ discharge products, although no spectroscopic results have been presented that could support it.

The catalytic effect toward cyclability should also be considered under the prism of selectivity in two possible mechanisms: (1) a catalyst that would lead to discharge products easier to recharge which would result in better cyclability and potential retention with cycles, and (2) a material that would selectively promote the reaction of superoxide radicals with Li$^+$ and not with the electrolyte species. However, in publications relating catalytic performance and battery cyclability, so far no ORR path has been presented that justifies cell improvement other than that noted by Debart et al. In Table 1 we summarize some results relative to the galvanostatic

TABLE 1 Catalytic Effect of Various Metal Oxides and Reduced Noble Metals on the Galvanostatic First Discharge of Li–O$_2$ Cells

Catalyst	Features	Support	Catalyst loading (wt%)	Electrolyte	Current (mA/g$_{carbon}$)	Potential (V$_{Li}$)	Capacity (mAh/g$_{carbon}$)	Ref.
Au	≈ 15 m^2/g$_{Au}$	Vulcan, supported	40	1 M LiClO$_4$ in propylene carbonate (PC)/DME	100	2.80	1400	9
CoFe$_2$O$_4$	1 to 5-μm particles	Super S, mixed	23	1 M LiPF$_6$ in PC	70	2.60	2000	42
Co$_3$O$_4$	1 to 5-μm particles	Super S, mixed	35	1 M LiPF$_6$ in PC	70	2.60	2000	42
CuO	1 to 5-μm particles	Super S, mixed	15	1 M LiPF$_6$ in PC	70	2.60	900	42
Fe$_2$O$_3$	1 to 5-μm particles	Super S, mixed	26	1 M LiPF$_6$ in PC	70	2.60	2700	42
	Unspecified	Super S, supported	26	1 M LiPF$_6$ in PC	70	2.60	2500	42
Fe$_3$O$_4$	1 to 5-μm particles	Super S, mixed	34	1 M LiPF$_6$ in PC	70	2.60	1200	42
La$_{0.8}$Sr$_{0.2}$MnO$_3$	1 to 5-μm particles	Super S, mixed	38	1 M LiPF$_6$ in PC	70	2.60	750	42
MnO$_2$	Electrolytic (EMD)	Super S, mixed	16	1 M LiPF$_6$ in PC	50	2.60	1000	43
	Electrolytic (EMD)	Super S, mixed	16	1 M LiPF$_6$ in PC	70	2.60	1000	42
	Electrolytic (EMD)	Super S, mixed	45	1 M LiPF$_6$ in PC	70	2.60	2700	46
	α-Phase nanowires	Super S, mixed	16	1 M LiPF$_6$ in PC	70	2.60	3000	8
	α-Phase nanoflakes	Multiwalled carbon nanotubes (MWCNTs), supported	40	1 M LiPF$_6$ in PC/EC/DME	70	2.70	1770	58
	α-Phase nanoflakes	MWCNTs, mixed	44	1 M LiPF$_6$ in PC/EC/DME	70	2.60	1260	58
	20 to 50 Nm nanoparticles	Norit, supported	45	1 M LiPF$_6$ in PC	70	2.60	4400	46
	20 to 50-nm nanoparticles	Super P, supported	45	1 M LiPF$_6$ in PC	70	2.50	3400	46
NiO	1 to 5-μm particles	Super S, mixed	14	1 M LiPF$_6$ in PC	70	2.60	1600	42
Pd	<25-nm nanoparticles	Norit, supported	40	1 M LiPF$_6$ in PC	70	2.65	1425	50
PdO	≈ 100-nm particles	Norit, supported	40	1 M LiPF$_6$ in PC	70	2.75	927	50
Pt	1 to 5-μm particles	Super S, supported	30	1 M LiPF$_6$ in PC	70	2.55	470	42
	Unspecified	Norit, supported	40	1 M LiPF$_6$ in PC	70	2.70	1027	50
PtAu	≈ 23 m^2/g$_{Au}$	Vulcan, supported	40	1 M LiClO$_4$ in PC/DME	100	2.60	1000	9
	4 to 10-nm nanoparticles	Vulcan, supported	40	1 M LiClO$_4$ in PC/DME	100	2.80	1500	9
Ru	Unspecified	Norit, supported	40	1 M LiPF$_6$ in PC	70	2.76	962	50
RuO$_2$	Unspecified	Norit, supported	40	1 M LiPF$_6$ in PC	70	2.89	743	50

first discharge of Li–O_2 cells in which various metal oxides and reduced noble metals are used as ORR catalysts.

3.2   Effect of Current Density, Catalyst Surface, and Loading

As reported in Section 3.1, the main indicator of the ORR activity of a catalyst is the overpotential decrease during discharge. It has been shown in several publications that the position of the potential plateau is dependent on the discharge rates employed. Therefore, different materials must always be compared under the same rate conditions. The trend is that as discharge rates increase, the discharge plateau is lowered. For example, Lu et al. [26] claim that both O_2 mass transport limitations in the electrolyte-filled pores of the electrode and low solid-state diffusion of Li^+ in the ORR products may be the cause of this trend. Nevertheless, in our opinion a kinetic catalyst effect plays a very important role and should not be overlooked. Figure 3

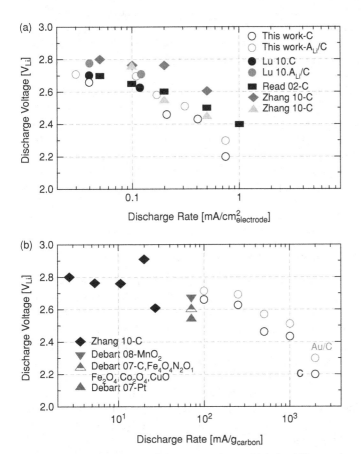

FIGURE 3   Comparison of discharge voltage of Li–O_2 cells at various rates, normalized to electrode geometric area (a) and carbon mass (b). For the papers cited in the figure, see the original paper. (From [26], with permission.)

shows the relation of discharge rates in either mA/g_{carbon} or mA/$cm^2_{electrode}$, and it can be seen that at all discharge rates the catalyzed cathodes present higher discharge voltages than those of noncatalyzed carbon-based cathodes. This is evidence of the catalytic effect of Au/C.

The effect of the catalyst surface and loading is directly related to the number of catalytic sites on the electrode surface and is of fundamental importance for every catalytic process. Catalysts that present higher dispersion present lower polarization, and at the same time it is believed that with a small particle size, the proportion of the interacting catalyst atoms increases and the electrochemical activity toward a reaction is enhanced. A study of the catalytic surface and the electrode activity has been carried out by Cheng and Scott [50]. Comparing a Pd with a PdO catalyst, they observe a great increase in the particle size during battery cycling. This catalyst agglomeration is considered to be the reason for the capacity fading reported. Unfortunately, no discharge data are given for other than the first cycle; nevertheless, it is clear that there is a great decrease in catalytic sites, due to the structural modification of the catalyst.

The effect of catalyst loading in the battery performance is still not clear, and studies on the optimization of the catalyst amount are perhaps still premature. Nevertheless, as has been shown in several published reports, it is something that must be considered when evaluating results. In the case of MnO_2-catalyzed composite electrodes, using loadings of 2.5 at% (16 wt%), no substantial overpotential drop was observed during discharge of the battery compared to the noncatalyzed cathodes [8,42]. However, in a later publication [47] using electrodes with 90 wt% MnO_2 material, the discharging voltage plateau was about 200 mV higher, although with much lower capacity, due to the lower porosity of the electrode. In addition, catalyst loading has to be studied under the prism of overall electrode energy density. As reported by Lu et al. [26], although the Au/C electrode that they employed presented considerably lower discharge overpotentials, its energy density was lower than that of noncatalyzed carbon electrodes.

3.3 Theoretical Origin of the Catalyst Effect

The design of an optimal catalyst for ORR is the key to improve the discharge performances. To foresee what would be the optimal catalyst, a better understanding of the theoretical origin of the ORR catalyst activity is of primary importance. For that reason, some groups already looked for a predictive tool for the best catalyst design. In a recent paper, the ORR activity of bulk catalyst surfaces of Pd, Pt, Ru, Au, and glassy carbon (GC) in DME electrolyte [48] has been claimed to be related to the oxygen adsorption energy on the surface. The activity order has been found to be Pd > Pt > Ru $\approx$ Au > GC, and this has been claimed to have its origin in the optimal adsorption energy between catalyst and oxygen. These energies are reported in a treatise on surface science by Hammer and Norskov [62] (Fig. 4) with the metals aligned exactly as in the periodic table. The Ru metal on the left shows the most intensive chemisorption, making the ORR catalysis more difficult. The best catalysts are Pd and Pt, since they have an intermediate value for the adsorption energy with oxygen. On the other side, there is Au, which shows an absorption energy too low

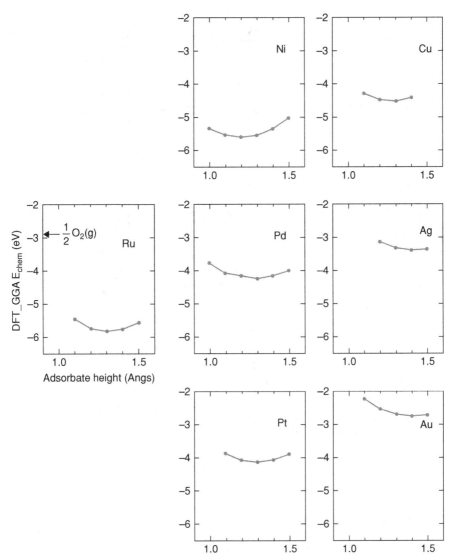

FIGURE 4   Oxygen chemisorption at many close-packed late-transition and noble metal surfaces (hexagonal close packed [hcp(0001)] for ruthenium, but face-centered cubic [fcc(111)] for the other metals). (From [62], with permission.)

to allow fast kinetics. This means that an oxygen–oxygen bond is stronger than a gold–oxygen bond and that the oxygen molecule will not be dissociated in contact with gold; nevertheless, the interaction between them remains. The value for GC is assumed to be similar to that of graphite, taken from reference [63]; the level of chemisorption with oxygen is found to be similar to that of Au. Hammer and Norskov noticed two trends in Figure 4: The farther to the left in the periodic table, the stronger is the chemisorption, and the farther down, the weaker is the interaction, meaning

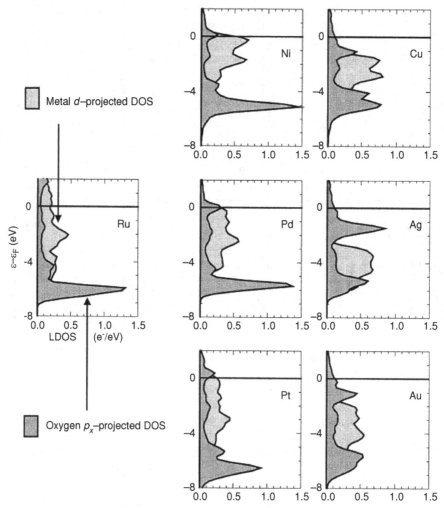

FIGURE 5 Local density of states projected onto the oxygen $2p_x$ state (dark-shaded area) for atomic oxygen, 1.3 Å above close-packed surfaces of late transition metals (cf. Fig. 4). The light-shaded areas give the metal d-projected DOS for the respective metal surfaces before the oxygen chemisorption. (From [62], with permission.)

that the $5d$ metals are the noblest. The final reason for the first observation resides in the interaction between the d orbitals of the metal with the p orbitals of the oxygen, as shown in Figure 5. Going from the right to the left in the periodic table, the d-band increases its energy, making the antibonding band moving above the Fermi level, thus less filled, and consequently, the bonding band more filled, which results in a stronger interaction. The second effect of weaker interaction down with the period is proportional to the oxygen–metal d coupling matrix element V_{ad}^2. This element is correlated with the overlap between the $2p$-states of oxygen and the d-states of the metal. From the Pauli principle, no two electrons can be in the same state; thus, the two states need to become orthogonal, and, in turn, this means that the higher the coupling matrix element, the stronger the repulsion (or the weaker the interaction). Calculation of V_{ad}^2 shows how it increases with the period.

This argument can be better related to the RDS of the ORR in aprotic electrolyte (i.e., the reduction of oxygen to superoxide ion radical). In the case of Ru, the bond with oxygen is so strong that the oxygen–oxygen bond is almost completely dissociated, making the RDS more difficult. In the case of Pt and Pd, the interaction is optimal and some states are available above the Fermi level to accept the electrons for the reduction. For Au and GC the oxygen is so weakly bonded that very few states are available above the Fermi level, making reduction more difficult.

Regarding OER catalysis, much more experimental work must be carried out on different types of catalysts in order to understand the trends in activity and correlate them with some theoretical parameter that could be used as a predictive tool.

3.4 Superoxide Stabilization with Large Cations

The electrochemistry of oxygen in aprotic solvents has been studied extensively using tetrabutylammonium (TBA) salts as the supporting electrolyte (e.g., see references 4,22, and 64–66). The generalized use of such compounds is justified by the large volume of the TBA cation; this, in contrast with Li^+, prevents the precipitation of solid oxides upon reaction with the superoxide radical, resulting from O_2 electroreduction [equation (3a)], and the subsequent poisoning of the working-electrode surface by this precipitate [3,24,66,67].

In a recent work by the Read group, Zhang et al. [61] explored the effect of this enhanced solubility of TBA oxides vs. Li oxides on the discharge of the Li–O_2 battery by adding various concentrations of a TBA salt to the electrolyte (which consisted of a 3 : 1 weight mixture of DME and PC, or of PC alone). Adding more than 50 mol% $TBASO_3CF_3$ to $LiSO_3CF_3$ resulted in a loss of discharge capacity, but a mere 5 mol% TBA in PC/DME increased the capacity from 732 mAh/g_{carbon} to 1068 mAh/g_{carbon} at a geometrical current density of 0.2 mA/cm^2 (the corresponding mass current density was not reported). The authors assigned this capacity improvement to the formation of more dense Li-(per)oxides (Li_2O_2, Li_2O), and to a larger Li_2O yield. The latter involves the exchange of four electrons per mole of Li_2O precipitate, against two electrons for each mole of Li_2O_2, and a correspondingly larger capacity. This capacity enhancement was expected to be promoted by the formation of a soluble TBA-intermediate (tetrabutylammonium-peroxide, TBA_2O_2) that would help to reduce the Li_2O_2 produced initially upon discharge into Li_2O, according to

$$Li_2O_2(s) + 2TBA^+(sol) \rightarrow TBA_2O_2(sol) + 2Li^+(sol) \quad (12)$$

$$TBA_2O_2(sol) + 4Li^+(sol) + 2e^- \rightarrow 2Li_2O(s) + 2TBA^+(sol) \quad (13)$$

Additionally, the discharge curves recorded in the electrolyte containing TBA started at a potential similar to that observed without tetrabutylammonium ($\approx 2.6\,V_{Li}$), but as the discharge proceeded, they featured a second potential plateau some 100 mV below this initial value (at $\approx 2.5\,V_{Li}$). While the authors related this lower voltage plateau to the TBA-assisted reduction of the initial Li_2O_2 into Li_2O, it is worth noting that this reaction has recently been ruled out of the ORR mechanism by Peng et al. [24]. However, these authors performed their electrochemical and Surface Enhanced Raman Spectroscopy (SERS) measurements in TBA-free acetonitrile (see Section 2.1).

Finally, Zhang's work showed that these effects were reproduced when TBA^+ was replaced with the N-methyl-N-butylpyrrolidinium cation (Pyr_{14}^+) in room-temperature ionic liquid $Pyr_{14}TFSI$. Unfortunately, the effects of the addition of these cations on the recharge potential and system's rechargability were not reported by the authors.

Interestingly, an improvement in Li-(per)oxide solubility upon inclusion of an additive in the electrolyte has recently been reported by De Giorgio et al. [68], who added tris(pentafluorophenyl) borane (TPFBP) to a solution of LiTFSI in $Pyr_{14}TFSI$. TPFBP can coordinate the anions resulting from O_2 electroreduction ($O_2^{\cdot-}$, O_2^{2-}). Its addition caused the disappearance of the peak corresponding to the oxidation of Li_2O_2 in a cyclic voltammogram (CV) recorded at 20 mV/s (and that was present on the CV at 200 mV/s). Moreover, this result agrees with the effects in actual Li–O_2 cells reported by Xu et al. [69]; in the latter paper, though, the addition of TPFBP did not result in an improvement in the discharge capacity, because the borate increased the viscosity of the electrolyte and decreased its contact angle with the surface of carbon. This hampered the creation of the triple-phase regions (O_2/electrolyte/carbon) required for the formation of Li precipitates and prevented a capacity improvement.

3.5 Capacity-Limiting Mechanisms

Beyond any doubt, the specific capacity of the oxygen electrode is one of the fundamental parameters that makes the Li–O_2 batteries attractive for their energy density [3,4]. This can be related to the kinetics only through the different catalyst selectivities possible, (i.e., how many electrons per mole of oxygen the catalyst could provide during discharge). Nevertheless, due to its importance, we also discuss this argument briefly.

The first parameter affecting the specific capacity is the electrode thickness; due to oxygen diffusion, limitations on excessive thickness result in decreased capacity [5,59]. In the absence of diffusion limitations, the growth of solid discharge products could be limited by the specific surface of the electrode material (usually, carbon) or by the electrode porosity. Trying to understand both influences, it is instructive to examine the possible reaction mechanisms during the growth of discharge deposits on the surface of carbon (or of a catalyst in the case of catalyzed cathodes). For the sake of simplicity, we first examine the limiting growth mechanisms when forming Li_2O_2, shown to occur during the first discharge in ether-based electrolytes [26,55,70]. In the initial phase of the reaction, lithium ions, oxygen, and electrons will react on the pristine carbon surface, forming a LiO_2 intermediate that disproportionates rapidly to oxygen and a solid peroxide layer (Fig. 6a and b). The growth of a solid discharge film on the carbon surface can occur by two conceptually different processes: at the film–solution interface, via the diffusion limitation of Li^+ and O_2 (see Fig. 6a); in this case, no catalytic effect of the underlying surface would be expected after the deposition of several monolayers; and (2) at the carbon–film interface, via electron-transport

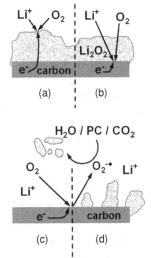

(a)   (b)

(c)   (d)

FIGURE 6 Possible reaction mechanisms of the discharge reaction at the cathode of a Li–O_2 cell in aprotic organic electrolytes. (a) ORR at the interface between deposited Li_2O_2 and the solvent, limited by the diffusion of lithium ions and molecular oxygen through Li_2O_2; (b) ORR at the interface between deposited Li_2O_2 and the carbon surface, limited by electron conduction through the Li_2O_2 deposit; reaction of superoxide ion radical with contaminants ($H_2O/PC/CO_2$) and formation of (c) soluble reaction products or (d) formation of a nonhomogeneous solid layer on the carbon surface. (From [73].)

limitation through the deposit (see Fig. 6b); in the second case, the catalyst still plays a role. The straightforward conclusion is that the capacity would be proportional to the specific surface area of the carbon. In this case, the discharge capacity–dependent deposit layer thickness on Vulcan electrodes can be estimated from the approximate external surface area of the electrodes: ≈ 100 m^2/g_{carbon} (assuming spherical carbon particles 30 nm in diameter with a density of ≈ 2 g/cm^3 [71]). Knowing the Li_2O_2 crystal structure [72], one monolayer of it provides ≈ 73 mAh/g_{carbon} (≈ 260 $\mu C/cm^2_{carbon}$). The charge for one monolayer of LiO_2 was estimated to be ≈ 200 $\mu C/cm^2_{carbon}$ [65], corresponding to ≈ 50 mAh/g_{carbon} per monolayer. In a recent paper [73] we correlated the discharge capacity in water-free DME (≈ 200 mAh/g_{carbon}) with the formation of only three or four monolayers of Li_2O_2 or LiO_2, respectively, able to stop the discharge reaction at a rate of 120 mA/g_{carbon}. On the other hand, in the presence of a reactive electrolyte (e.g., PC) or a contaminant such as water, $O_2^{\cdot -}$ reacts with them preferentially, leading to two possible scenarios. In the first, the reaction products could be slightly soluble in the electrolyte and diffuse into the bulk of the solution, leaving the carbon surface free (Fig. 6c); evidence for this is given in a recent article by Mitchell et al. [57], where the formation of Li_2O_2 particles disconnected from the carbon substrate of the electrode was observed. The second scenario occurs when the eventual solid product formed on the electrode grows in the form of a cracked layer (Fig. 6d) that exposes the underlying cathode somewhat to the electrolyte. The latter is similar to the case of the oxygen reduction reaction in a proton-exchange membrane fuel cell at $-20°C$, where the ORR kinetics of platinum are observed throughout the discharge process despite the formation of thick water–ice layers on the surface of the carbon-supported platinum catalyst [74]. Since solid-state diffusion of O_2 through water–ice or through a Li_2O_2 deposit would be expected to be slow at room temperature, a buildup of more than three monolayers would only be possible if the deposit formed were substantially "cracked" rather than being a smooth film (Fig. 6d). In all these cases, the specific capacity would not be related to the specific surface but to the total porosity of the electrode.

3.6   Electrolyte Reactivity and Alternative Reaction Mechanisms in ORR and OER

Under aprotic conditions the $O_2^{\cdot-}$ is a strong nucleophile and a moderated one-electron oxidant with a closed coordination sphere; its reactivity is also dependent on its level of solvation in the aprotic solvent. ORR kinetics in aprotic media have been studied extensively in the past in the scope of superoxide stabilization and to drive reactions on organic substrates, using such solvents as dimethyl sulfoxide, dimethyl-formamide, and, especially, acetonitrile [21,22,64,75–77]. In the Li–O$_2$ battery field, the study is again directed toward $O_2^{\cdot-}$ stabilization, although under the perspective of electrochemical reaction with the Li$^+$ cations. As has already been studied in the field of Li-ion batteries, all organic electrolytes have potential windows within which the electrolyte corrosion currents can be considered insignificant [78]. Nevertheless, the presence of oxygen radicals in a Li–O$_2$ cell transforms the chemistry of the system, affecting the electrolyte stability window. Therefore, the compatibility between organic solvent and $O_2^{\cdot-}$ has to be studied in more detail. Catalyst evaluation can be carried out only when the electrolyte decomposition is not a dominant process and the reaction kinetics are toward the desirable products. Several recent publications on this subject present the problem of using the most common organic solvents in Li batteries, trying to decipher the solvent decomposition kinetics that seem to represent the prevailing process [19,55,79–81].

Electrolyte stability during charge is at the moment not studied as extensively as during discharge. The lack of O$_2$ radicals means that the electrolyte decomposition is strictly potential dependent and as long as the upper potential voltage is not too high, the electrolyte corrosion currents can be considered insignificant. Nevertheless, the corrosion currents for each organic solvent should be evaluated individually for each composite electrode. As the electrolyte decomposition reaction is catalytic, factors such as electrode surface area, number of active sites, and catalyst activity are parameters that would strongly influence the electrolyte stability and could lead to high electrolyte degradation.

The main problem in studying ORR or OER catalysts seems to lie in the cathode during battery discharge, when lithium cations and oxygen radicals coexist in the positive electrode. Of course, studying the cathode separately from the anode is only possible when a reference electrode is incorporated in the cell; furthermore, the two sides of the cell should be separated by a stable Li-conducting ceramic material, preventing diffusion of the decomposition products from the anode to the cathode. To date, neither of these two prerequisites has been presented in the literature.

On the cathode side during discharge, the two mechanisms through which the superoxide ion radical may react with organic substrates are (1) proton/hydrogen extraction via nucleophilic attack and/or (2) electron transfer [82,83]. In a recent publication by Bryantsev et al., [79] the stability of the most commonly used solvents is predicted using density functional theory. Their results are in good agreement with published experimental data. The use of PC as a solvent is a good example of the difficulties of catalyst evaluation in a system that employs a reactive electrolyte. This solvent is used widely due to its low vapor pressure and the fact that it forms a steady solid electrolyte interphase (SEI) vs. lithium [84]. Nevertheless, as has recently been demonstrated, this organic solvent is unstable against oxygen reduction products formed in the air cathode of Li–O_2 batteries [55,56,81]. Evidence from Fourier transform infrared (FTIR) and nuclear maganetic resource (NMR) spectroscopic data indicate that the main product of the discharge is not Li_2O_2 but several lithium–organic species [81]. Mass spectrometry results support these data, with the main evolutionary product detected during battery charging being CO_2 instead of O_2 [55,81]. With this in mind, it is easy to understand the change in the ORR activity of Pt and Au, depending on the type of electrolyte used (PC/DME or pure DME), as reported by Lu et al. [9,48].

The most stable organic solvents for Li-ORR products, according to the density functional theory (DFT) studies, seem to be the glymes [79]. Although they are not considered stable after a few cycles [80] and do not form a steady SEI vs. Li [85], they represent the most reliable electrolyte systems at the moment and should therefore be used for cathode catalyst evaluation.

4   OXYGEN EVOLUTION REACTION KINETICS

Earlier we have shown that the reduction of oxygen upon battery discharge typically occurs some 300 to 500 mV below the thermodynamic potential values (i.e., between ≈ 2.6 and ≈ 2.4 V_{Li}; see Section 3.1), subsequent electrooxidation of the ORR products upon recharge on bare carbon typically requires operative voltages above 4.0 V_{Li}. Examples of this large overpotential on carbon-based materials with largely different surface areas and morphologies include ≈ 4.3 V_{Li} using carbon nanofibers (and recharging at a current density of 43 mA/g_{carbon}) [57], ≈ 4.7 V_{Li} with Super S carbon (45 m^2/g, 70 mA/g_{carbon}) [42], and ≈ 4.4 V_{Li} with Vulcan XC-72 carbon (240 m^2/g, 150 mA/g_{carbon}) [4].

Of course, this large OER overpotential also results in a remarkable loss of efficiency for the battery pack and can additionally lead to the electrooxidation of the cell's electrolyte, ultimately affecting its cyclability. Li–O_2 cells with acceptable round-trip efficiency and durability will therefore require appropriate catalysts for this OER; it is thence understandable that an increasing number of works in the literature deal with the development of such materials and an understanding of the variables that govern their catalytic activity.

4.1   Oxide Catalysts

The use of metal oxides as ORR and OER catalysts in Li–O_2 cells was introduced by Read [25], who prepared electrodes by mixing λ-MnO_2 and Super P carbon and observed a first recharge plateau at ≈ 4.0 V_{Li} (imposing a geometrical current density of 0.1 mA/cm^2; the current normalized per mass of material was not reported). This seminal work was followed by several papers from Bruce's group. In the first of these publications, Ogasawara et al. [43] mixed electrolytic manganese dioxide with Super S carbon black and ball-milled Li_2O_2, observing a stable potential of ≈ 4.3 V_{Li} upon charging at a rate of 10 mA/g_{carbon}. X-ray diffraction showed that this process involved the disappearance of the peroxide, while in situ mass spectrometry confirmed the evolution of oxygen upon charging, due to oxidation of the initial Li_2O_2 into Li^+ and O_2 [equation (1)].

In their next paper, Bruce's group presented the results from the screening of several oxide-based catalysts [42]. The recharge potentials obtained by mixing 1 to 5-μm particles of Fe_3O_4, CuO, or $CoFe_2O_4$ with Super S carbon were similar to those observed in Ogasawara et al.'s previous work [43] (i.e., ≈ 4.3 V_{Li}, yet now charging at a higher rate of 70 mA/g_{carbon}), but when the catalyst of choice was Co_3O_4, this potential fell to ≈ 4.0 V_{Li}. Next, Débart et al. [8] compared the cell performance of several manganese oxides (including various manganese dioxide phases and morphologies, plus Mn_2O_3 and Mn_3O_4) mixed with Super P. The best results were obtained using α-MnO_2 nanowires, which again catalyzed the recharge at a steady potential of ≈ 4.0 V_{Li}. Similar recharging potential has also been reported by Garsuch et al. [86] using a commercial MnO_2 mixed with Ketjen black carbon (cycled at a current density of 67 mA/g_{carbon}).

This positive catalytic effect of manganese dioxides on a cell's recharge potential motivated several other works featuring MnO_2-based catalysts. Recently, Li et al. [87] prepared MnO_2 nanoflakes supported on multiwalled carbon nanotubes and observed recharging potentials of only ≈ 3.85 and ≈ 3.90 V_{Li} at charging rates of 70 and 180 mA/g_{carbon}, respectively. Ishihara's group has also published two works using α-MnO_2 as the cathode catalyst [47,49]; in their case, though, 12 wt% palladium was also added to the mixture of manganese oxide and carbon. The presence of this noble metal brought the recharging potential down by ≈ 100 mV: from the ≈ 3.7 V_{Li} observed with MnO_2 alone to the ≈ 3.6 V_{Li} measured after Pd was added to the cathode (with both charging potentials registered at a rate of ≈ 26 mA/g_{carbon}).

Cheng and Scott have also published two papers using manganese oxide as an ORR/OER catalyst [46,50]. In the first of these works [46], the authors prepared 20 to 50-nm particles of MnO_x supported on two different carbon blacks (Norit and Super P) and one acetylene black. While the capacity upon discharge increased with the porosity of the support, the average recharging potential of ≈ 4.1 V_{Li} (at 70 mA/g_{carbon}) was similar for all three carbons, as one would expect from the purely catalytic effect of the manganese oxide. In their second publication [50], MnO_2 was only part of a comparison between various noble metals and their oxides, including Pt, Pd, PdO, Ru, and RuO_2. It is worth noting that PdO and RuO_2 featured average recharging potentials of ≈ 3.7 and ≈ 4.0 V_{Li} (again upon recharge at 70 mA/g_{carbon}), respectively. Focusing on Pd and PdO, the paper concluded that the oxide was more adequate than its reduced counterpart because it suffered from fever aggregation issues, even if it delivered a lower capacity in the first discharge, and thus featured better capacity retention upon cycling.

Interestingly, direct charging of electrodes prefilled with Li-peroxide first presented by Ogasawara et al. [43] has recently been extended to some of the metal-oxide catalysts previously featured in reference 42, in a collaboration between the Tarascon and Bruce groups [27]. After mixing with Super P and Li_2O_2, the metal oxides manage to catalyze the oxidation of the Li-peroxide at potentials 200 to 700 mV lower than those required to recharge the reduction products in discharged cells [42] (Fig. 7). This difference between Li_2O_2 oxidation and actual recharge potentials

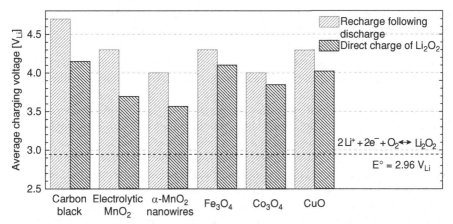

FIGURE 7 Comparison of the average charging voltages recorded with various metal oxide catalysts, upon recharge of cells that had previously been discharged, or when electrodes prefilled with Li_2O_2 are directly charged. The average recharging potentials were extracted from the works of Débart et al. [8,42]. The voltages for the direct oxidation of Li_2O_2 were compiled from a paper by Giordani et al. [27], and for Fe_3O_4, from work by Xu et al. [69]. Besides of the latter work, all cell tests were performed using a solution of 1 M $LiPF_6$ in propylene carbonate as the electrolyte, and (re)charging at a rate of 70 mA/g_{carbon} (as compared to 1 M LiTFSI in a 1 : 1 weight mixture of PC and EC, and a charging rate of 31 mA/g_{carbon} in Xu et al's work).

suggests that the species formed in the initial discharge of a $Li–O_2$ cell do not consist exclusively of Li_2O_2, and must include solid products that are harder to oxidize than this peroxide. Indeed, all of these catalyst-related studies were performed using carbonate-based electrolytes; as outlined above (see Section 3.6), this electrolyte decomposes upon reaction with the superoxide radical formed during cell discharge to produce lithium carboxylates and Li_2CO_3 [81]. For electrodes prefilled with the latter carbonate and also containing electrolytic MnO_2, the potential required for its oxidation upon charge at 70 mA/g_{carbon} spanned between ≈ 3.7 and ≈ 4.2 V_{Li} [81] (as compared to a minimum of 4.2 V_{Li} in reference 88), while Li-propyl dicarbonate, Li-formate, and Li-acetate were oxidized at ≈ 3.6, ≈ 3.7, and ≈ 4.0 V_{Li}, respectively. This distribution of initially unexpected discharge products, each with its corresponding recharge potential, could explain the mixed voltages higher than the 4.0 V_{Li} typically required for cell recharge (Fig. 7).

4.2 Nonnoble Metal Catalysts

Metal oxides may be the OER catalysts of choice in many current $Li–O_2$ studies, but the first work on these novel energy-storage devices featured a nonnoble metal catalyst, based on a heat-treated cobalt N_4-macrocycle. Indeed, the cathode electrodes in the seminal paper by Abrahams and Jiang [2] consisted of Co-phthalocyanine impregnated on Chevron acetylene black and heat-treated in argon at $800°C$. The resulting material displayed a recharging potential of ≈ 3.7 V_{Li}, compared to ≈ 4.0 V_{Li} for the noncatalyzed carbon-based electrode (in both cases upon charging at a geometrical current density of 0.05 mA/cm^2, the corresponding carbon-mass-normalized current density was not reported).

More recently, a catalyst based on heat-treated Cu- and Fe-phthalocyanines has also been featured in two works by Read's group [38,61]. Unfortunately, neither of the papers put an emphasis on the recharge behavior of this material, deciding instead to focus exclusively on its ORR properties. Nevertheless, the authors did observe that this Fe–Cu material catalyzed the chemical disproportionation of hydrogen peroxide (i.e. $H_2O_2 \rightarrow H_2O + \frac{1}{2}O_2$) upon immersion of the powder in an aqueous solution of H_2O_2 [61]. This observation is in good agreement with previous reports on the catalytic activity of resembling materials for this chemical reaction in aqueous medium [89] and would tie in with the work of Giordani et al. [27]. In that paper the authors drew a clear correlation between the H_2O_2-disproportionation activities of several metal oxides and their subsequent performance during charging of cells prefilled with Li_2O_2. Considering this positive correlation and Abraham and Jiang's work, more attention should be paid in the future to the possible catalytic properties of non–noble metal catalysts for the OER in Li–O$_2$ batteries.

4.3 Noble Metal Catalysts

Finally, noble metal catalysts have also been the subject of a number of recent publications. Most of these works were conducted by Shao-Horn's group [3,4,9,26,48], which again focused primarily on the ORR behavior of these noble materials (see Section 3.1). However, in the first of their papers, Lu et al. [4] did compare the OER activity of Au and Pt nanoparticles supported on Vulcan carbon (in both cases with metal loadings of 40 wt%, well above the 10–20 wt% used in most catalyst-related work). After discharging the cells down to 2.0 V_{Li}, the Pt-based catalyst displayed the best recharge performance (with a mean potential of ≈ 3.8 V_{Li} at ≈ 200 mA/g$_{carbon}$), followed by Au and Vulcan carbon (which recharged at ≈ 4.2 and ≈ 4.4 V_{Li}, respectively, at the same current density). These OER-activity trends were confirmed further in direct-charge tests of electrodes prefilled with Li_2O_2, where potentiostatic oxidation at 4.0 V_{Li} yielded net currents of ≈ 1000 mA/g$_{carbon}$ for 40 wt% Pt/C and ≈ 100 mA/g$_{carbon}$ for 40 wt% Au/C (vs. a negligible current < 10 mA/g$_{carbon}$ for noncatalyzed Vulcan).

Following this study and subsequent work on the intrinsic ORR activity of Au, Pt, and glassy carbon surfaces [3], Lu et al. brought together the best ORR and OER catalysts (Au and Pt, respectively) by preparing Pt–Au nanoparticles supported on Vulcan carbon [9]. The resulting catalyst featured an average recharging potential of ≈ 3.6 V_{Li} at 100 mA/g$_{carbon}$, similar to that obtained with Pt nanoparticles alone and 400 mV below the ≈ 4.0 V_{Li} reported for Au nanoparticles. This result, along with the discharge potential (matching that of Au/C, as discussed in Section 3.1), confirmed the bifunctionality of these carbon-supported nanoparticles, where the Au atoms are responsible for the ORR, and subsequent recharge is catalyzed by Pt.

More recently, Cheng and Scott [50] demonstrated recharging potentials of ≈ 3.7, ≈ 3.8, and ≈ 3.9 V_{Li} for Pd, Pt, and Ru particles, respectively, supported on Norit carbon (again using loadings of ≈ 40 wt% metal/C, while recharging at a rate of 70 mA/g_{carbon}). Focusing on Pd and PdO catalysts, while the round-trip efficiency and initial capacity of Pd were better than that of PdO, the particles of Pd tended to agglomerate or get dissolved upon cycling. This resulted in fast fading of the initial capacity, making the authors suggest that the oxides, less affected by these effects, were more adequate for these applications.

Before closing this section, it is worth noting that all of these studies were performed in carbonate-based electrolytes that do not exclusively yield the Li_2O_x expected initially upon cell discharge (see Section 3.6). It is therefore possible that the OER-activity trends reported therein should be different for other (hopefully more stable) electrolytes. That has indeed been the case for the ORR-activity trends reported by Lu et al. [48] in their most recent work using 1,2-dimethoxyethane (DME) as the electrolyte. These authors found Pt to be more ORR-active than Au (compared to the opposite behavior when the measurements were performed in a mixture of this DME with PC [4,9]). To summarize the results of OER catalysis, Table 2 indicates the average recharge potential (first cycle) of Li–O_2 batteries using various metal oxides and reduced noble metals, and Table 3 shows the average charging potentials in catalyzed cathodes prefilled with Li_2O_2.

5 CONCLUSIONS

In this chapter, dealing with the oxygen electrode kinetics on Li–Air cells, we began by discussing the features of various types of Li–air batteries, their thermodynamics, and their kinetic mechanisms. The presence of water or contaminants from the environment makes such batteries more difficult to recharge and more complicated in design. Thus, we focus on the aprotic Li–O_2 cell type, since this is the most studied and the most likely to be rechargeable, provided that the contaminations from the ambient are avoided. For this type of cell the kinetic mechanisms for ORR and OER have recently been clarified for a gold surface in acetonitrile, but much more work is needed to understand the kinetic path for all the other types of catalysts and electrolytes. A predictive tool for ORR catalytic activity has recently been established in the case of noble metals and carbon using DME as a solvent. Nevertheless, the most interesting catalysts for ORR are metal oxides and nonnoble metals, both for their higher activity and their low cost. In general, from the evidence reported in the literature, the most likely product formed after discharge in the oxygen electrode is lithium peroxide; this also seems the most easily recharged. Unfortunately, electrolyte reactivity with the discharge products plays a crucial role, as early as on the ORR kinetic path, making it very difficult to draw conclusions as to pure ORR kinetic activities. For this reason, an electrolyte with very low reactivity toward ORR products has to be found, and then more work is needed to clarify the limitations in rate capability and cyclability with such an ideal electrolyte for a Li–O_2 system. Also shown in this chapter is the strong dependence of the specific capacity on the choice of electrolyte, most of all when the latter is reactive (e.g., PC), and on the contaminations from the environment. For this reason and for all the very different experimental conditions and setups used in the literature, it is very difficult to draw clear and reproducible correlations between discharge-specific capacities and used materials. Attention must be

TABLE 2 Effect of Various Metal Oxides and Reduced Noble Metals on the Average Recharge Potential (First Cycle) of Li–O$_2$ Batteries

Catalyst	Features	Support	Catalyst loading (wt%)	Electrolyte	Current (mA/g$_{carbon}$)	Potential (V$_{Li}$)	Ref.
Au	≈ 15 m^2/g$_{Au}$	Vulcan, supported	40	1 M LiClO$_4$ in PC/DME	100	4.10	9
CoFe$_2$O$_4$	1 to 5-μm particles	Super S, mixed	23	1 M LiPF$_6$ in PC	70	4.30	42
Co$_3$O$_4$	1 to 5-μm particles	Super S, mixed	35	1 M LiPF$_6$ in PC	70	4.00	42
CuO	1 to 5-μm particles	Super S, mixed	15	1 M LiPF$_6$ in PC	70	4.30	42
Fe$_3$O$_4$	1 to 5-μm particles	Super S, mixed	37	1 M LiPF$_6$ in PC	70	4.30	42
MnO$_2$	Electrolytic (EMD)	Super S, mixed	16	1 M LiPF$_6$ in PC	70	4.30	43
	Electrolytic (EMD)	Super S, mixed	16	1 M LiPF$_6$ in PC	70	4.30	42
	Electrolytic (EMD)	Norit, mixed	45	1 M LiPF$_6$ in PC	70	4.10	46
	α-Phase nanowires	Super P, mixed	16	1 M LiPF$_6$ in PC	70	4.00	8
	α-Phase nanoflakes	MWCNTs, supported	40	1 M LiPF$_6$ in PC/EC/DME	70	3.85	58
	α-Phase nanoflakes	MWCNTs, mixed	44	1 M LiPF$_6$ in PC/EC/DME	70	4.00	58
	20 to 50-nm nanoparticles	Norit, supported	45	1 M LiPF$_6$ in PC	70	4.20	46
	20 to 50-nm nanoparticles	Super P, supported	45	1 M LiPF$_6$ in PC	70	4.00	46
Pd	< 25-nm nanoparticles	Norit, supported	40	1 M LiPF$_6$ in PC	70	3.90	50
PdO	≈ 100-nm particles	Norit, supported	40	1 M LiPF$_6$ in PC	70	3.70	50
Pt	Unspecified	Norit, supported	40	1 M LiPF$_6$ in PC	70	4.10	50
	≈ 23 m^2/g$_{Au}$	Vulcan, supported	40	1 M LiClO$_4$ in PC/DME	100	3.60	9
PtAu	4 to 10-nm nanoparticles	Vulcan, supported	40	1 M LiClO$_4$ in PC/DME	100	3.60	9
Ru	Unspecified	Norit, supported	40	1 M LiPF$_6$ in PC	70	4.00	50
RuO$_2$	Unspecified	Norit, supported	40	1 M LiPF$_6$ in PC	70	4.10	50

TABLE 3 Average Charging Potentials in Catalyzed Cathodes Prefilled with Li_2O_2

Catalyst[a]	Features	Support	Catalyst loading (wt%)	Electrolyte	Current (mA/g_{carbon})	Potential (V_{Li})	Ref.
Au*	≈ 13 m^2/g_{Au}	Vulcan, supported	40	1 M $LiClO_4$ in PC/DME	100	4.00	4
Co_3O_4	High-surface-area nanoparticles	Super P, mixed	63	1 M $LiPF_6$ in PC	70	3.85	27
CuO	Mesoporous	Super P, mixed	63	1 M $LiPF_6$ in PC	70	4.00	27
Fe_2O_3	α-Phase, mesoporous	Super P, mixed	63	1 M $LiPF_6$ in PC	70	4.15	27
Fe_3O_4	< 50-nm nanoparticles	Super P, mixed	18	1 M LiTFSI in PC/EC	31	4.10	69
MnO_2	Electrolytic (EMD)	Super S, mixed	29	1 M $LiPF_6$ in PC	10	4.30	43
	Electrolytic (EMD)	Super P, mixed	63	1 M $LiPF_6$ in PC	70	3.70	27
	α-Phase nanowires	Super P, mixed	63	1 M $LiPF_6$ in PC	70	3.60	27
	α-Phase, bulk	Super P, mixed	63	1 M $LiPF_6$ in PC	70	3.65	27
NiO	Mesoporous	Super P, mixed	63	1 M $LiPF_6$ in PC	70	4.15	27
Pt*	≈ 80 m^2/g_{Au}	Vulcan, supported	40	1M $LiClO_4$ in PC/DME	> 200	4.00	4

[a]The charging experiments with an asterisk were performed potentiostatically, conversely to the galvanostatic procedure followed in all other work.

paid to control all the experimental conditions to improve reproducibility and separate the different variables that affect the working performances. Much more research must be carried out on OER catalysts, since OER is the electrochemical reaction that presents the highest overpotential to be decreased (nearly 1 V, compared to less than 0.3 V for ORR), with the simultaneous beneficial secondary effect of decreasing the electrolyte decomposition during charge. Additional theoretical studies are needed to understand some structural properties, also to be used as predictive tools for OER kinetic activity.

REFERENCES

1. F. T. Wagner, B. Lakshmanan, and M. F. Mathias, *J. Phys. Chem. Lett.*, **1**, 2204 (2010).
2. K. M. Abraham and Z. A. Jiang, *J. Electrochem. Soc.*, **143**, 1 (1996).
3. Y.-C. Lu, H. A. Gasteiger, E. Crumlin, J. R. McGuire, and Y. Shao-Horn, *J. Electrochem. Soc.*, **157**, A1016 (2010).
4. Y.-C. Lu, H. A. Gasteiger, M. C. Parent, V. Chiloyan, and Y. Shao-Horn, *Electrochem. Solid-State. Lett.*, **13**, A69 (2010).
5. S. D. Beattie, D. M. Manolescu, and S. L. Blair, *J. Electrochem. Soc.*, **156**, A44 (2009).
6. T. Kuboki, T. Okuyama, T. Ohsaki, and N. Takami, *J. Power Sources*, **146**, 766 (2005).
7. X.-H. Yang, P. He, and Y.-y. Xia, *Electrochem. Commun.*, **11**, 1127 (2009).
8. A. Débart, A. J. Paterson, J. Bao, and P. G. Bruce, *Angew. Chem. Int. Ed.*, **47**, 4521 (2008).
9. Y.-C. Lu, Z. Xu, H. A. Gasteiger, S. Chen, K. Hamad-Schifferli, and Y. Shao-Horn, *J. Am. Chem. Soc.*, **132**, 12170 (2010).
10. G. Girishkumar, B. McCloskey, A. C. Luntz, S. Swanson, and W. Wilcke, *J. Phys. Chem. Lett.*, **1**, 2193 (2010).
11. A. Meitav and E. Peled, *J. Electroanal. Chem.*, **134**, 49 (1982).
12. E. Peled, in *Lithium Batteries.* J.-P. Gabano, Ed., Academic Press, New York, 1983, pp. 43–72.
13. D. Rahner, S. Machill, and K. Siury, *J. Power Sources*, **68**, 69 (1997).
14. M.-K. Song, S. Park, F. M. Alamgir, J. Cho, and M. Liu, *Mater. Sci. Eng. R*, **72**, 203 (2011).
15. J.-S. Lee, S. Tai Kim, R. Cao, N.-S. Choi, M. Liu, K. T. Lee, and J. Cho, *Adv. Energy Mater.*, **1**, 34 (2011).
16. C. O'Laoire, in *Chemistry Dissertations*, Paper 17, Northeastern University, Boston, 2010.
17. M. W. Chase, Jr., *J. Phys. Chem. Ref. Data*, Monograph 9, 1506–1510 (1998).
18. C, O'Laoire, S. Mukerjee, K. M. Abraham, E. J. Plichta, and M. A. Hendrickson, *J. Phys. Chem. C*, **113**, 20127 (2009).
19. C. O'Laoire, S. Mukerjee, K. M. Abraham, E. J. Plichta, and M. A. Hendrickson, *J. Phys. Chem. C*, **114**, 9178 (2010).
20. B. Kumar, J. Kumar, R. Leese, J. P. Fellner, S. J. Rodrigues, and K. M. Abraham, *J. Electrochem. Soc.*, **157**, A50 (2009).
21. D. T. Sawyer, in *Oxygen Chemistry*, Oxford University Press, New York, 1991, pp. 19–51.
22. D. T. Sawyer and J. L. Roberts, *J. Electroanal. Chem.*, **12**, 90 (1966).
23. J. Hassoun, F. Croce, M. Armand, and B. Scrosati, *Angew. Chem. Int. Ed.*, **50**, 2999 (2011).
24. Z. Peng, S. A. Freunberger, L. J. Hardwick, Y. Chen, V. Giordani, F. Bardé, P. Novák, D. Graham, J.-M. Tarascon, and P. G. Bruce, *Angew. Chem. Int. Ed.*, **50**, 6351 (2011).
25. J. Read, *J. Electrochem. Soc.*, **149**, A1190 (2002).
26. Y.-C. Lu, D. G. Kwabi, K. P. C. Yao, J. R. Harding, J. Zhou, L. Zuin, and Y. Shao-Horn, *Energy Envlron. Sci.*, **4**, 2999 (2011).
27. V. Giordani, S. A. Freunberger, P. G. Bruce, J.-M. Tarascon, and D. Larcher, *Electrochem. Solid-State. Lett.*, **13**, A180 (2010).
28. R. P. Hamlen and T. B. Atwater, in *Handbook of Batteries*, 3rd ed., D. Linden and T. B. Reddy, Eds., McGraw-Hill, New York, 2002, pp. 38.46–38.48.

29. S. J. Visco, B. D. Katz, Y. S. Nimon, and L. C. De Longhe, U.S. Patent 72,82,295, 2007.
30. S. J. Visco, E. Nimon, and L. C. De Longhe, in *Encyclopedia of Electrochemical Power Sources*, J. Garche, Ed., Elsevier, Amsterdam, 2009, pp. 376–382.
31. Y. Wang and H. Zhou, *J. Power Sources*, **195**, 358 (2010).
32. T. Zhang, N. Imanishi, Y. Shimonishi, A. Hirano, J. Xie, Y. Takeda, O. Yamamoto, and N. Sammes, *J. Electrochem. Soc.*, **157**, A214 (2010).
33. C. H. Hamann, A. Hamnett, and W. Vielstich, in *Electrochemistry*, Wiley-VCH, Weinheim, Germany, 2007, p. 346.
34. B. B. Blizanac, C. A. Lucas, M. E. Galllgher, M. Arenz, P. N. Ross, and N. M. Markovic, *J. Phys. Chem. B*, **108**, 625 (2004).
35. G. Jürmann, D. J. Schiffrin, and K. Tammeveski, *Electrochim. Acta*, **53**, 390 (2007).
36. M. Pourbaix, *Atlas of Electrochemical Equilibria in Aqueous Solutions*, National Association of Corrosion Engineers (NACE), Houston, TX, and Centre Belge d'Etude de la Corrosion (CEBELCOR), Brussels, Belgium, 1974.
37. R. Jasinski, *Nature (London)*, **201**, 1212 (1964).
38. X. Ren, S. S. Zhang, D. T. Tran, and J. Read, *J. Mater. Chem.*, **21**, 10118 (2011).
39. M. Piana, M. Boccia, A. Filpi, E. Flammia, H. A. Miller, M. Orsini, F. Salusti, S. Santiccioli, F. Ciardelli, and A. Pucci, *J. Power Sources*, **195**, 5875 (2010).
40. S. Catanorchi and M. Piana, Patent WO 2009/124905 A1, 2009.
41. M. Piana, S. Catanorchi, and H. A. Gasteiger, *ECS Trans.*, **16**, 2045 (2008).
42. A. Débart, J. Bao, G. Armstrong, and P. G. Bruce, *J. Power Sources*, **174**, 1177 (2007).
43. T. Ogasawara, A. Debart, M. Holzapfel, P. Novak, and P. G. Bruce, *J. Am. Chem. Soc.*, **128**, 1390 (2006).
44. H. Minowa, M. Hayashi, M. Takahashi, and T. Shodai, *J. Electrochem. Soc. Jpn.*, **78**, 353 (2010).
45. C. S. Johnson, D. W. Dees, M. F. Mansuetto, M. M. Thackeray, D. R. Vissers, D. Argyriou, C. K. Loong, and L. Christensen, *J. Power Sources*, **68**, 570 (1997).
46. H. Cheng and K. Scott, *J. Power Sources*, **195**, 1370 (2010).
47. A. K. Thapa and T. Ishihara, *J. Power Sources*, **196**, 7016 (2011).
48. Y.-C. Lu, H. A. Gasteiger, and Y. Shao-Horn, *J. Am. Chem. Soc.*, **133**, 19048 (2011).
49. A. K. Thapa, Y. Hidaka, H. Hagiwara, S. Ida, and T. Ishihara, *J. Electrochem. Soc.*, **158**, A1483 (2011).
50. H. Cheng and K. Scott, *Appl. Catal. B*, **108–109**, 140 (2011).
51. R. E. Williford and J.-G. Zhang, *J. Power Sources*, **194**, 1164 (2009).
52. M. Eswaran, N. Munichandraiah, and L. G. Scanlon, *Electrochem. Solid-State Lett.*, **13**, A121 (2010).
53. F. Mizuno, S. Nakanishi, Y. Kotani, S. Yokoishi, and H. Iba, *J. Electrochem. Soc. Japan*, **78**, 403 (2010).
54. C. O'Laoire, S. Mukerjee, E. J. Plichta, M. A. Hendrickson, and K. M. Abraham, *J. Electrochem. Soc.*, **158**, A302 (2011).
55. B. D. McCloskey, D. S. Bethune, R. M. Shelby, G. Girishkumar, and A. C. Luntz, *J. Phys. Chem. Lett.*, **2**, 1161 (2011).
56. J. Xiao, J. Hu, D. Wang, D. Hu, W. Xu, G. L. Graff, Z. Nie, J. Liu, and J.-G. Zhang, *J. Power Sources*, **196**, 5674 (2011).
57. R. R. Mitchell, B. M. Gallant, C. V. Thompson, and Y. Shao-Horn, *Energy Environ. Sci.*, **4**, 2952 (2011).
58. Y. Li, J. Wang, X. Li, J. Liu, D. Geng, J. Yang, R. Li, and X. Sun, *Electrochem. Commun.*, **13**, 668 (2011).
59. G. Q. Zhang, J. P. Zheng, R. Liang, C. Zhang, B. Wang, M. Hendrickson, and E. J. Plichta, *J. Electrochem. Soc.*, **157**, A953 (2010).
60. Y. Li, J. Wang, X. Li, D. Geng, R. Li, and X. Sun, *Chem. Commun.*, **47**, 9438 (2011).
61. S. S. Zhang, X. Ren, and J. Read, *Electrochim. Acta*, **56**, 4544 (2011).
62. B. Hammer and J. K. Nørskov, in *Advances in Catalysis*, Vol. 45, H. K. Bruce and C. Gates, Ed., Elsevier, San Diego, CA, 2000, pp. 71–129.
63. D. C. Sorescu, K. D. Jordan, and P. Avouris, *J. Phys. Chem. B*, **105**, 11227 (2001).
64. D. T. Sawyer, G. Chiericato, C. T. Angelis, E. J. Nanni, and T. Tsuchiya, *Anal. Chem.*, **54**, 1720 (1982).

65. D. Aurbach, M. Daroux, P. Faguy, and E. Yeager, *J. Electroanal. Chem.*, **297**, 225 (1991).
66. C. O'Laoire, S. Mukerjee, K. M. Abraham, E. J. Plichta, and M. A. Hendrickson, *J. Phys. Chem. C*, **113**, 20127 (2009).
67. Y.-C. Lu, H. A. Gasteiger, and Y. Shao-Horn, *Electrochem. Solid-State. Lett.*, **14**, A70 (2011).
68. F. De Giorgio, F. Soavi, and M. Mastragostino, *Electrochem. Commun.*, **13**, 1090 (2011).
69. W. Xu, V. V. Viswanathan, D. Wang, S. A. Towne, J. Xiao, Z. Nie, D. Hu, and J.-G. Zhang, *J. Power Sources*, **196**, 3894 (2011).
70. G. Cohn, D. D. Macdonald, and Y. Ein-Eli, *ChemSusChem*, **4**, 1124 (2011).
71. K. Kinoshita, in *Carbon: Electrochemical and Physicochemical Properties*, K. Kinoshita, Ed., Wiley, New York, 1988.
72. L. G. Cota and P. De la Mora, *Acta Crystalloyr.*, **B61**, 133 (2005).
73. S. Meini, M. Piana, N. Tsiouvaras, H. A. Garsuch, and H. A. Gasteiger, *Electrochem. Solid-State. Lett.*, **15**, A45 (2012).
74. E. L. Thompson, J. Jorne, W. B. Gu, and H. A. Gasteiger, *J. Electrochem. Soc.*, **155**, B625 (2008).
75. T. A. Lorenzola, B. A. Lopez, and M. C. Giordano, *J. Electrochem. Soc.*, **130**, 1359 (1983).
76. R. Dietz, A. E. J. Forno, B. E. Larcombe, and M. E. Peover, *J. Chem. Soc. B*, **6**, 816 (1970).
77. F. B. Magno, G. and M. M. Andreuzzi Sedea, *Electroanal. Chem.*, **97**, 85 (1979).
78. F. Ossola, G. Pistoia, R. Seeber, and P. Ugo, *Electrochim. Acta*, **33**, 47 (1988).
79. V. S. Bryantsev, V. Giordani, W. Walker, M. Blanco, S. Zecevic, K. Sasaki, J. Uddin, D. Addison, and G. V. Chase, *J. Phys. Chem. A*, **115**, 12399 (2011).
80. S. A. Freunberger, Y. Chen, N. E. Drewett, L. J. Hardwick, F. Bardé, and P. G. Bruce, *Angew. Chem. Int. Ed.*, **123**, 8768 (2011).
81. S. A. Freunberger, Y. Chen, Z. Peng, J. M. Griffin, L. J. Hardwick, F. Bardé, P. Novák, and P. G. Bruce, *J. Am. Chem. Soc.*, **133**, 8040 (2011).
82. D. T. Sawyer and J. S. Valentine, *Acc. Chem. Res.*, **14**, 393 (1981).
83. A. A. Frimer and I. Rosenthal, *Photochem. Photobiol.*, **28**, 711 (1978).
84. E. Peled, *J. Power Sources*, **9**, 253 (1983).
85. K. Xu, *Chem. Rev.*, **104**, 4303 (2004).
86. A. Garsuch, D. M. Badine, K. Leitner, L. H. S. Gasparotto, N. Borisenko, F. Endres, M. Vracar, J. Janek, and R. Oesten, *Z. Phys. Chem.*, **225**, 1 (2011).
87. J. Li, N. Wang, Y. Zhao, Y. Ding, and L. Guan, *Electrochem. Commun.*, **13**, 698 (2011).
88. A. K. Thapa, K. Saimen, and T. Ishihara, *Electrochem. Solid-State. Lett.*, **13**, A165 (2010).
89. F. Jaouen and J. P. Dodelet, *J. Phys. Chem. C*, **113**, 15422 (2009).

CHAPTER *12*

LITHIUM-ION BATTERIES AND SUPERCAPACITORS FOR USE IN HYBRID ELECTRIC VEHICLES

Catia Arbizzani, Libero Damen, Mariachiara Lazzari,
Francesca Soavi, and Marina Mastragostino

1 INTRODUCTION

The worldwide demand for clean, low-fuel-consuming transport is promoting the development of safe high-energy and power electrochemical storage and conversion systems. The market success of clean transportation by electric (EV) and hybrid electric (HEV) vehicles requires high-efficiency, safe, low-cost electrochemical energy storage and conversion systems. The performance requirements for these systems depend on the level of power train hybridization and on the driving cycle [1,2].

Mass-market penetration of pure EVs with a driving range of at least 500 km calls for batteries able to deliver 500 Wh/kg at the battery pack level. This target is quite far from what is presently featured by lithium-ion batteries, the best-performing batteries on the market, which are expected to approach 200 Wh/kg in the near future by material optimization. New battery chemistries are under study for EV applications, with great attention to safety and costs [3,4]. Batteries are currently used in light electric transport involving a short driving range and in new-generation HEVs with a high level of hybridization, where they power the electric engine allowing the internal combustion one to operate at its most efficient point [1]. Due to their capability to store and deliver charge in a few seconds or more, electrochemical double-layer carbon supercapacitors are presently used only for applications having

Lithium Batteries: Advanced Technologies and Applications, First Edition.
Edited by Bruno Scrosati, K. M. Abraham, Walter van Schalkwijk, and Jusef Hassoun.

high peak to average power demands and for smoothing the strong, short-time power solicitations required in transport (e.g., start/stop, regenerative breaking) [5].

The targets set by the U.S. Advanced Battery Consortium (USABC) and the U.S. Department of Energy (DOE) for power-assisted HEVs (i.e., full HEVs with the highest level of hybridization) are a pulse power of at least 625 W/kg for 10 s over more than 3×10^5 shallow cycles (25 Wh/cycle) and 7.5 Wh/kg of total available energy [2]. Hence, today lithium-ion batteries are considered to be the best candidates for HEV applications, with some concerns about safety. The batteries that employ a carbon-coated lithium iron phosphate cathode, which demonstrates both good thermal stability and specific capacity, are the most promising for safe, large-format modules that are required for power assist in HEVs [6,7]. However, conventional lithium-ion batteries combine high-energy materials with flammable organic solvents and hence can suffer premature failure if subjected to abusive conditions because of spontaneous heat-evolving reactions, which may lead to fires and explosions [8].

Electrochemical double-layer carbon supercapacitors (EDLCs), with positive and negative carbon electrodes charged and discharged by physical processes, are intrinsically more tolerant than batteries to such abusive conditions, and, for the high energy efficiencies even above 95%, can easily dissipate the amount of heat released in the charge–discharge cycles. EDLCs satisfy the USABC power and cycle-life requirements for power-assisted HEVs, but even the best-performing EDLCs on the market, which operate with organic electrolytes and a maximum cell voltage (V_{max}) of 2.7 V, do not meet the specific energy target [5,9].

Ionic liquids (ILs) play a key role in the development of both safe lithium-ion batteries and high-energy EDLCs for HEV applications. The IL class includes a variety of room-temperature molten salts, and pyrrolidinium-based ILs feature a favorable combination of such physicochemical properties as low vapor pressure, high thermal stability, good ionic conductivity, and a wide electrochemical stability window. ILs are also often classified as nonflammable materials, although a more appropriate description of many ILs would be nonvolatile (up to the decomposition temperature) class IIIB combustible materials (flash points above 200°C) [10–16].

Here we compare the performance of a graphite/LiFePO$_4$ battery and that of an EDLC assembled with electrolytes based on N-butyl-N-methylpyrrolidinium bis-(trifluoromethanesulfonyl)imide (PYR$_{14}$TFSI) and N-methoxyethyl-N-methylpyrrolidinium bis(trifluoromethanesulfonyl)imide (PYR$_{1(2O1)}$TFSI) ILs, respectively, tested in view of HEV applications. Table 1 provides some specifications of the systems under study [17,18]. The battery was assembled with a PYR$_{14}$TFSI–LiTFSI 0.4 m electrolyte solution added with 10% w/w of vinylene carbonate (VC) for solid electrolyte interphase (SEI) formation on the graphite anode. PYR$_{14}$TFSI was selected as an electrolyte component because it is thermally stable over 350°C even under an oxygen atmosphere and is more difficult to ignite and burn than is a conventional organic electrolyte [19].

The EDLC featured PYR$_{1(2O1)}$TFSI IL as electrolyte and was assembled without additional solvent, thanks to the low freezing point of the IL ($< -95°C$). PYR$_{1(2O1)}$TFSI also features a 5.0 V-wide electrochemical stability, which allowed us to reach a cell voltage of 3.8 V by and asymmetric configuration with positive and

TABLE 1   Specifications of Battery and EDLC Under Study and Parameters of Their SCT and HPPC Tests[a]

	Battery: Graphite/ LiFePO$_4$	Supercapacitor: EDLC
Electrolyte	90% PYR$_{14}$TFSI–LiTFSI 0.4 m –VC 10% w/w	PYR$_{1(2O1)}$TFSI
σ at 60°C	9.8 mS/cm	8.4 mS/cm
$m_{total\ composite}$	10.4 mg/cm^2	16.8 mg/cm^2
$V_{max} \div V_{min}$	3.6–2.0 V	3.7–2.035 V
	Static Capacity Test (SCT)	
I_{SCT} at 60°C	1C 0.4 mA/cm^2; 19 mA/g$_{module}$	5C 0.9 mA/cm^2; 27 mA/g$_{module}$
$Q_{100\%DOD}$	19 mAh/g$_{module}$	5.0 mAh/g$_{module}$
$E_{100\%DOD}$	51 Wh/kg$_{module}$	15 Wh/kg$_{module}$
	Hybrid Pulse-Power Characterization (HPPC) Test–10 s	
I_{dis} at 60°C	5C	34 C
$I_{reg} = 75\%\ I_{dis}$	3.45 mA/cm^2; 165 mA/g$_{module}$	5 mA/cm^2; 150 mA/g$_{module}$
DOD$_{min} \div$ DOD$_{max}$	10–70%	7–84%

[a]Electrolyte compositions and conductivities (σ), total positive and negative composite electrode mass ($m_{total\ composite}$), maximum and minimum cutoff voltages ($V_{max} \div V_{min}$), SCT (I_{SCT}) discharge current, specific capacities ($Q_{100\%\ DOD}$), and specific energies ($E_{100\%\ DOD}$) delivered upon discharges with 100% depth of discharge (DOD), HPPC discharge (I_{dis}), and regenerative (I_{reg}) pulse currents and DOD window [17,18].

negative electrodes of different carbon loadings [20]. Note that the high cell voltage was achievable because in EDLCs the electrode potential varies along the capacitive charge process and is limited only by IL oxidation or reduction. In contrast, in battery the cell voltage is set by the lithium insertion potentials of the LiFePO$_4$ and graphite electrodes, and for the graphite/LiFePO$_4$ battery is about 3.2 V. Table 1 shows that at 60°C, which is a high-temperature condition of HEVs, the inherent conductivity of PYR$_{1(2O1)}$TFSI closely approaches that of the PYR$_{14}$TFSI–LiTFSI 0.4 m 90%–VC 10% solution.

The IL-based battery and EDLC were tested by protocols that were developed to simulate the dynamic functioning of the systems in power-assisted full HEV, where, typically, the battery is used during acceleration for a short time and kept within a DOD range (never approaching the fully charged or fully discharged state) by regenerative braking or the engine. These benchmarks include the static capacity test (SCT), which provides the total available energy of the system and the hybrid pulse-power characterization (HPPC), which gives the dynamic power and energy capabilities at various states of charge and depths of discharge (DODs). The tests together provide the dynamic-power capability over the available energy of the systems [21–23].

2   EXPERIMENTAL PROCESS

The battery was assembled as in reference 17 with 0.4 m LiTFSI (3M)–PYR$_{14}$TFSI (Solvent Innovations)–10% w/w VC (Fluka). LiFePO$_4$ was synthesized as in reference 7, and the LiFePO$_4$ electrodes (0.61 cm^2, composite mass $= 5.4$ mg/cm^2) were prepared by lamination on a carbon-coated aluminum grid (Lamart) of a 80 wt% LiFePO$_4$, 15 wt% carbon-conducting additive (SuperP, MMM Carbon Co.), and 5 wt% pTFE composite.

The battery anode (0.61 cm^2, composite mass $= 5.0$ mg/cm^2) featured Graphite Timrex KS-15 (TIMCAL)/SuperP/PVDF $= 70 : 10 : 20$ wt% composition and a copper foil current collector.

The EDLC was assembled as in reference 18, with PYR$_{1(2O1)}$TFSI (Evonik Industries) and composite electrodes (0.64 cm^2) with 95% w/w carbon–5% w/w pTFE and carbon-coated aluminum current collectors. The carbon was PICACTIF SUPERCAP BP10 (Pica) treated at 1050°C in Ar for 2 h; the carbon displayed a total pore volume of 1.2 cm^3/g with a pore size distribution centered at 2.7 nm. The positive and negative electrodes featured carbon-binder composite loadings of 10.5 and 6.3 mg/cm^2, respectively.

The battery and the EDLC featured a Swagelok-type configuration and were assembled in a dry box (MBraun Labmaster 130, H$_2$O and O$_2$ < 1 ppm) using a fiberglass filter as a separator (Durieux, 200 μm thick when pressed); the separator and electrodes were soaked under vacuum with the IL before assembly. The electrochemical tests were performed in a Thermoblock (FALC) oven at 60°C using a Perkin-Elmer VMP multichannel potentiostat. The conductivity measurements were performed by a Radiometer Analytical CDM210 conductivity meter.

3 RESULTS AND DISCUSSION

The characterization of the IL-based battery and EDLC for HEV applications was performed following the standards stated by the DOE [21,22]. The protocol that we adopted includes the SCT and the HPPC in the temperature range -30 to $+60°C$ and here we report the results obtained at 60°C. According to Stewart et al., [23], the specific parameters evaluated by such tests and reported in the following refer to a total module mass (m_{module}), which is twice the total composite electrode mass (i.e., $m_{module} = 2m_{total\ composite}$), so as to include the other components' weights.

The SCT is carried out to evaluate the available energy of systems at various DODs. It was performed by a deep galvanostatic discharge from the V_{max} voltage down to the minimum cell voltage V_{min} at currents of the same order of magnitude for the battery and EDLC (i.e., 0.4 and 0.9 mA/cm^2, corresponding to 19 and 27 mA/g$_{module}$, respectively). Table 1 reports the SCT cutoff voltages and discharge rates (I_{SCT}) used for the battery and EDLC tests, and Figure 1 compares the corresponding plots of the cell discharge voltage and of the specific cumulative energy removed during discharge vs. DOD. The battery was tested at a 1C rate (i.e., at fast current regimes) and featured a flat discharge voltage plateau at about 3.2 V, which keeps constant up to 85% DOD. In contrast, as a result of the capacitive response of the EDLC, the supercapacitor voltage decreased linearly with discharge depth from $V_{max} = 3.7$ at DOD $= 0\%$ down to a V_{min} of 2.035 V, which corresponded to 55% V_{max} and that, according to the manuals, marked 100% of usable

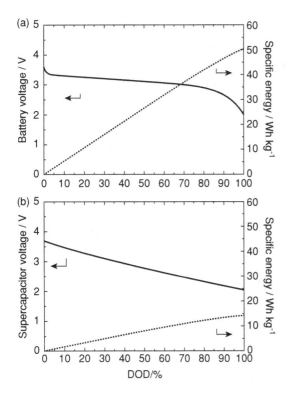

FIGURE 1 SCT cell voltage profile at 60°C of the battery at a 1C rate (a) and of the EDLC at a 5C rate (b). The dotted lines indicate the relationships between cumulative energy and DOD.

DOD. The EDLC discharge current corresponded to a 5C rate, which is very slow for supercapacitors that work on a time scale of seconds. Table 1 also shows the experimental specific capacities ($Q_{100\% \text{ DOD}}$) and the specific energies ($E_{100\% \text{ DOD}}$) delivered upon discharge with 100% DOD. The high specific capacitance of the EDLC, of 13 F/g_{module} (corresponding to a single carbon electrode capacitance of about 100 F/g) brings about a $Q_{100\% \text{ DOD}}$ that is one-fourth that of the battery. Given that more than 60% of the battery discharge takes place at cell voltages higher than 3 V, the battery can deliver a total energy more than three times higher than that of the EDLC.

The HPPC, which provides the dynamic power and energy capabilities over the $V_{max} \div V_{min}$ voltage range by a test profile that incorporates both discharge and regenerative pulses, was performed as follows. The tests begin with a deep charge at a given cell potential, followed by a rest period. Then a sequence that consists of a 10-s discharge pulse at the I_{dis} current, a 40-s (for the battery) or 5-s (for the EDLC) rest period step, a 10-s regenerative pulse at the $I_{reg} = 75\%I_{dis}$ current, and a 1C (for the battery) or 5C (for the EDLC) discharge with 10% DOD is repeated until the V_{min} cutoff potential is reached. Different rest periods were used for the battery and the EDLC to take into account the self-discharge of the latter, which is inherently higher than that of the former. Table 1 reports the HPPC cutoff voltages, the discharge rates (I_{dis}) used for the battery and the EDLC, which are comparable, and the DOD window

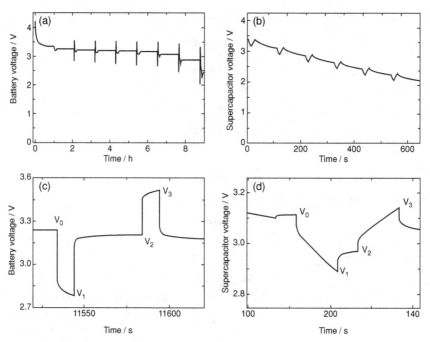

FIGURE 2 HPPC cell voltage profile at 60°C of the battery [(a) and (c)] and of the EDLC [(b) and (d)].

(DOD$_{min}$ ÷ DOD$_{max}$), where the pulse sequence is applied. Figure 2 shows the HPPC cell voltage profiles of the two systems at 60°C. The magnification of these profiles in Figure 2c and d highlights the main differences between a battery and an EDLC under discharge and regenerative pulses. The cell voltage variation of the battery is affected primarily by the ohmic drop or rise at the beginning of the pulses, which in turn is originated mainly by the electrode discharge–charge resistances (including electronic and ionic contributions) and is higher than that featured by a battery with the same electrode materials but with a conventional electrolyte [7]. By contrast, the EDLC voltage variation on the pulse is affected principally by the linear decrease or increase of the cell voltage over time, which is an intrinsic characteristic of the capacitive discharge–charge process of the supercapacitor.

Such differences reflect the discharge (R_{dis}) and regenerative (R_{reg}) pulse resistances, which are calculated by

$$R_{dis} = \frac{V_1 - V_0}{I_{dis}} \tag{1}$$

$$R_{reg} = \frac{V_3 - V_2}{I_{reg}} \tag{2}$$

and that are plotted for each pulse (i.e., for each DOD) in Fig. 3. The resistances of the battery are more than double those of the EDLC. Furthermore, the formers increase with DOD, along with the increase in electrode resistance with the state of discharge.

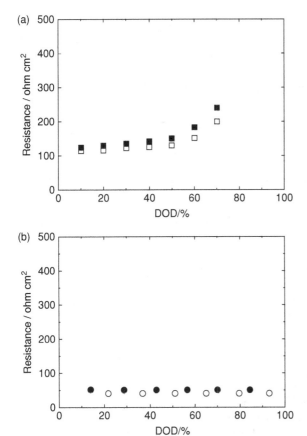

FIGURE 3 R_{dis} and R_{reg} values at different DOD values of the battery (a) and the EDLC (b) evaluated from 10-s HPPC pulses at 60°C.

By contrast, the resistances of EDLC remain constant with DOD, and, being affected mainly by the capacitive cell potential variation, depend on pulse time: the shorter the pulse time, the lower the EDLC R_{dis} and R_{reg} values. R_{dis} and R_{reg} are used to evaluate the specific discharge (P_{dis}) and regenerative (P_{reg}) pulse-power capabilities that are shown in Figure 4 vs. the cumulative energy:

$$P_{dis} = \frac{V_{min}(V_0 - V_{min})}{R_{dis} m_{module}} \tag{3}$$

$$P_{reg} = \frac{V_{max}(V_{max} - V_2)}{R_{reg} m_{module}} \tag{4}$$

Figure 4 shows that both the battery and the EDLC can fulfill the DOE–USABC pulse power capability target for HEV (i.e., $P_{dis} = 625$ W/kg over 10-s pulses with $P_{reg} = 80\%\ P_{dis}$). The battery and the EDLC surpass the pulse power goal in the DOD ranges 10 to 55% and 8 to 67%, respectively, and these correspond to the states

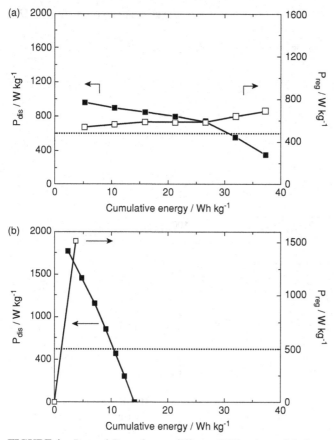

FIGURE 4 P_{dis} and P_{reg} values at different DOD values of the battery (a) and the EDLC (b) evaluated from 10-s HPPC pulses at 60°C; dotted lines indicate the DOE–USABC target for power-assisted HEV.

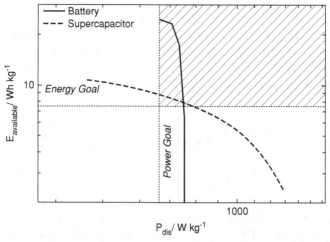

FIGURE 5 Available energy ($E_{available}$) vs. 10-s-discharge pulse-power capability (P_{dis}) of the battery and the EDLC at 60°C.

of charge in which the systems should be maintained during their use in power-assist mode. The corresponding available energies between these DODs are 25 Wh/kg for the battery and 9.2 Wh/kg for the EDLC, above the target of 7.5 Wh/kg.

Figure 5 shows plots of the amount of energy available for a given pulse power vs. the pulse power for the battery and the EDLC. They represent the energy (or power) available over the operating region where a specified power (or energy) demand can be met. The shaded zone, which identifies where energy and power targets are satisfied simultaneously, crosses the performance of the two systems under study. This demonstrates that both an IL-based battery and an EDLC can fulfill the power-assist HEV requisites. In the shaded area of Figure 5, although the usable P_{dis} values are almost the same for the two systems, the $E_{available}$ values span

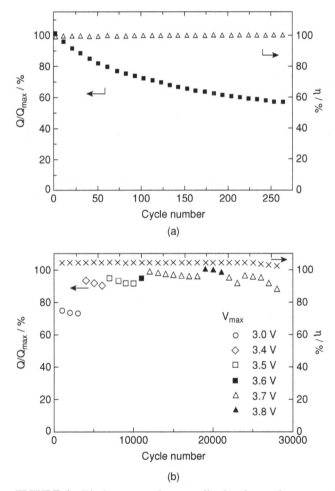

(a)

(b)

FIGURE 6   Discharge capacity normalized to the maximum capacity and efficiency vs. cycle number at 60°C of the battery cycled at 0.4 mA/cm² (1C rate) (a) and of the EDLC cycled at 20 mA/cm² (110 C rate) (b).

a range that is wider for the battery than for the supercapacitor. Hence, although the IL-based battery performs worst than that featuring the same electrode materials but a conventional electrolyte [7], Figure 5 shows that the former is more suitable than the IL-based EDLC for power-assist use in HEVs. However, the battery demonstrates a lower performance retention than that of the EDLC upon cycling in conventional mode (deep galvanostatic charge–discharge). Figure 6 compares at the same operating time ($\sim$20 days) the capacity of the systems normalized to the corresponding highest capacity (Q_{max}) and shows that whereas the EDLC is stable over 27,000 cycles, the battery fades to 45% of the initial capacity after 275 cycles and then seems to stabilize at such a value. However, even after 275 cycles, the cycled battery surpasses the pulse power goal in the range 10 to 45% DOD with an available energy of 9 Wh/kg which is similar to that of the EDLC and is still above the DOE target.

4 CONCLUSIONS

This study demonstrates that the use of ILs as electrolyte components of batteries and supercapacitors is a powerful strategy to develop systems for power assist in full HEVs. Ionic liquids are more difficult to ignite than are conventional solvents; hence, in batteries, the former increase safety and are beneficial to the evolution of a large battery market. In EDLCs, ILs of wide electrochemical stability are allowed to reach high cell voltages and high specific energies, which make supercapacitors potential competitors with lithium-ion batteries for power-assisted HEVs, with the added advantage of inherently higher safety and cycle life. Therefore, whereas in batteries the use of ILs mainly promotes safety, in supercapacitors such use also demonstrates a great advantage in terms of energy.

REFERENCES

1. E. Karden, S. Ploumen, B. Fricke, T. Miller, and K. Snyder, *J. Power Sources*, **168**, 2–11 (2007).
2. http://www.uscar.org.
3. J. B. Goodenough and Y. Kim, *Chem. Mater.*, **22**, 587–603 (2010).
4. M.-K. Song, S. Park, F. M. Alamgir, J. Cho, and M. Liu, *Mater. Sci. Eng. R*, **72**, 203–252 (2011).
5. J. R. Miller and A. F. Burke, *Electrochem. Soc. Interface*, **17**, 53–57 (2008).
6. A. K. Padhi, K. S. Nanjundaswamy, and J. B. Goodenough, *J. Electrochem. Soc.*, **144**, 1188–1194 (1997).
7. S. Beninati, L. Damen, and M. Mastragostino, *J. Power Sources*, **194**, 1094–1098 (2009).
8. P. G. Balakrishnan, R. Ramesh, and T. Prem Kumar, *J. Power Sources*, **155**, 401–414 (2006).
9. A. F. Burke, *Electrochim. Acta*, **53**, 1083–1091 (2007).
10. G. B. Appetecchi, M. Montanino, A. Balducci, S. F. Lux, M. Winter, and S. Passerini, *J. Power Sources*, **192**, 599–605 (2009).
11. A. Lewandoski and A. Swiderska-Mocek, *J. Power Sources*, **194**, 601–609 (2009).
12. M. Mastragostino and F. Soavi, Electrochemical Capacitors: Ionic Liquid Electrolytes, in J. Garche, C. K. Dyer, P. Moseley, Z. Ogumi, D. Rand, and B. Scrosati, Eds., *Encyclopedia of Electrochemical Power Sources*, Vol. 1, Elsevier, Amsterdam, 2009, pp. 649–657.
13. G. B. Appetecchi, M. Montanino, M. Carewska, M. Moreno, F. Alessandrini, and S. Passerini, *Electrochim. Acta*, **56**, 1300–1307 (2011).

14. M. Lazzari, C. Arbizzani, F. Soavi, and M. Mastragostino, EDLCs Based on Solvent-Free Ionic Liquids, in F. Beguin, E. Frackowiak, and M. Lu, Eds., *Supercapacitors: Materials, Systems and Applications*, WILEY-VCH, Weineheim, 2013, pp. 289–306.

15. D. M. Fox, W. H. Awad, J. W. Gilman, P. H. Maupin, H. C. De Long, and P. C. Trulove, *Green Chem.*, **5**, 724–727 (2003).

16. M. Smiglak, M. Reichert, J. D. Holbrey, J. S. Wilkes, L. Sun, J. S. Thrasher, K. Kirichenko, S. Singh, A. R. Katritzky, and R. D. Rogers, *Chem. Commun.*, 2554–2556 (2006).

17. L. Damen, M. Lazzari, and M. Mastragostino, *J. Power Sources*, **196**, 8692–8695 (2011).

18. M. Lazzari, F. Soavi, and M. Mastragostino, *J. Electrochem. Soc.*, **156**, A661–A666 (2009).

19. C. Arbizzani, G. Gabrielli, and M. Mastragostino, *J. Power Sources*, **196**, 4801–4805 (2011).

20. C. Arbizzani, M. Biso, D. Cericola, M. Lazzari, F. Soavi, and M. Mastragostino, *J. Power Sources*, **185**, 1575–1579 (2008).

21. INEEL, *FreedomCAR Battery Test Manual for Power-Assist Hybrid Electric Vehicles*, prepared for the U.S. Department of Energy, 2003.

22. INEEL, *FreedomCAR Ultracapacitor Test Manual*, prepared for the U.S. Department of Energy, 2004.

23. S. G. Stewart, V. Srinivasan, and J. Newman, *J. Electrochem. Soc.*, **155**, A664–A671 (2008).

CHAPTER 13

$Li_4Ti_5O_{12}$ FOR HIGH-POWER, LONG-LIFE, AND SAFE LITHIUM-ION BATTERIES

Zonghai Chen, I. Belharouak, Yang-Kook Sun, and Khalil Amine

1   INTRODUCTION

Among the currently available energy storage technologies, the lithium-ion battery has the highest energy density and hence has received intense attention from both the academic community and industry for use as the power source in hybrid electric vehicles (HEVs), plug-in hybrid electric vehicles (PHEVs), and full electric vehicles (EVs) [1,2]. However, the large-scale deployment of lithium-ion batteries to power HEVs, PHEVs, and EVs is hindered significantly by several major technological barriers: high cost, insufficient life, intrinsically poor safety characteristics, and poor low-temperature performance ($< -20°C$). In particular, the cost of the battery pack to power vehicles will be the major deciding factor to entice consumers to purchase PHEVs, HEVs, or full EVs. In the case of full EVs, the cost of the battery pack can be compensated for significantly by elimination of the internal combustion engine. However, energy-demanding full-EV application requires a very high-performance battery to store enough energy for long driving distances (> 100 miles). The energy density of such a battery pack should be at least double that of state-of-the-art lithium-ion technology so that a battery of reasonable size can be manufactured to meet the

Lithium Batteries: Advanced Technologies and Applications, First Edition.
Edited by Bruno Scrosati, K. M. Abraham, Walter van Schalkwijk, and Jusef Hassoun.

high energy demand of full EVs. An alternative to the full EV is the PHEV, in which a battery with a high energy density and high power as well as an electric motor will be added to a vehicle with a small internal combustion engine. In this case, the battery pack provides energy for full electric driving for a reasonable distance (10 to 40 miles). When the battery pack is drained to a certain level, the vehicle will operate in the internal combustion mode, relying on gasoline for energy, until the battery is recharged. In contrast to the full EV, the cost of the internal combustion engine remains and hence the PHEV cost can be substantially higher than that of conventional vehicles using internal combustion engines. It is physically and economically more practical to develop lithium-ion batteries for HEVs, which require only a small high-power battery to store or deliver a small amount of energy frequently during braking, accelerating, and startup. After about a decade of continuous development, several cell chemistries have been reported [3,4] (e.g., graphite/LiNi$_{0.8}$Co$_{0.15}$Al$_{0.05}$O$_2$, graphite/LiFePO$_4$, graphite/Li$_{1+x}$Mn$_{2-x}$O$_4$, and Li$_4$Ti$_5$O$_{12}$/Li$_{1+x}$Mn$_{2-x}$O$_4$) that can meet the power, life, and cost requirements for the HEV battery [5], whereas a battery technology to meet the requirements of PHEVs and EVs is still under development.

The safety issue is another major barrier that hinders the deployment of lithium-ion batteries for the electrification of automobiles. In a fully charged graphite/LiMO$_2$ (M = transition metal) cell, the cathode is composed of an oxidative transition metal oxide, and the anode is mainly LiC$_6$, whose chemical reactivity similar is to that of lithium metal. Sandwiched between them is a nonaqueous electrolyte, typically a solution of LiPF$_6$ in a mixture solvent of ethylene carbonate and ethyl methyl carbonate, which can be both oxidized and reduced in the wide working potential window of lithium-ion batteries. Kinetically triggering the reaction between the cathode and the electrolyte or between the anode and the electrolyte generally leads to the release of a large amount of heat during a short period of time, potentially leading to fire or explosion of the battery (also called *thermal runaway*). Lots of blame has been cast on the instability of the delithiated cathode, which starts to decompose at above 200°C and generates a large amount of heat, sometimes larger than that generated from an equivalent amount of lithiated graphite [6,7]. However, more attention should be paid to the anode side from a safety perspective, because the solid electrolyte interphase (SEI) on the graphite surface tends to decompose at a temperature as low as 60°C [8]. The SEI layer is an organic–inorganic composite thin film that is formed during the initial insertion of lithium into graphite at about 0.8 V and that suppresses the continuous chemical reaction between lithiated graphite and the electrolyte [9]. When the battery is exposed to a temperature above its critical temperature, the SEI layer will decompose and expose lithiated graphite directly to the nonaqueous electrolyte, resulting in a continuous exothermal reaction between the lithiated graphite and the electrolyte [10]. The continuous heat flow from the reaction can slowly raise the internal temperature of the battery and eventually trigger a major reaction of the cathode at above 200°C. Therefore, the thermal stability of the SEI is critical to the safety of lithium-ion batteries using graphite as the anode.

An alternative anode material, Li$_4$Ti$_5$O$_{12}$, for lithium-ion batteries operates at about 1.55 V [11] (i.e., above 0.8 V vs. Li$^+$/Li), and a stable SEI layer is not required. It has been consistently reported that Li$_4$Ti$_5$O$_{12}$ can offer a significant safety advantage over a graphite anode, and many researchers are attempting to develop high power, long life, and extremely safe lithium-ion chemistries with Li$_4$Ti$_5$O$_{12}$ anodes for HEV applications.

2 SYNTHESIS OF Li$_4$Ti$_5$O$_{12}$

The synthesis and characterization of Li$_{3+y}$Ti$_{6-y}$O$_{12}$ spinels ($0 \leq y \leq 1$) have been under investigation since the 1970s, due to their superconductivity at a relatively high transition temperature [12]. The reversible insertion and removal of lithium into Li$_4$Ti$_5$O$_{12}$ was reported about a decade later [11]. Solid-state synthesis is generally used for the preparation of Li$_4$Ti$_5$O$_{12}$ because of its simplicity, low cost, and well-grown crystals at a relatively high sintering temperature (800 to 1000°C). TiO$_2$ is widely used as the titanium source for this approach, and the typical lithium source can be either LiOH or Li$_2$CO$_3$; TiO$_2$ is mixed with an equivalent amount of lithium source, and the mixture is then sintered at high temperature (800 to 1000°C) for 8 to 12 h [13–15]. Research efforts have shown that the Li$_4$Ti$_5$O$_{12}$ obtained from this solid-state synthesis was an electronic insulator with large, well-grown primary particles. Lithium cells using Li$_4$Ti$_5$O$_{12}$ anodes, whose theoretical capacity is 175 mAh/g, generally delivered a specific capacity of about 100 mAh/g and had poor rate capability [11]. It was believed that the electrochemical performance of Li$_4$Ti$_5$O$_{12}$ could be improved dramatically by carefully controlling the particle size of the final active materials to shorten the diffusion length of the lithium electrons inside the particle. Aldon et al. [16] reported that this goal can be accomplished easily by a sol–gel method. In this method, both soluble titanium salt and lithium salt are added to a solvent such as ethanol to form a precursor gel, and the solvent is then evaporated to obtain nanosized precursor particles, which are then annealed at 800 to 1000°C to obtain the final Li$_4$Ti$_5$O$_{12}$ powder. It has been reported that nanometer-sized materials can approach the theoretical specific capacity and result in good rate capability [4]. However, the energy density for batteries using nanosized Li$_4$Ti$_5$O$_{12}$ anode materials remains problematic, for two reasons. First, Li$_4$Ti$_5$O$_{12}$ operates at 1.55 V vs. Li$^+$/Li, which is about 1.5 V higher than graphitic anodes. Second, high porosity and low loading density are expected for electrodes using nanometer-sized materials and can lead to further reduction in battery volumetric energy density.

A practical approach to increasing the loading density of the Li$_4$Ti$_5$O$_{12}$ electrode is to synthesize materials consisting of micrometer-sized secondary particles composed of nanometer-sized primary particles (see Fig. 1b for an exemplary scanning electron microscopy image illustrating this morphology) [4]; the micrometer-sized particles improve the tapping density and the load density of the electrodes, while the nanometer-sized primary particles maintain all the benefits associated with nano Li$_4$Ti$_5$O$_{12}$ powders. In this approach, TiCl$_4$ or other soluble titanium salt is used as the starting material to prepare the porous TiO$_2$·2H$_2$O precursor. During high-temperature sintering, the porous structure in the precursor is maintained, and the removal of the crystalline water in the precursor during material synthesis helps to maintain a percolated tunnel structure inside the particle; this tunnel structure will uptake nonaqueous electrolyte after the cell is assembled and will increase the accessibility of Li$_4$Ti$_5$O$_{12}$ toward lithium ions.

Further modification of the synthesis method is to add a carbon source such as pitch to the starting material: for example, a mixture of mesoporous TiO_2, $LiOH$, and pitch [17]. During the sintering process, the added carbon source maintains the reducing environment, decomposes at a high temperature, and generates a carbon coating layer on the surface of the final $Li_4Ti_5O_{12}$; the carbon coating can then enhance the electronic conductivity of the materials and further improve the rate capability of lithium-ion batteries using carbon-coated $Li_4Ti_5O_{12}$ anodes.

3 STRUCTURAL INSIGHT OF Li₄Ti₅O₁₂

The $Li_4Ti_5O_{12}$ anode material has a spinel structure of space group $Fd\bar{3}m$, whose unit cell contains eight formula units of $(Li)^{8a}(Li_{1/3}Ti_{5/3})^{16d}O_4^{32e}$, in which the 8a tetrahedral sites are fully occupied by lithium, the 16d octahedral sites are shared between lithium and titanium with an atomic ratio of $1:6$, and the 32e sites are filled with oxygen atoms (see Fig. 1a). One formula unit of $Li_4Ti_5O_{12}$ has the capability of electrochemical and reversible uptake of up to three lithium atoms to deliver a theoretical capacity of 175 mAh/g, converting $Li_7Ti_5O_{12}$ to a rock salt structure (see Fig. 1b). The phase transformation from the spinel to the rock salt structure results in only slight shrinkage of the lattice parameter, from 8.3595 Å to 8.3538 Å, with only a 0.2% change in the cell volume. The $Li_4Ti_5O_{12}$ material is widely considered as zero strain for lithium insertion, and an extremely long cycle life for lithium insertion and removal is expected because its small lattice change can help to maintain the structural stability of $Li_{4+x}Ti_5O_{12}$ and the mechanical integrity of the binder/carbon black matrix that holds the $Li_{4+x}Ti_5O_{12}$ particles together with a good electron-conducting pathway to the current collector.

The electrochemical insertion of lithium atoms into $Li_4Ti_5O_{12}$ is a two-phase process that forms a solid solution between $Li_4Ti_5O_{12}$ and $Li_7Ti_5O_{12}$ and that presents as a long voltage plateau at 1.55 V vs. Li^+/Li. Worth mentioning is that three lithium atoms in $Li_4Ti_5O_{12}$ occupy 8a tetrahedral sites and that no lithium is present in the 8a sites of $Li_7Ti_5O_{12}$. The three 8a lithium atoms migrate to the 16c sites during the insertion process, and the newly inserted lithium atoms fill the remaining 16c sites [16,18,19]. The migration of lithium atoms from 8a to 16c sites is driven by

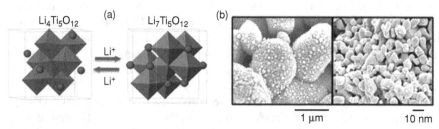

FIGURE 1 (a) Structure of $Li_4Ti_5O_{12}$ and $Li_7Ti_5O_{12}$ showing no volume change after charge and discharge; (b) scanning electron microscopy under low and high magnification of nanostructured $Li_4Ti_5O_{12}$ with micrometer-sized secondary particles and nanometer-sized primary particles. (From [4].)

the coulombic repulsion between the inserted lithium atoms in the 16c sites, whose distance to the nearby 8a sites is only 1.81 Å. The kinetics of lithium migration between 8a and 16c sites, which is determined by the energy barrier between them, can hinder the insertion process. The thermal migration of lithium from 8a to 16c sites in Li$_4$Ti$_5$O$_{12}$ without extra inserted lithium (or no repulsion force) was observed at about 700°C [18]. This high-energy barrier for lithium migration can lead to the cooccupation of 8a and 16c sites when fast insertion occurs. Therefore, reduced reversible capacity and low rate capability are generally reported for micrometer sized Li$_4$Ti$_5$O$_{12}$, in which the cooccupation of 8a and 16c sites is not energetically favored. Borghols et al. [19] suggested that the boundary effect of materials with a large specific surface area can help to relax the repulsion force and make it easier for cooccupation of 8a and 16c sites in the near-surface region. This suggestion agrees well with the strategies to develop nanosized or nanostructured materials to maximize the electrochemical performance.

An indirect support for the hypothesis of 8a- and 16c-site cooccupation is that the chemical lithiation of Li$_4$Ti$_5$O$_{12}$, which is much faster than the electrochemical process, resulted in only a small degree of 8a-site lithium migration into 16c sites, and the majority of the lithium inserted was trapped irreversibly in 48f sites [16]. It was also reported that more lithium could be inserted into Li$_7$Ti$_5$O$_{12}$ electrochemically when a potential below 1.0 V vs. Li$^+$/Li was applied [18, 20]. However, a large hysteresis effect was observed in the voltage profile within the window below 1.5 V, probably due to the repulsion force between lithium in the 16c and 8a sites.

4 SUPERIOR ELECTROCHEMICAL PERFORMANCE OF Li$_4$Ti$_5$O$_{12}$-BASED LITHIUM-ION CHEMISTRY

In a practical lithium-ion cell, the capacity of the anode is generally not less than that of the cathode to avoid the lithium plating on the anode side when the cell is fully charged. Therefore, the reversible capacity of the cell is determined by the amount of accessible lithium in the cathode material; any side reactions that can consume either lithium or charge/electron will contribute directly to the capacity loss of the cell [21,22]. One of the sources that consumes lithium continuously is the decomposition or formation of the SEI layer on the graphite surface at elevated temperatures ($\geq$ 60°C) [10,22]. This consumption will become worse when the graphitic anode is paired with a lithium transition metal oxide/phosphate cathode because the migration of a trace amount of transition metal from the cathode side to the anode side could have a detrimental impact on the thermal stability of the SEI layer and accelerate a side reaction between the lithiated anode and the nonaqueous electrolyte, increasing the capacity fade at elevated temperatures [23]. Alternatively, Li$_4$Ti$_5$O$_{12}$ operates at a higher potential ($\sim$1.55 V), and an SEI layer is not required. Therefore, the consumption of valuable lithium to form the SEI layer as well as the negative impact of migrated transition metals can be reduced dramatically. In addition, the intrinsic zero strain during lithium insertion and removal is another major contributor to the superior electrochemical performance of Li$_4$Ti$_5$O$_{12}$ over graphitic anodes.

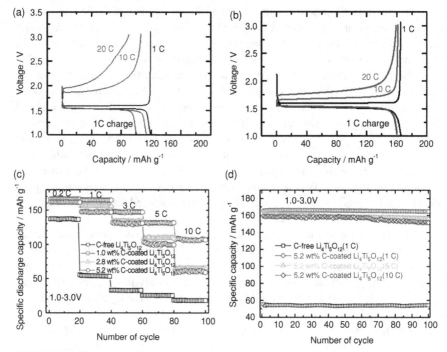

FIGURE 2   Charge and discharge curves of (a) micrometer-sized Li$_4$Ti$_5$O$_{12}$ and (b) nanostructured Li$_4$Ti$_5$O$_{12}$ with micrometer-sized secondary particles and nanometer-sized primary particles. The test was carried out in a half-cell. Initially, half-cells were charged and discharged at the 0.2C rate, and they were then charged at the 1C rate and discharged at different rates. (From [4].) (c) Rate capability of carbon-free and carbon-coated Li$_4$Ti$_5$O$_{12}$ from the 0.2C rate (0.34 mA/g) to the 10C rate (1.7 A/g); (d) cyclability of 5.2 wt% carbon-coated Li$_4$Ti$_5$O$_{12}$ from the 1C rate (0.17 A/g) to the 10C rate (1.7 A/g) in comparison with carbon-free Li$_4$Ti$_5$O$_{12}$ cycled at the 1C rate (0.15 A/g). (From [17].)

Figure 2a shows the charge–discharge profile of a half-cell with a conventional micrometer-sized Li$_4$Ti$_5$O$_{-12}$ electrode. The cell was discharged at the 1C rate to reach full insertion of lithium into the working electrode material. Then the cell was charged to 3.0 V using different rates (1C, 10C, and 20C). At the 1C rate, the cell delivered a specific capacity of about 120 mAh/g. The charge capacity decreased to about 80 mAh/g at 20C with a large hysteresis loop in the voltage profile, due to the difficulty in lithium migration between the 8a and 16c sites in bulk materials (see the discussion above). Figure 2b shows the charge–discharge profile of a half-cell using specially designed Li$_4$Ti$_5$O$_{12}$ with micrometer-sized secondary particles and nanometer-sized primary particles. The testing procedure was identical to the one used for the test results shown in Figure 2a. The nanostructured Li$_4$Ti$_5$O$_{12}$ delivered a very high reversible capacity (>160 mAh/g for all three rates), and the capacity was very high when the cell was charged at the 20C rate.

Figure 2c shows the discharge capacity vs. cycle number of half-cells using carbon-free and carbon-coated Li$_4$Ti$_5$O$_{12}$ tested with different charge and discharge rates. The benefit of carbon coating on the electrochemical performance of $Li_4Ti_5O_{12}$ is clearly evident. The discharge capacity of carbon-free $Li_4Ti_5O_{12}$ dropped dramatically with increasing discharge rate. The capacity was very low (18 mAh/g) at the 10C rate. For the carbon-coated $Li_4Ti_5O_{12}$, the rate capability depended strongly on the amount of coated carbon, especially at high rates. The increase in the amount of coated carbon on $Li_4Ti_5O_{12}$ particles led to a noticeable enhancement in rate capability. If the capacity obtained at C/5 is used as a standard, the capacity retention at the 10C rate is 37%, 40%, and 65% for anodes with 1.0, 2.8, and 5.2 wt% carbon coating, respectively. Figure 2d shows the capacity retention of a half-cell with the $Li_4Ti_5O_{12}$ electrode having 5.2 wt% coated carbon tested at the 1, 5, and 10C rates over 100 cycles, as well as another half-cell using carbon-free $Li_4Ti_5O_{12}$ cycled at the 1C rate as a control. The reversible capacity for the carbon-coated $Li_4Ti_5O_{12}$ was about 160 mAh/g throughout the entire cycling test, even though the applied current was increased to the 10C rate (1.7 A/g). Meanwhile, the carbon-free $Li_4Ti_5O_{12}$ had a low reversible capacity of 54 mAh/g at the 1C rate. These test results indicate that the carbon-coated $Li_4Ti_5O_{12}$ is sufficiently stable for high-rate cycling.

The performance advantage of nanosized or nanostructured $Li_4Ti_5O_{12}$ over graphitic anodes can be clearly demonstrated when they are paired with lithium transition metal oxides to form a full cell configuration. Figure 3a compares the capacity retention of two full cells using $Li_{1+x}Mn_{2-x}O_4$ (LMO) as the cathode material over extended cycling tests. The anode was either carbon or nanostructured $Li_4Ti_5O_{12}$ with micrometer-sized secondary particles and nanometer-sized primary particles (MSNP-LTO). To accelerate the cycle life test, the cells were charged and discharged at high current density (5C) and high temperature (55°C). The MSNP-LTO/LMO cell showed no capacity fade after 1000 cycles under these aggressive test conditions, while the carbon/LMO cell lost more than 25% capacity under the same test conditions. In the carbon/LMO cell, the capacity fade was attributed to manganese migration from the cathode to the anode side that compromised the integrity of the SEI layer. In contrast, no SEI layer was required for the MSNP-LTO electrode because of its high stability in the charged state and its high operating voltage (1.5 V vs. Li^+/Li). Figure 3b compares the power fade vs. aging time of MSNP-LTO/LMO and carbon/LMO cells. Both cells were aged at 60% state of charge and 55°C for 8 to 10 weeks.

The carbon/LMO cell showed poor calendar life, with over 38% power fade after only 8 weeks, caused mainly by the interfacial impedance rise at the carbon anode. In contrast, the MSNP-LTO/LMO cell showed a small initial power fade (less than 4%) during the first 4 weeks. Thereafter, the power fade stabilized at about 7% after 10 weeks. Similar results were also reported for $Li_4Ti_5O_{12}$-based fuel cells having other cathode materials with higher energy density than $Li_{1+x}Mn_{2-x}O_4$. For example, Jung et al. [17] investigated the electrochemical performance of a full cell using carbon-coated $Li_4Ti_5O_{12}$ as the anode and $Li[Ni_{0.45}Co_{0.1}Mn_{1.45}]O_4$ as the cathode (a 4.7-V high-voltage spinel) and reported that when cycled at 1C for 500 cycles at room temperature the cell maintained an excellent capacity retention of 85.4%.

Figure 3c compares the area-specific impedance (ASI) for MSNP-LTO/LMO, micrometer-sized LTO/LMO, and carbon/LMO cells that were cycled using the hybrid

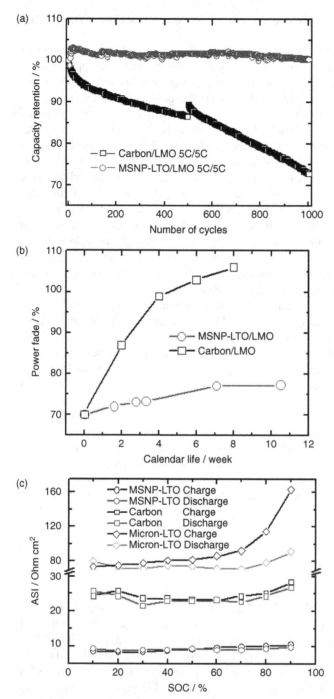

FIGURE 3 (a) Capacity retention during cycling at 55°C of MSNP-LTO/LMO and carbon/LMO cells; both cells were charged and discharged at the 5C rate. (b) Power fade vs. aging time at 55°C of MSNP-LTO/LMO and carbon/LMO cells. The power of each cell was determined from the hybrid pulse power characterization test. (c) Area-specific impedance vs. state of charge of MSNP-LTO/LMO, carbon/LMO, and micrometer-sized LTO/LMO cells. (From [4].)

power performance characteristic (HPPC) testing procedure proposed by the Partnership of Next Generation Vehicles (PNGV). Hybrid electric vehicles require very high charge and discharge pulse power for vehicle acceleration and regenerative braking. To meet the 25-kW pulse power requirement in HEVs, the ASI of electrodes in the cell must be less than 35 $\Omega \cdot cm^2$. These values were then used to determine the energy swings within the limits of the pulse power capability required. For the MSNP-LTO/LMO cell, the ASI values were around 9 $\Omega \cdot cm^2$, about one-eighth of the ASI values for the cell with LTO anodes composed of 8-μm particles (80 $\Omega \cdot cm^2$). Moreover, the ASI values for the MSNP-LTO/LMO cell were one-fourth of the upper ASI limit to power HEVs and were almost one-third of the ASIs for the conventional carbon/LMO system. In addition, the pulse-power test of the cells at $-30°C$ showed that the MSNP-LTO/LMO could meet the 5-kW requirement during the cold cranking test, whereas the carbon/LMO cell failed.

5   UNMATCHED SAFETY CHARACTERISTICS OF LITHIUM-ION BATTERIES USING Li$_4$Ti$_5$O$_{12}$

Figure 4 compares the different scanning calorimetry profiles of lithiated carbon and lithiated LTO with the presence of the nonaqueous electrolyte [24]. The results show that the fully lithiated graphite anode generated much more heat than the fully lithiated Li$_{6.9}$Ti$_5$O$_{12}$ anode (2750 vs. 383 J/g, respectively). In the case of lithiated graphite, the thermal degradation of the SEI layer occurred at around 100°C, leading to a wide exothermal reaction because of the continuous degradation of the SEI layer and lithium removal from the graphitic layers. This exothermal reaction was much reduced for the Li$_{6.9}$Ti$_5$O$_{12}$, because the SEI layer may not be formed upon lithiation of Li$_4$Ti$_5$O$_{12}$ at the higher operating potential. In summary, the thermal characteristics of Li$_{4+x}$Ti$_5$O$_{12}$ phases are significantly better than those of lithiated graphite (i.e., higher onset temperature and much less generated heat, as shown in the accumulated heat curve in Figure 4b).

   Lu et al. [13] reported that substantial heat could also be generated during normal high-rate operation of lithium-ion batteries. This heat generation can be even worse in a large battery pack, where the thermal management system will be less efficient at removing heat from the battery than that for a small pack. Figure 5a and b show the heat profiles of LTO/Li$_{1+x}$Mn$_{2-x}$O$_4$ and mesocarbon microbead (MCMB)/Li$_{1+x}$Mn$_{2-x}$O$_4$ cells during charge at 0.1 mA (C/10) and 1.0 mA (1C), respectively. The heat flow was measured by isothermal microcalorimetry. For both charge rates, the heat variation of the MCMB/Li$_{1.156}$Mn$_{1.844}$O$_4$ cell is larger than that of the LTO/Li$_{1.156}$Mn$_{1.844}$O$_4$ cell. Figure 5c shows the total heat generation of LTO/Li$_{1.156}$Mn$_{1.844}$O$_4$ and MCMB/Li$_{1.156}$Mn$_{1.844}$O$_4$ cells during the charge as a function of the charge rates. Clearly, the total heat generation of the cell with the LTO anode is smaller than that with the MCMB anode. This result indicates that the LTO/Li$_{1.156}$Mn$_{1.844}$O$_4$ cell has less heat generation and better thermal safety.

   Wu et al. [4] also demonstrated the unmatched safety characteristics of LTO-based lithium batteries in large format pouch cells after nail penetration and overcharge abuse, the most severe abuse conditions for lithium-ion batteries.

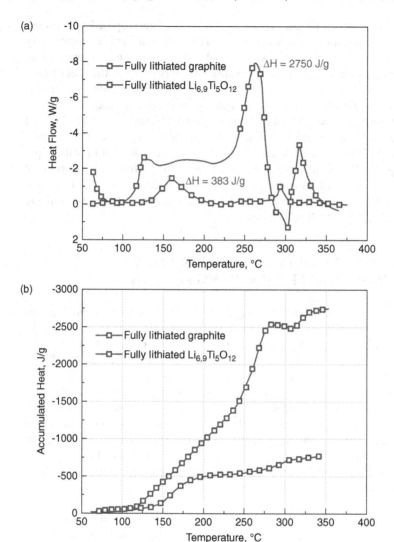

FIGURE 4 (a) Differential scanning calorimetry curves of fully lithiated graphite and fully lithiated $Li_4Ti_5O_{12}$; [24]. (b) corresponding accumulated heat curves. (From [24].)

Figure 6a and b show the nail penetration test results for a carbon/LMO cell and an MSNP-LTO/LMO cell, respectively. Both cells were fully charged before testing. Once the nail penetrated the carbon/LMO cell, the cell temperature increased dramatically to 150°C, caused by internal shorting and reaction of the nonaqueous electrolyte with the lithiated carbon anode.

In contrast, the MSNP-LTO/LMO cell showed only a 5°C increase in the cell temperature after nail penetration, clearly indicating outstanding abuse tolerance. Other battery systems based on carbon anodes and $LiNi_{0.8}Co_{0.15}Al_{0.05}O_2$ or $LiFePO_4$ cathodes also show larger thermal events than the MSNP-LTO/LMO system.

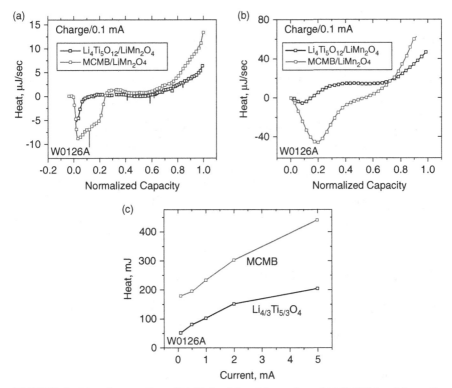

FIGURE 5   Heat flow profiles of $Li_4Ti_5O_{12}/Li_{1.156}Mn_{1.844}O_4$ and MCMB/$Li_{1.156}Mn_{1.844}O_4$ cells as a function of the normalized capacity at (a) 0.1 mA and (b) 1 mA charge rates; (c) total heat generation of $Li_4Ti_5O_{12}/Li_{1.156}Mn_{1.844}O_4$ and MCMB/$Li_{1.156}Mn_{1.844}O_4$ cells during the charge processes as a function of the charge rate. (From [13].)

Figure 6c shows the temperature profile of a 4-V carbon/LMO cell that was overcharged to 12 V. After about 45 min, the cell temperature increased significantly and reached over 500°C, followed by an explosion. When the 2.5-V MSNP-LTO/LMO cell was overcharged to 15 V, the cell temperature after overcharge reached only 80°C, with no explosion or thermal runaway (Fig. 6d). In fact, the cell was able to operate normally after this severe overcharge test.

6   CLOSING REMARKS

In brief, a synthesis route to prepare nanostructured $Li_4Ti_5O_{12}$ has been well established. Bulk $Li_4Ti_5O_{12}$ with micrometer-sized particles has low specific capacity and poor rate capability, due to the difficulty of lithium migration from the 8a to 16c sites in bulk materials. This issue can be resolved by taking advantage of the boundary effect of nanostructured materials. It has been consistently reported that lithium-ion batteries using nanostructured $Li_4Ti_5O_{12}$ as the anode had extremely long cycle and calendar

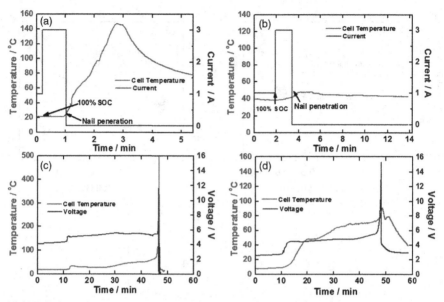

FIGURE 6　Nail penetration test showing the surface temperature of fully charged (a) carbon/LMO and (b) MSNP-LTO/LMO cells; overcharge test showing the surface temperature and voltage of (c) carbon/LMO and (d) MSNP-LTO/LMO cells. (From [4].)

lives, high power capability, and unmatched safety characteristics. As a quick example, Table 1 compares the performance of a battery pack using $Li_{1+x}Mn_{2-x}O_4$ as the cathode and $Li_4Ti_5O_{12}$ with micrometer-sized secondary particles and nanometer-sized primary particles as the anode against the technical requirements for hybrid electric vehicles proposed by FreedomCAR [5]. The data clearly show that this exemplary lithium-ion chemistry has already exceeded all the technical requirements

TABLE 1　Gap Analysis Comparing the Performance of an MSNP-LTO/LMO Battery Pack and HEV Battery Pack in Terms of PNGV Requirements

	HEV battery requirement	MSNP-LTO/LMO battery pack performance based on BSF 112
Pulse discharge power (10 s)	25 kW	32.66 kW
Peak regenerative peak power (10 s)	20 kW, 55-Wh pulse	26.13 kW
Total available energy	0.3 kWh at 1C rate	0.43 kWh
Minimum round-trip energy efficiency	90% 25-Wh cycle	97%
Cold cranking power at −30°C	5 kW	5 kW
Maximum weight	40 kg	12 kg
Maximum volume	32 L	8.9 L
Maximum allowable self-discharge rate	50 Wh/day	1 Wh/day
Equipment operation temperature range	−30 to +52°C	−30 to +55°C

needed. Furthermore, it has recently been reported that $Li_4Ti_5O_{12}$ is also an ideal candidate as the anode electrolyte for extremely large-scale and low-cost viscous flow batteries, due to its long life and excellent safety performance [25]. However, $Li_4Ti_5O_{12}$ has an intrinsic gassing issue associated with lithiated $Li_{4+x}Ti_5O_{12}$ that is rarely mentioned in the literature. The lithiated $Li_{4+x}Ti_5O_{12}$ has a tendency to react with nonaqueous electrolyte and generate gas when aged at elevated temperature. Resolving the gassing issue would enable successful deployment of the $Li_4Ti_5O_{12}$ anode.

Acknowledgments

This research was funded by the U.S. Department of Energy, FreedomCAR, and the Vehicle Technologies Office. Argonne National Laboratory is operated for the U.S. Department of Energy by UChicago Argonne, LLC, under contract DE-AC02-06CH11357. This work was also supported by a Human Resources Development of the Korea Institute of Energy Technology Evaluation and Planning grant funded by the Korean Government's Ministry of Knowledge Economy (No. 20104010100560). The authors also acknowledge EnerDel for fruitful collaboration.

REFERENCES

1. J. B. Goodenough, H. D. Abruña, and M. V. Buchanan, *Basic Research Needs for Electric Energy Storage: Report of the Basic Energy Sciences Workshop on Electrical Energy Storage*, U.S. Department of Energy, Office of Basic Energy Sciences, Washington, DC, 2007.
2. G. Jeong, Y. U. Kim, H. Kim, Y. J. Kim, and H. J. Sohn, *Energy Environ. Sci.*, **4**(6), 1986–2002 (2011).
3. S. Y. Chung, J. T. Bloking, and Y. M. Chiang, *Nat. Mater.*, **1**(2), 123–128 (2002).
4. K. Amine, I. Belharouak, Z. H. Chen, T. Tran, H. Yumoto, N. Ota, S. T. Myung, and Y. K. Sun, *Adv. Mater.*, **22**(28), 3052–3057 (2010).
5. *Partnership of New Generation Vehicle Battery Test Manual*, revision 3, U.S. Department of Energy, (2001).
6. H. J. Bang, H. Joachin, H. Yang, K. Amine, and J. Prakash, *J. Electrochem. Soc.*, **153**(4), A731–A737 (2006).
7. I. Belharouak, W. Q. Lu, D. Vissers, and K. Amine, *Electrochem. Commun.*, **8**(2), 329–335 (2006).
8. M. Holzapfel, F. Alloin, and R. Yazami, *Electrochim. Acta*, **49**(4), 581–589 (2004).
9. S. Flandrois and B. Simon, *Carbon*, **37**(2), 165–180 (1999).
10. Z. H. Chen, Y. Qin, Y. Ren, W. Q. Lu, C. Orendorff, E. P. Roth, and K. Amine, *Energy Environ. Sci.*, **4**(10), 4023–4030 (2011).
11. K. M. Colbow, J. R. Dahn, and R. R. Haering, *J. Power Sources*, **26**(3–4), 397–402 (1989).
12. A. Moodenba, D. C. Johnston, and R. Viswanat, *Mater. Res. Bull.*, **9**(12), 1671–1676 (1974).
13. W. Lu, I. Belharouak, J. Liu, and K. Amine, *J Power Sources*, **174**(2), 673–677 (2007).
14. H. M. Wu, I. Belharouak, H. Deng, A. Abouimrane, Y. K. Sun, and K. Amine, *J. Electrochem. Soc.*, **156**(12), A1047–A1050 (2009).
15. J. W. Jiang, J. Chen, and J. R. Dahn, *J. Electrochem. Soc.*, **151**(12), A2082–A2087 (2004).
16. L. Aldon, P. Kubiak, M. Womes, J. C. Jumas, J. Olivier-Fourcade, J. L. Tirado, J. I. Corredor, and C. P. Vicente, *Chem. Mater.*, **16**(26), 5721–5725 (2004).
17. H. G. Jung, S. T. Myung, C. S. Yoon, S. B. Son, K. H. Oh, K. Amine, B. Scrosati, and Y. K. Sun, *Energy Environ. Sci.*, **4**(4), 1345–1351 (2011).

18. A. Laumann, H. Boysen, M. Bremholm, K. T. Fehr, M. Hoelzel, and M. Holzapfel, *Chem. Mater.*, **23**(11), 2753–2759 (2011).

19. W. J. H. Borghols, M. Wagemaker, U. Lafont, E. M. Kelder, and F. M. Mulder, *J. Am. Chem. Soc.*, **131**(49), 17786–17792 (2009).

20. W. Lu, I. Belharouak, J. Liu, and K. Amine, *J. Electrochem. Soc.*, **154**(2), A114–A118 (2007).

21. A. J. Smith, J. C. Burns, D. Xiong, and J. R. Dahn, *J. Electrochem. Soc.*, **158**(10), A1136–A1142 (2011).

22. D. J. Xiong, J. C. Burns, A. J. Smith, N. Sinha, and J. R. Dahn, *J. Electrochem. Soc.*, **158**(12), A1431–A1435 (2011).

23. K. Amine, Z. H. Chen, Z. Zhang, J. Liu, W. Q. Lu, Y. Qin, J. Lu, L. Curtis, and Y. K. Sun, *J. Mater. Chem.*, **21**(44), 17754–17759 (2011).

24. I. Belharouak, Y. K. Sun, W. Lu, and K. Amine, *J. Electrochem. Soc.*, **154**(12), A1083–A1087 (2007).

25. M. Duduta, B. Ho, V. C. Wood, P. Limthongkul, V. E. Brunini, W. C. Carter, and Y. M. Chiang, *Adv. Energy Mater.*, **1**(4), 511–516 (2011).

CHAPTER 14

SAFE LITHIIUM RECHARGEABLE BATTERIES BASED ON IONIC LIQUIDS

A. Guerfi, A. Vijh, and K. Zaghib

1 INTRODUCTION

Lithium-ion batteries have become the dominant discrete power sources for a variety of applications in electronic devices, mobile telephones, laptop computers, and a large variety of other portable instruments and appliances. They provide a very high energy density, are light and compact, and exhibit excellent cyclability and reliability. The current commercial Li-ion batteries are based on organic liquids (e.g., ethylene carbonate, dimethyl carbonate) that have a high dielectric constant and thus are good solvents for salts; they also show a fairly large electrochemical window of stability.

However, these organic solvents have high vapor pressures, and in case of accidental battery shorts or thermal runaway, can lead to fires and explosions. Some

Lithium Batteries: Advanced Technologies and Applications, First Edition.
Edited by Bruno Scrosati, K. M. Abraham, Walter van Schalkwijk, and Jusef Hassoun.
© 2013 John Wiley & Sons, Inc. Published 2013 by John Wiley & Sons, Inc.

of these dramatic accidents have occurred from time to time, leading to recalls of millions of batteries and creating concern, even panic, among consumers. Such safety issues become paramount in large Li-ion batteries of interest in electric cars, especially if charge and discharge is carried out at high rates. Thus, safety has become a central issue in the development of this technology.

A major avenue for creating a safe Li-ion battery is to replace organic solvents or at least to diminish their flammability and high vapor pressure. The approach adopted at Hydro-Québec, through its research institute, is to use ionic liquids as solvents in these batteries; we have also discovered some solvent mixture compositions of organic solvents and ionic liquids which retain the best characteristics of both constituents and provide safe battery electrolytes.

The objective of the present review is to summarize the work on rechargeable Li-ion batteries carried out and continued at Hydro-Québec (HQ); this work is placed in context by mentioning work on this subject published by workers elsewhere. This is not, however, intended to be an exhaustive review of lithium rechargeable batteries in ionic liquids. The use of ionic liquids in Li-ion batteries has been surveyed relatively recently [1].

2   IONIC LIQUIDS

Ionic liquids are molten salts comprised of multiatomic organic ions, which have a low melting point, ideally much lower than room temperature; thus, these salts are liquids at temperatures at which batteries are required to operate. They constitute the solvent part of the electrolyte in the battery. Ionic liquids (ILs) exhibit several characteristics that make them attractive in Li-ion batteries. Some of these features are [2]:

- Nonflammability
- Low vapor pressure
- Low toxicity and high environmental compatibility
- Large potential window/electrochemical stability
- High thermal stability
- Acceptably high conductivity
- Material compatibility with the anodes and cathodes during charge and discharge

In battery work, ionic liquids also possess some disadvantages: high cost, high viscosity, and high contact angle (and thence poor wetting) with some electrode materials and configurations.

Why do ionic liquids salts have low melting temperatures? It is because they invariably comprise large, nonsymmetric ions possessing a univalent charge (Fig. 1). A very low charge/radius ratio of both the cation and the anion leads to very low lattice energy of the salts, with a consequent low melting point and very low cation–anion electrostatic forces. Such cations and anions are also difficult to discharge on

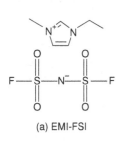

1-ethyl-3-mthyleimidazolium
bis(fluorosulfonyl)imide

N-Methyl-N-propylpyrrolidinium
bis(fluorosulfonyl)imide

(a) EMI-FSI

(d) Py13-FSI

FIGURE 1 Typical ionic liquid properties that make their use attractive in safe lithium rechargeable batteries.

the electrodes, at least at lower potentials, so that a large window of electrochemical stability is obtained. Since these ionic liquids are essentially ionic salts that have undergone a phase change at room temperature (i.e., melted), they have little vapor pressure, in contrast to covalently bonded liquids such as ethylene carbonate, propylene carbonate, or water.

3 LITHIUM-METAL RECHARGEABLE BATTERIES IN IONIC LIQUIDS

Rechargeable batteries based on lithium metal are sought because they provide higher theoretical energy density than that of the alternatives. The key limitation in commercializing such batteries is the growth of dendrites during cycling: this poses a potential hazard and leads to reduced cycle lifetime. Several strategies have been attempted to suppress dendritic growth. Use of solid electrolytes to act as mechanical barriers, or choosing electrolytes that provide a suitable passivation layer, called the solid electrolyte interphase (SEI), are the more popular strategies. SEI layers are obtained in high-dielectric-constant aprotic organic electrolyte compositions based on compounds such as ethylene carbonate (EC) and diethyl carbonate (DEC) mixtures.

These electrolytes have high vapor pressures and are flammable, thus potentially hazardous. To overcome these problems, use of ILs has been proposed by a number of workers [3]. To enhance lithium-ion transport in ionic liquids of interest in batteries, addition of zwitterions to the electrolyte has been proposed [4]; zwitterionic compounds are designed to tether together the anion and cation constituting the battery ionic liquid.

The mechanism of the enhancement of Li-ion transport in ionic liquids by zwitterions is not known, except for the speculation that they prevent migration of the ionic liquid under the influence of an electric field [4]; it is more likely that they provide "bridges" of charges for the diffusing lithium ions, somewhat similar to proton conduction in ice [5]. Bryne et al. [4] present some work on lithium-metal cells.

Owing to the aforementioned difficulties in cycling lithium-metal anodes, they are used mostly in primary batteries; the same is true of lithium-metal alloys [6].

Commercial rechargeable lithium batteries invariably use graphite intercalation compounds as the active anode materials rather than lithium-metal electrodes; the latter are used in some batteries using solid polymer electrolytes, although some unresolved interfacial problems still exist in them.

The use of lithium-alloy anodes presents additional problems: There is usually a large volume change in the electrode on charge–discharge cycles, leading to their decrepitation or crumbling [7]; these problems can perhaps be reduced by the use of nanostructured or microstructured electrodes [8]. Nanomaterials would, however, have a tendency to undergo sintering during cycling.

Research work on rechargeable lithium metal–ionic liquid batteries at the Institut de Recherche d'Hydro-Québec (IREQ) has been done primarily in collaboration with the Central Research Institute of Electrical Power Industry (CRIEPI) of Japan. In this joint IREQ–CRIEPI work, charge–discharge curves have been reported for the system [9]:

$$Li/Py13(FSI)–0.7 \text{ M LiFSI}/LiFePO_4$$

| Anode | Ionic liquid electrolyte | Cathode |

In the half-cell studies, the ionic liquid showed lower first coulombic efficiency, 80% compared to EC–DEC at 93% with the graphite anode. However, the reversible capacity was higher than 360 mAh/g for all cells at a C/24 rate. On the cathode side, a lower reversible capacity, 143 mAh/g, was obtained with Py13(FSI)–LiFSI; however, a comparable reversible capacity was found in EC–DEC and EMI(FSI)–LiFSI. Impregnation of the IL in the cathode was enhanced by using vacuum and heating at 60°C. With these conditions, the reversible capacity improved to 160 mAh/g in the cathode at C/24. The power decreases when polymer is added. At a 1C rate, the capacity was 106 mAh/g, compared to 140 mAh/g without the polymer; however, at a 10C rate, the capacity is comparable.

In other work from IREQ–CRIEPI cooperation, studies on a graphite/lithium metal cell using an $FSI(N(SO_2F)_2{}^-)$ anion were reported [10]. The cells showed stable impedance and thermal behavior and good capacity retention over 150 cycles. This work was an extension of work published previously [11].

A number of investigators have also reported on the behavior of lithium metal, as either an anode or a cathode or both, in a variety of ionic liquid–based electrolytes: these studies have been tabulated in a recent survey [1]. This work is not reviewed here, as our purpose is to provide a brief look at the HQ work on safe rechargeable lithium batteries involving ionic liquids.

4 LITHIUM INTERCALATION RECHARGEABLE BATTERIES INVOLVING IONIC LIQUIDS: HQ WORK

Here we review the work on Li-ion rechargeable batteries containing electrolytes based on ionic liquids (either entirely or partially) in our work at HQ. Work on rechargeable lithium batteries in ionic liquids done elsewhere has been summarized recently [1] and is not surveyed again here. It is important to note that the work to be reviewed pertains mostly to the electrochemical behavior of anodic and cathodic half-cells in which lithium intercalation–deintercalation is examined; at the experimental level, these half-cells can be investigated only by incorporating them in a battery configuration against a counterelectrode whose behavior is well understood (e.g., a lithium-metal electrode in the present case).

4.1 Graphite Anodes and LiFePO₄ Cathodes: Comparison of Behavior in Some Ionic Liquids and Conventional EC–DEC Electrolytes

Ambient-temperature ionic liquid based on bis(fluorosulfonyl)imide (FSI) as the anion and 1-ethyl-3-methyleimidazolium (EMI) or N-methyl-N-propylpyrrolidinium (Py13) as the cation were investigated with a natural graphite anode and a LiFePO₄ cathode in lithium cells. The electrochemical performance was compared to the conventional solvent EC–DEC with 1 M LiPF₆ or 1 M LiFSI. The ionic liquid showed lower first coulombic efficiency at 80% compared to EC–DEC at 93%. The impedance spectroscopy measurement showed higher resistance of the diffusion part, which increased in the order EC–DEC–LiFSI < EC/DEC–LiPF₆ < Py13(FSI)–LiFSI = EMI(FSI)–LiFSI; however, a comparable reversible capacity was found in EC–DEC and EMI(FSI)–LiFSI. The high viscosity of the ILs suggested using different conditions, such as vacuum and 60°C, to improve the impregnation of ILs in the electrodes, and the reversible capacity was improved to 160 mAh/g at C/24. The high rate capability of LiFePO₄ was evaluated in polymer–IL and compared to the pure IL cell configuration. The power performance was decreased when the polymer is added; at C/10 only 126 mAh/g was delivered compared to 155 mAh/g.

Graphite Anode   The impedance measurement is shown in Figure 2; a comparable interface resistance is observed with graphite anode vs. Li, between ionic liquid and reference electrolyte EC/DEC–LiPF₆ at 80 Ω. However, EC/DEC–LiFSI-based salt showed a lower interface impedance at 65 Ω. In the diffusion part, the ionic liquids show higher resistance, which increases in the order EC/DEC–LFSI < EC/DEC–LiPF₆ < Py13(FSI)–LiFSI = EMI(FSI)–LiFSI = 20 Ω. Due to the high viscosity of the IL, the diffusion resistance is consequently higher.

Figure 3 shows the first two cycles of discharge–charge curves of anode graphite (a) in EC/DEC–LiPF₆, (b) in EC/DEC–LiFSI, (c) in EMI(FSI), and (d) in Py13(FSI). These charge–discharge cycles were obtained at a C/24 rate between 0 and 2.5 V at ambient temperature. For the standard cell (a), a reversible capacity very close to theoretical capacity was obtained at the second discharge of 365 mAh/g, with high coulombic efficiency in the first cycle (CE1) at 92.7%; these data reflect the performance of our electrodes in the standard electrolyte that was used as a reference for comparison in this study.

When LiPF₆ is replaced by LiFSI salt (b), the anode shows excellent performance with reversible capacity close to the theoretical capacity of 369 mAh/g and 93% of coulombic efficiency in the first cycle. The LiFSI salt has a positive effect on the formation of coherent passive layer on the graphite.

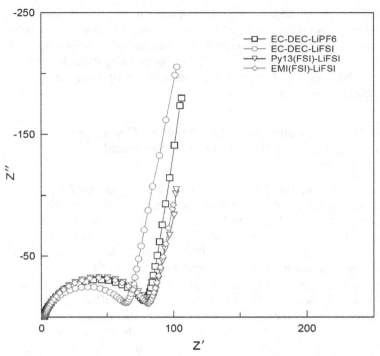

FIGURE 2 Impedance spectra of Li/graphite anodes before cycling in various electrolytes.

In (c), the cell with IL based on EMI–FSI shows 362 mAh/g as the reversible capacity but a coulombic efficiency of only 80.5%. However, for the ionic liquid Py13 (d), the reversible capacity is close to theoretical with 367 mAh/g, while the first coulombic efficiency was 80%. All these data explain well that LiFSI salt in FSI-based ionic liquid is suitable for use in the anode graphite side without any secondary reaction. In Table 1 we summarize the first electrochemical data for the graphite anode for the first cycle. The coulombic efficiency in the second cycle (CE2) still doesn't reach 100%, which is probably associated with some side reactions, and then the passivation layer on the graphite cannot be established during the first cycles (Fig. 3b).

The cyclability of the cells was evaluated at 1C when discharging and at C/4 when charging. The discharge capacities of the graphite with various electrolytes are shown in Figure 4. The capacity evolution was quite stable for an organic electrolyte during cycling. The reversible discharge capacity was slightly lower in $LiFP_6$ than in LiFSI salt. On the other hand, when the electrolyte is an ionic liquid, the discharge capacity still increased during the first five cycles for Py13(FSI) and 10 cycles for EMI(FSI). This result indicates that the IL Py13(FSI) electrolyte needs more cycles to form a passivation layer on graphite surface particles.

For the IL EMI(FSI), the capacity increases can be attributed to the improvement in wettability in the electrode bulk with lithium intercalation and deintercalation in graphite. However, Ishikawa et al. [12] reported quite stable cyclability of the

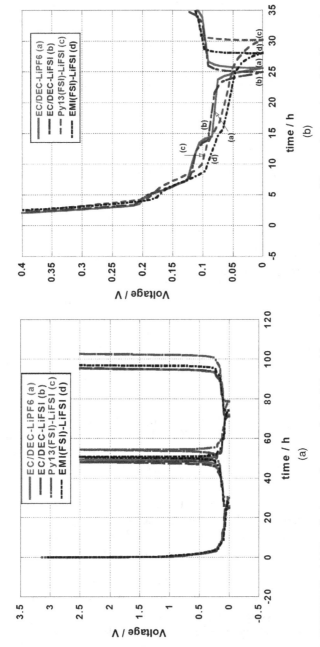

FIGURE 3 (a) First discharge–charge cycles for Li/graphite anodes in various electrolytes; (b) expanded version of the first cycle between 0.5 and 0 V.

297

TABLE 1   First Electrochemical Characteristics of the Graphite Anode

Electrolyte	First discharge (mAh/g)	CE1 (%)	Reversible capacity (mAh/g)	CE2 (%)
EC/DEC–1 M LiPF$_6$	398	92.7	365	100
EC/DEC–1 M LiFSI	382	93.0	369	100
Py13–FSI + 0.7 M LiFSI	468	80	367	98.3
EMI–FSI + 0.7 M LiFSI	432	80.5	362	97.6

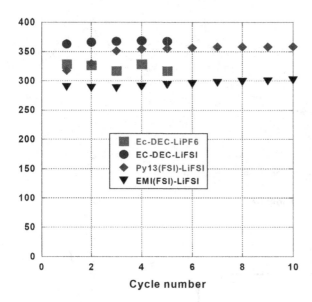

FIGURE 4   Cycling behavior of Li/graphite anodes in various electrolytes represented as discharged capacity in mAh/g of the electrolytes indicated.

Li/graphite anode with ionic liquid EMI(FSI) with 0.8 M LiTFSI salt, and they show a stable capacity of 360 mAh/g at C/5. These data probably suggest the combined stabilizing effect of FSI cation in IL and the additive salt type on the anode graphite to the SEI layer and thus quite high reversible capacity.

LiFePO$_4$ Cathode   In the cells, the LiFePO$_4$ cathode side exhibited interface resistance behavior different from that of the graphite anode side (Fig. 5). The highest interface resistance with a reference electrolyte was obtained at 240 Ω. The ionic liquids both show a lower interface resistance: 54 and 64 Ω, respectively, for Py13(FSI) and EMI(FSI). This result can probably be explained by the fact that the cathode bulk was not completely wetted in the ionic liquid configuration cells when the cathode was simply dipped in IL under normal conditions. Perhaps some LiFePO$_4$ particles did not have any passivation layer yet and thus did not contribute to the total resistance of the cathode interface. For the cathode, the electrochemical performance of the Li/LiFePO$_4$ cells was examined with ionic liquids and compared to that in a conventional organic solvent. Figure 6 shows the first charge–discharge curves at C/24 between 4 and 2.5 V.

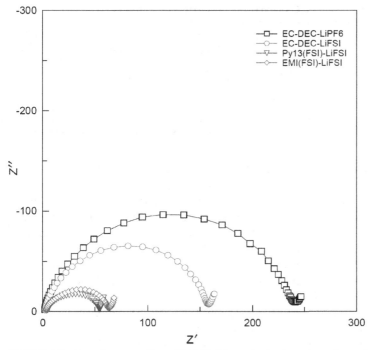

FIGURE 5   Impedance profiles of Li/LiFePO$_4$ cathodes in various electrolytes.

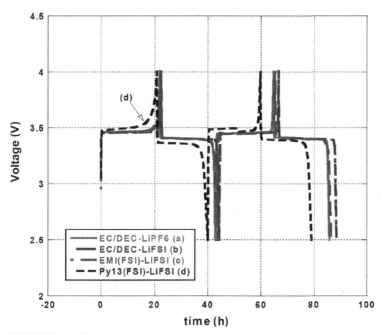

FIGURE 6   First charge–discharge cycle of Li/LiFePO$_4$ cells in various electrolytes.

TABLE 2   First Electrochemical Characteristics of the LiFePO$_4$ Cathode

Electrolyte	First discharge (mAh/g)	CE1 (%)	Reversible capacity (mAh/g)	CE2 (%)
EC/DEC–1 M LiPF$_6$	158.2	97.5	158.0	98.0
EC/DEC–1 M LiFSI	156.5	98.05	156.5	98.0
Py13–FSI + 0.7 M LiFSI	151.3	93.0	143.3	98.3
EMI–FSI + 0.7 M LiFSI	164.0	95.0	160	97.0

The reversible capacity with EC/DEC–LiPF6, curve (a), was 158 mAh/g, with 97.5% as coulombic efficiency in the first cycle (CE1). In a charge–discharge curve (b) with LiFSI salt, the reversible capacity was quite comparable to that of EC/DEC–LiPF$_6$ electrolyte at 156.5 and 98% in the first-cycle coulombic efficiency. With the ionic liquid Py13–FSI, curve (c), a lower reversible capacity of 143 mAh/g was obtained with only 93% coulombic efficiency. However, with ionic liquid EMI(FS), curve (d), higher reversible capacity and coulombic efficiency, respectively, with 160 mAh/g and 95% were obtained. The reversible capacity and the coulombic efficiency are summarized in Table 2.

The high viscosity of Py13(FSI), twice that of EMI(FSI) can make the lithium extraction from a LiFePO$_4$ structure more difficult even at a low rate such as C/24. This result is clearly described on the charge curve with not-well-defined curvature at the end of the charging plateau. In the LiFePO$_4$ cells, the viscosity perhaps affects the performance, also due to the carbon coating on the surface of LiFePO$_4$ particles. When an electrolyte such as an ionic liquid has a high viscosity, the wettability of the carbon layer is more difficult, due to its large surface area. Then the lithium ion cannot migrate easily across this layer, particularly in the first cycles. Moreover, the viscosity can also prohibit the wettability of all the electrodes in depth, both anode and cathode because of the quasi three-dimensional fractal nature of the electrodes.

The cycling data at 1C for discharge and C/4 for charge are presented in Figure 7. The highest capacity was obtained with EMI(FSI) at 145 mAh/g compared to EC/DEC–LiPF$_6$ and EC/DEC–LiFSI, which give around 137 and 139, respectively. However, Py13(FSI) shows a capacity decrease to 105 mAh/g. The coulombic efficiency for all the electrolytes found stabilized at 100% in the second cycle, except for the Py13(EMI), for which the coulombic efficiency reaches 100% only in the sixth cycle. Since ionic conductivity is inversely related to the viscosity, the Py13(FSI) tends not to yield good capacity at 1C compared to other electrolytes. Thus, it is necessary to evaluate the cathode in these electrolytes with different increasing discharge rates in order to push the power limitation of these electrolytes. To investigate the power performance of LiFePO$_4$ in different electrolytes, a rate capability test was employed. The rate dependence of the various electrolytes is summarized in Figure 8 in Ragone plots. The discharge current was varied at different rates, whereas the charge current was maintained constant at C/4.

The organic solvent EC/DEC with LiFSI salt shows a quite reasonable performance at a high rate; for example, at a 15C rate, high discharge capacities such as 105 mAh/g were delivered by a cell with LiFSI. For rates above 20C, the capacity

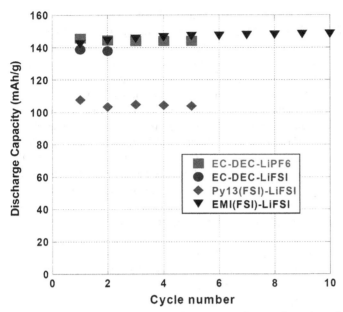

FIGURE 7 Cycling behavior of Li/LiFePO$_4$ cells in various electrolytes.

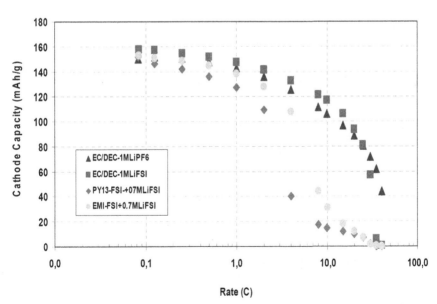

FIGURE 8 Rate capability of Li/LiFePO$_4$ cells at different discharge rates in various electrolytes.

starts diverging from the cell with LiPF$_6$ as a salt. With ionic liquids when higher viscosity is used, EMI(FSI) shows a comparable performance upto a 1C rate; at a rate above 4C, the capacity dropped to 45 mAh/g. For a more viscous electrolyte such as Py13(FSI), the C-rate performance was lower and the gap with other cells starts at a C/2 rate. At 4C, the capacity dropped to a low value of 40 mAh/g.

From these data it is clear that the power performance depends on the viscosity and the ionic conductivity. Furthermore, the LiFSI salt has shown a good result for both organic and ionic electrolytes. Matsumoto et al. [13] have compared the performance at a high rate of the Li/LiFSI–EMI/LiCoO$_2$ with FSI or TFSI as the anion. He confirmed the higher performance of EMI when an FSI anion is used. He found that the capacity retention ratio 1/0.1C for EMI(FSI) and Py13(FSI) cells was 93% and 87%, respectively. However, when a TFSI anion was used, the EMI(TFSI) cell showed a lower capacity retention ratio 1/0.1C of only 43%. Furthermore, the performance of these ionic liquids depends not only on the viscosity and conductivity, but also on the anion and on the salt additive types.

For further tests we have selected the Py13(FSI) IL based on its higher safety level as reported by Wang et al. [14]. To improve the wettability of the cathodes with the ionic liquid, the LiFePO$_4$ cathode was pretreated by immersing it in Py13(FSI)–LiFSI IL, then putting it under vacuum at 60°C for 8 h. This cathode was then evaluated with a lithium metal cell.

The first charge–discharge cycles of the pretreated cell, realized at 25°C, are shown in Figure 9 compared to a cell prepared without vacuum at 25°C. The

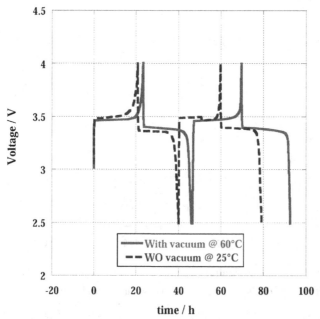

FIGURE 9 First charge–discharge of Li/LiFePO$_4$ cells with Py13(FSI)–LiFSI with and without vacuum.

coulombic efficiency was improved to 100% in the first and second cycles compared to 92.6% and 98.3%. The impedance spectra of these cells before and after cycling are shown in Figure 10. The interface resistance is higher when the cathode is pre-treated under vacuum: 70 Ω vs. 50 Ω. However, after two cycles at C/24, the interface impedances are lower in the pretreated cathodes.

The reversible capacity was increased by 14%, to 160 mAh/g. The cycling life at 1C shows a stable capacity at 140 mAh/g, with an improvement of 36% (Fig. 11). The viscosity limitation of the IL Py13(FSI) was twice that of EMI(FSI). The high rate capability shows a small increase in the power performance up to the 2C rate (Fig. 12). Above 2C, a high jump in capacity was delivered: from 40 mAh/g to 80 mAh/g. However, for the higher rates, the capacity dropped comparably to the same level as in cells in which the cathode wetting was not improved with vacuum pretreatment, as described above.

4.2 Lithium-Ion Ionic Liquid/Gel-Polymer Battery Systems

In the quest for new electrolyte compositions for rechargeable Li-ion batteries which are highly safe even at high charge–discharge rates or under extremely abusive conditions, we have also examined mixtures of ionic liquids with a suitable polymer in order to form a gel [15]. The effect of gel-polymer media was investigated with Py13(FSI) ionic liquid. The gel polymer was prepared by mixing of 5 wt% of poly-mer with Py13(FSI)–0.7 M LiFSI and 1000 ppm of the thermal initiator Perkadox (Akzo Nobel, United States). The polymer used was made from an ether-based low-molecular-weight cross-linkable polymer precursor (TA210, Daiichi Kogyo Seiyaku, Japan). The chemical formula has a trifunction consisting of poly(alkylene oxide) main chain with acrylate chain ends. This polymer has demonstrated good electro-chemical stability and high compatibility with high-voltage cathodes such as $LiCoO_2$. The electrodes were immersed in the mixture (polymer + IL + initiator) and then heated at 60°C under vacuum, a step necessary to help the (polymer + IL) electrolyte to penetrate the pores deeply across the electrode thickness. Then the cell was heated at 60°C for 1 h to form the gel polymer.

Li/LiFePO₄ Cells with Py13(FSI)–0.7 M LiFSI plus Polymer The effect of poly-mer addition to the IL was analyzed by in situ impedance spectroscopy. Figure 13 shows the charge–discharge data for the first cycle and the associated impedance spec-tra at different states of charge of the Li/FePO₄ cells at (a) IL [Py13(FSI)–LiFSI], (b) IL + 1% polymer, and (c) IL + 5% polymer; the first coulombic efficiency was found to be 97%, 100%, and 99%, respectively, with these cell configurations. The reversible capacity was close to 159 mAh/g for all cells, which is comparable to that of a cell with a standard electrolyte.

Figure 14a shows the interface resistance R_i for IL cells as well as that for cells to which 1 or 5 wt% polymer is added. The R_i value is increased when 5% polymer is added to the IL cell. However, when only 1% of polymer is added, R_i is lower than in the cell with IL alone at different percentages of depth of discharge (DOD). The average R_i values are maintained constant during the LiFePO₄-FePO₄ transformation with a higher R_i value at 90% DOD followed by lower R_i at monophase FePO₄.

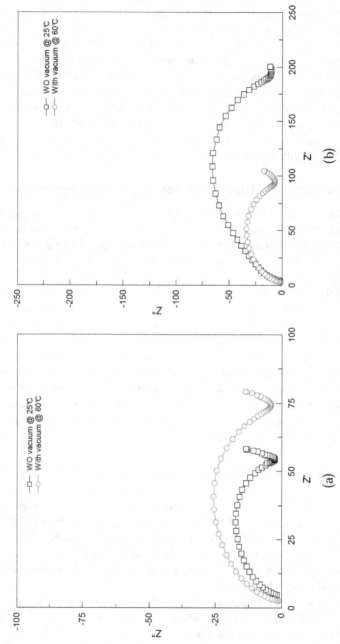

FIGURE 10 Impedance of Li/LiFePO$_4$ cells with Py13(FSI)–LiFSI: (a) freshly assembled cells, and (b) after two cycles.

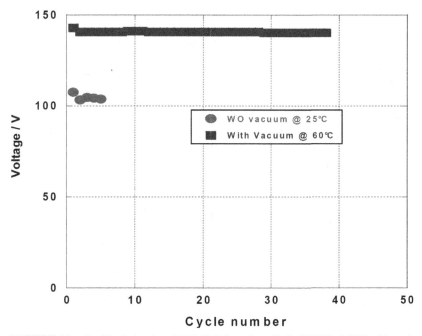

FIGURE 11   Cycling behavior of Li/LiFePO$_4$ cells with Py13(FSI)–LiFSI with and without vacuum.

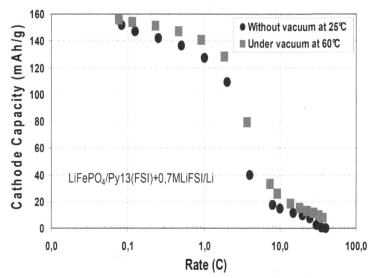

FIGURE 12   Rate capability of Li/LiFePO$_4$ cells with Py13(FSI)–LiFSI with and without vacuum.

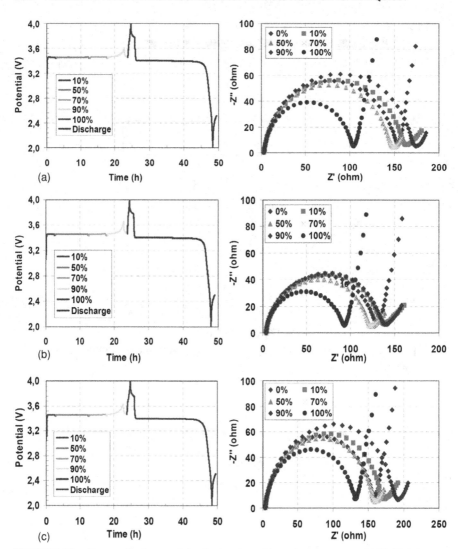

FIGURE 13 Charge–discharge of the first cycle and impedance spectra of Li/LiFePO$_4$ cells with Py13(FSI)-0.7MLiFSI: (a) 0% polymer; (b) 1% polymer; (c) 5% polymer additive.

This behavior of the R_i variation is still present, independent of polymer addition. Therefore, from these results we can assume that the addition of a small amount of polymer (1%) improves the interface of the cathode in cells with IL. This small amount of polymer can play an important role in making a stable solid–electrolyte interphase (SEI) layer, better than when the IL alone is used. The addition of more polymer in the IL cell (5%) shows a contrary effect, with an increase in the interface resistance that is even higher than that of an IL without additive. We may infer that the addition of 1% polymer stabilizes the SEI and then reduces the interface resistance by forming a thin passivation film. A higher polymer content in the IL forms a thicker resistive

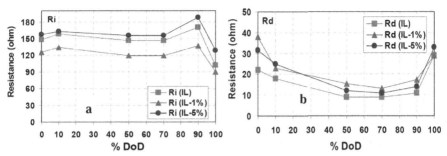

FIGURE 14   Interface resistance R_i (a) and diffusion resistance R_d; (b) of $Li/LiFePO_4$ cells with (IL + 0% polymer), (IL + 1% polymer), and (IL + 5% polymer) as a function of the state of charge.

layer, which can lead to an increase in the R_i. The polymer would be expected to form a stable thin layer that is permeable to Li ions on both the anode and the cathode. In the half-cells, the polymer would passivate the lithium metal, decreasing its reactivity and reducing its surface energy.

The diffusion resistance is plotted in Figure 14b as a function of % DOD with different amounts of polymer in the IL. The same R_d behavior may be noted at different states of charge with or without polymer additions. The highest value of R_d is obtained with the monophase iron phosphate: at 0% DOD with $LiFeO_4$ phase and at 100% DOD with $FePO_4$ phase. The diffusion process is easier when the diphase $LiFePO_4$–$FePO_4$ coexists, particularly in the range 50 to 70% DOD.

To investigate whether there is further stabilization after cycling, cells were cycled at C/24 for three cycles. The impedance measurements were taken at the fully delithiated state (100% DOD). Figure 15 shows the impedance curves of the cells: (a) standard electrolyte (EC/DEC–1 M $LiPF_6$), (b) IL (Py13–FSI), (c) (IL + 1% polymer), and (d) (IL + 5% polymer). The stabilization can be reached for the standard electrolyte (Fig. 15a) after the first cycle at 100 Ω of the total resistance (R_t) compared to the IL cell, in which the stabilization can only be achieved after three cycles, with $R_t = 194$ Ω and $R_i = 156$ Ω (Fig. 15b) compared to the standard electrolyte (Fig. 15a) with $R_t = 105$ Ω and $R_i = 37$ Ω. When a small amount of polymer is added (1%), stabilization still occurred at the third cycle but at $R_t = 163$ Ω and $R_i = 124$ Ω (Fig. 15). The addition of a higher polymer content (5%) (Fig. 15d) shows that the interface is not well stabilized, and R_t is 200 Ω with 166 Ω associated with interface resistance. Also to be noted is the fact that the initial ohmic resistance in Fig. 15 was found to be higher in the standard organic electrolyte; 23 Ω. This resistance may be attributed to the electrolyte and/or to the organic reaction film formed by interaction of the organic electrolyte with the lithium-metal electrode. In the presence of IL, however, this ohmic resistance becomes vanishingly small, as would be expected for these high-conducting media, which also have no tendency to form resistive films at the surface, which, if anything, would be expected to be covered by a conducting ionic salt layer.

The cells were cycled at a C/4 rate for long-cycle-life aging (Fig. 16) and the reversible capacity was found stable in all cases at 149, 152, and 148 mAh/g,

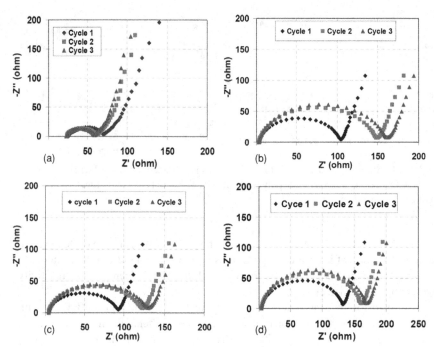

FIGURE 15 Impedance curves of the Li/LiFePO₄ cells: (a) standard electrolyte (EC/DEC–1 M LiPF6); (b) (IL + 0% polymer); (c) (IL + 1% polymer); (d) (IL + 5% polymer).

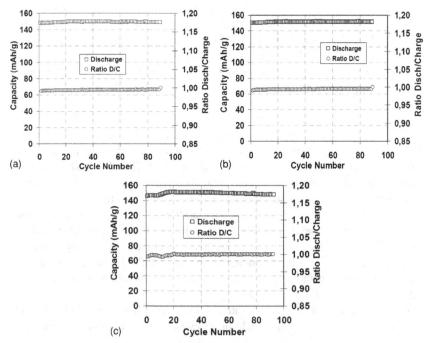

FIGURE 16 Cycling behavior of Li/LiFePO₄ cells in various electrolytes.

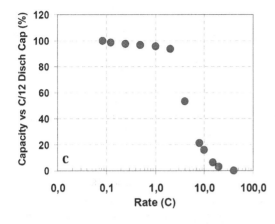

FIGURE 17 Rate capability of Li/graphite cell with Py13(FSI)–0.7 M LiFSI + 5% polymer.

respectively for (a) IL, (b) (IL + 1 Wt% polymer), and (c) (IL + 5 Wt% polymer). The coulombic efficiency remains constant during the cycling life with 99.6% for all cells. However, we noted abnormal behavior with 5% polymer cell. In the first 20 cycles, the capacity increases and the coulombic efficiency fluctuates before the cell reaches a stable state. Moreover, this result confirms the stabilization of both electrode interfaces of LiFePO$_4$, and lithium electrodes with IL, which probably suppresses the dendrite formation on the lithium metal.

The rate capability of the cell, Li/(IL + 5 Wt% polymer)/LiFePo$_4$, is shown in Figure 17. The cell delivers the full capacity until around the C/2 rate, and then the discharged capacity begins to decrease to reaching 82% at 2C. At a 4C rate, the capacity drops sharply to 47% and continues dropping as the discharge rate is increased. This drop in the capacity above the 2C rate perhaps indicates that the limiting process lies in the ionic liquid. It is well known that a very high concentration of ions in the molten salt can lead to an "overpopulation" of ions that can cause phenomena such as ion-pair formation and "salting out," in the sense that the ion–ion distance is short, thus disabling the "free" ion to make a full contribution to the conduction. Such a situation does not, of course, exist in a standard organic electrolyte in which the solvent molecules are abundantly available to solvate the lithium ions, which then contribute to conduction [15].

Lithium-Ion Cell with IL and Polymer Gel: Graphite Anode and LiFePO$_4$ Cathode   We have attempted to make a full Li-ion battery with LiFePO$_4$ as the cathode and graphite as the anode, by using Py13(FSI)–0.7 M LiFSI as the electrolyte, to which we have added 5 Wt% polymer. Before making the Li-ion cell, the graphite anode was also evaluated separately in a half-cell with the same electrolyte composition (IL + 5 Wt% polymer). Figure 18 shows the cycling behavior at C/4 of the graphite anode. In the first few cycles, a small fading of about 3% was noted and then the cell recovered, with good reversible capacity stabilization at 342 mAh/g. Our choice for this amount of polymer in the IL–polymer mixture is dictated by the fact that the desired electrode–gel interface that was the goal here can only be formed at polymer concentration of 5% or above.

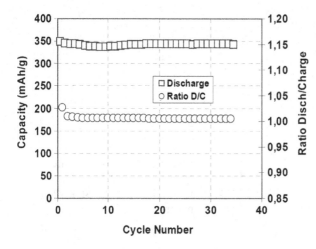

FIGURE 18 Cycling behavior of Li/graphite cell with Py13(FSI)–0.7 M LiFSI + 5% polymer.

The cells have an active surface area of 104 cm^2 (Fig. 19) and an installed capacity of 38 mAh. The first charge–discharge cycles at C/24 show an increase in the coulombic efficiency (CE) from 68.4% in the first cycle to 96.5% in the third cycle (Fig. 20). The low first CE cycle in the Li-ion cell is more closely related to the graphite anode, which has only 80% CE, in the first cycle. Thus, further improvements should be aimed at the graphite anode side. The reversible capacity was 113 mAh/g. After three cycles, the CE recovers to 96.5%, but the battery still needs more cycles to be efficient. The power capability was evaluated for this IL Li-ion battery based on the discharge capacity obtained at C/12. The charge was maintained in the constant regime at C/6 and the discharge regime ranged from C/12 to 40C. A stable capacity,

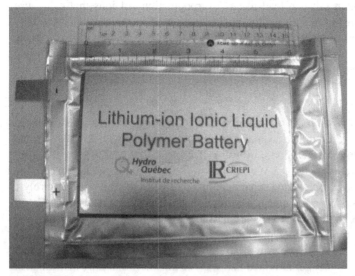

FIGURE 19 Li-ion flat aluminum bag cell with an active surface area of 104 cm^2.

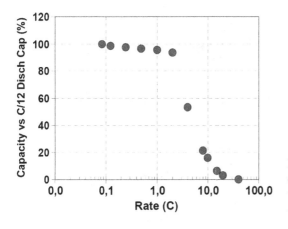

FIGURE 20   First charge–discharge cycles of an Li-ion cell with Py13(FSI)–0.7 M LiFSI + 5% polymer.

independent of the discharge rates, was obtained below and up to a 2C rate, as shown in Figure 21. The capacity starts decreasing at rates above 2C, with 54% of the rated capacity still delivered at 4C. As we have explained above, this limitation arises from the high concentration of ions in IL, which induce a lower free pathway of ions in the media, owing to ion pairing and other factors.

On the long cycling life, the cell was cycled at the C/4 rate between 4 and 2 V. Figure 22 shows the behavior of the discharged capacity normalized to the initial discharged capacity as a function of the cycle number. After 30 cycles, the capacity of the cell fades by 10%. This capacity fade, represented by 0.33% per cycle, is slightly higher than that in the Li-ion cell with a standard electrolyte.

Based on the cycling data of the separate half-cells (Figs. 16 and 18), we cannot attribute this fading to any of the half-cells; both electrodes separately have shown a good stability with a Py13–FSI–polymer electrolyte. But when we assemble the full Li-ion cell, the source of lithium ions is limited; thus, the graphite anode half-cell (80%) consumes more lithium ion to form its SEI on the graphite anode than does the standard electrolyte. However, this SEI layer needs a few cycles until it is fully

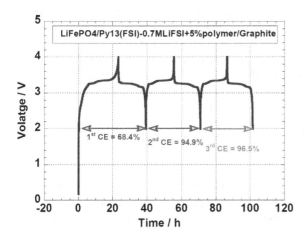

FIGURE 21   Rate capability of a Li-ion cell with Py13(FSI)–0.7 M LiFSI + 5% polymer.

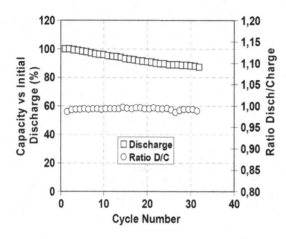

FIGURE 22 Cycling behavior of a Li-ion cell with Py13(FSI)–0.7 M LiFSI + 5% polymer.

stabilized. Hence, the balancing process between the anode and cathode capacities in the IL media can be one of the problems causing capacity fading. The large active surface area of (104 cm^2) of the electrodes raises a technological difficulty related to the wettability of the all the active material of both electrodes with an IL–polymer electrolyte. Notwithstanding this fact, we demonstrate that a Li-ion battery having LiFePO$_4$ and graphite as the cathode and anode, respectively, with an ionic liquid as the electrolyte is achievable. Some applications in which high power is not a requirement can benefit from this technology. However, many technical parameters, such as electrode porosity, separator porosity, and the wetting process, should be improved during further technical development.

4.3 Electrolytes with Enhanced Safety and Electrochemical Performance for Lithium-Ion Batteries: Mixtures of Ionic Liquids and Organic Aprotic Solvents

On the anode side, ionic liquids give rise to the unstable formation of SEI layer on the graphite material. Decomposition of ionic liquids on the anode graphite material has restricted their application in Li-ion batteries. Many studies have focused on this to improve the stability of the SEI layer. The addition of vinylene carbonate (VC) to the IL based on TFSI can suppress the reduction of this IL [16–18].

Here we review HQ work carried out to explore ways to increase the conductivity of battery electrolytes based on ILs but without compromising their stability, especially their nonflammability. This led us to examine mixtures of ILs with organic electrolytes usually employed in commercial Li-ion batteries: ethylene carbonate (EC) and diethylene carbonate (DEC). The properties of these mixtures are described below [19]. The formulas of some of the ionic liquids used in these mixtures are given in Figure 23.

Properties of Mixtures of Ionic Liquids and Aprotic Organic Solvents We have measured the effect of the salt concentration on the ionic conductivity in various ILs (i.e., EMImTFSI, PMImTFSI, HMImTFSI). The high viscosity and relative low

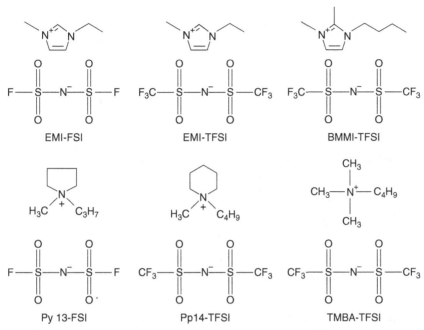

FIGURE 23 Chemical formulas of some ionic liquids used in mixed electrolytes.

conductivity of ILs are compounded by the fact that their conductivity *decreases* when a salt (e.g., LiTFSI) is added to make an electrolyte solution (Fig. 24). This behavior is contrary to that of conventional aqueous and nonaqueous solvents. One way to overcome this difficulty is to add an organic solvent such as EC and DEC. Many interesting properties have been observed in this new work on these mixtures.

First, we measured the viscosity of the mixed and pure electrolytes at temperatures from −20 to +80°C (Fig. 25). The viscosity varies inversely with the temperature. As the temperature increases, the viscosity decreases. When the temperature is changed from −20 to +80°C, the viscosity decreases 30 to 35 times for mixed and pure electrolytes, respectively. The addition of IL in the electrolyte increases the viscosity of the mixture; before reaching a 1 : 1 ratio by volume, the variation is very slow, which is probably due to the domination of the organic electrolyte aspect. In Figure 26 we present the relative variation in viscosity, Δ [$\Delta = (V_x - V_{50})/V_{50}$], of the mixture with a percentage of IL, where V_x is the viscosity at x% IL and V_{50} is the viscosity when 50% IL is added. From this representation we can observe clearly the effect on the viscosity of the mixture of IL addition to the organic electrolyte. A very small variation in the viscosity when the IL is less than 50% and a rapid increase in Δ after the IL ratio increases to above 50% in the electrolyte mixture are observed for all temperatures. This behavior indicates the possible unavailability of sufficient organic solvent for the solvation of ions, leading to some ion-pair formation as in a quasilattice.

To clarify the effect of ionic liquids on battery safety, flammability tests of electrolytes with and without IL were carried out. Figure 27 shows direct flame being

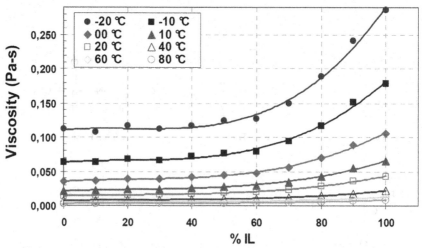

FIGURE 24 Conductivity of different ionic liquid–TFSI as a function of LiTFSI salt concentration.

applied to pure electrolytes and their mixtures. The electrolyte flammability occurred with a pure organic electrolyte in the first second of ignition; in contrast to this behavior, pure IL did not show any combustion even after more than 20 s of exposure to flame. As soon as IL is added to the electrolyte, the flame exposure time increases before a flammability level is reached. After adding 40% IL to the electrolyte, no

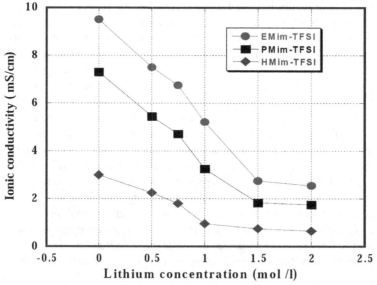

FIGURE 25 Viscosity at various temperatures as a function of ionic liquid content in an EC/DEC–VC–LiPF$_6$ electrolyte.

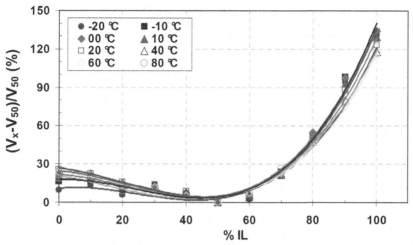

FIGURE 26 Relative variation Δ of viscosity as a function of ionic liquid content in an EC/DEC–VC–LiPF$_6$ electrolyte.

flammability is observed and the ignition period is increased. This result shows that safety can be improved in an Li-ion battery when the percentage of IL added in the organic electrolyte is at least 40%.

The nonflammability characteristic of IL-based electrolytes, so crucial for the safety of batteries, is maintained in these mixtures provided that at least 40% of the volume of the mixture is constituted by the IL (Fig. 27 and Table 3). When

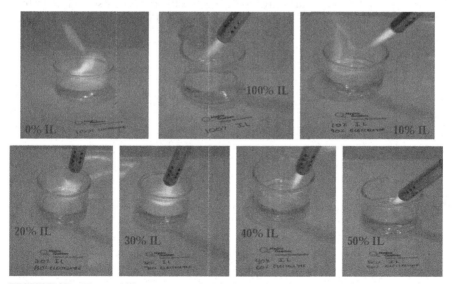

FIGURE 27 Flammability tests on various EC/DEC–VC compositions mixed with an EMI–TFSI electrolyte.

TABLE 3 Viscosity, Ionic Conductivity, and Flammability as a Function of the Percentage of IL in EC–DEC–VC–LiPF$_6$ Electrolytes

Ionic liquid (5)	At 25°C		
	Viscosity (Pa · s)	Conductivity (mS/cm)	Flammability
0% EMI–TFSI	12.1	8.50	Yes
10% EMI–TFSI	12.7	9.45	Yes
20% EMI–TFSI	13.7	9.31	Yes
30% EMI–TFSI	14.1	9.41	Yes
40% EMI–TFSI	14.9	10.09	No
50% EMI–TFSI	16.3	10.11	No
60% EMI–TFSI	17	10.45	No
70% EMI–TFSI	19.9	10.26	No
80% EMI–TFSI	24.5	10.13	No
90% EMI–TFSI	30.5	9.78	No
100% EMI–TFSI	36.3	8.61	No

the conductivity and viscosity of IL/EC–DEC mixtures are examined at different compositions, very interesting trends arise (Fig. 28). When the IL in mixtures is increased from 0 to around 60%, the conductivity rises, as one would expect if a salt (IL) is being added increasingly to the conventional solvent EC–EDC; additions of IL also increase the viscosity, but not so steeply. When there is about 60% IL in the mixture, further increases in IL lead to decreases in conductivity, as if the organic solvent left is in sufficient to "solvate" all the new "salt" (IL) being added and a mixture's behavior approaches that of, and is being dominated by, the behavior pattern of pure IL; correspondingly, not unexpectedly, there is a steep rise in the viscosity of the mixtures with an increasing IL fraction (Fig. 28).

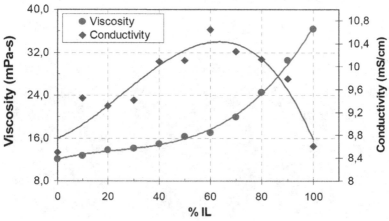

FIGURE 28 Conductivity and viscosity of the various EC/DEC–VC–1 M LiPF$_6$ compositions mixed with an EMI–TFSI electrolyte.

From the point of view of batteries, a remarkable *optimum* range has been discovered when the proportion of IL in the mixture is 40 to 60%: high conductivity, low viscosity, and nonflammability (Figs. 27 and 28 and Table 3). These are the principal desirable properties of a battery electrolyte. The expected inverse correlation between conductivity and viscosity (Table 3) of the electrolytes does not follow the mixtures' trend because the organic solvent has covalent bonding and few ions (by autoionization) and hence low conductivity, whereas the ionic liquids are by definition full of ions and therefore have higher conductivity than that of the organic solvents. Although viscosity increases when IL is added to a mixture, the conductivity shows a maximum at 60% IL and then goes down with increasing IL (see Fig. 28 and related comments).

We have also investigated in situ measurement of evaporation temperatures or the vapor pressure of some mixture electrolytes inside the SEM using a Peltier device and scanning the temperature range from −20 to +180°C. Figure 29 shows a scanning electron microscope (SEM) image of drops of IL, organic electrolyte, and their mixture (30 and 50% IL). For the pure IL drop, no decrease in the drop size was observed until the highest temperature limit of the device, 182°C, was reached. On the other hand, the organic electrolyte starts evaporation at around 50°C. When IL is added, the mixture drops show a strong direct relationship between the IL content in the electrolyte and evaporation temperature—hence the evaporation temperature, which is related to the flammability, can be increased by adding some IL to the organic electrolyte. When 30% IL is added to the electrolyte, the evaporation temperature increases to about 125°C, and it reaches a higher temperature of 176°C when the IL concentration in the organic solvent is 1 : 1. Therefore, the evaporation temperature of the mixture, related to the flammability, can be increased by 252% with 50% IL in the organic solvent.

The thermal properties of an electrolyte are central to battery safety. Thermal gravimetric analysis of all electrolytes with IL concentrations of 0%, 30%, 50%, 70%,

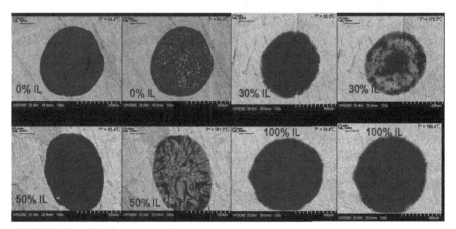

FIGURE 29   SEM images from in situ analysis of temperature evaporation of a mixed electrolyte with various IL contents.

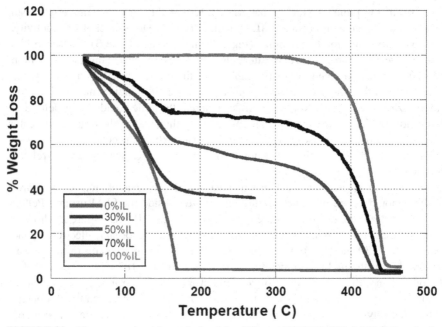

FIGURE 30 Thermogravimetric analysis of the different EC/DEC–VC–1 M LiPF₆ mixed with various EMI/TFSI ratios.

and 100% is shown in Figure 30. Up to 300°C, there is essentially no weight loss for pure IL, and less than a 3% loss was measured at 350°C, which explains the thermal stability of IL and confirms that only a small amount of volatile species is released. In contrast, for the organic electrolyte, the weight loss was 30% when the electrolyte is heated from 25 to 100°C and 80% at 160°C. By adding IL in the organic electrolyte, the thermal stability of the mixed electrolyte doesnot show a notable improvement up to 100°C. Electrolytes having 30%, 50%, and 70% IL show losses of 23%, 15%, and 10%, respectively, at 100°C. This result probably confirms in situ SEM analysis where the electrolyte drops were found to evaporate at different stages, depending on the proportion of IL in the mixtures. Although the IL addition improves the thermal stability of the organic electrolyte, the mixed electrolyte doesnot form a new phase, however, this first weight loss from 25 to 100°C is attributed to the decomposition of salt and organic solvent in the mixtures.

Anode and Cathode Behavior

Anode As reported in our previous work [2], graphite cannot be intercalated when pure IL based on the TFSI anion is used. Discharge–charge data for the first cycles of graphite electrode at a C/24 rate with pure and mixture electrolytes are shown in Figure 31. As is well known, intercalation in the graphite usually occurs below 250 mV; however for pure IL, no intercalation is observed. Only some exfoliate phenomena were noted, and a low coulombic efficiency is obtained. This contrasts

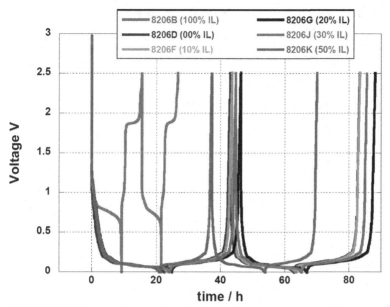

FIGURE 31   First-cycle behavior of Li/graphite of Li/EC–DEC–VC–1 M LiPF$_6$/graphite cells with EC/DEC–VC–LiPF$_6$ with various EMI/TFSI ratios.

with the results we found with pure IL based on an FSI anion, where the anion plays a key role in stabilizing the SEI layer and getting Li-ion impermeable film [19]. When organic solvent is added to the IL electrolyte in the ratio 1 : 1, a passivation SEI layer is obtained at 800 mV, followed by an intercalation compound formation. This is due to EC-based organic solvent decomposition during the reduction in the range 800 to 250 mV. Hence, a more stable passivation layer is built on the graphite particles. The coulombic efficiencies and reversible capacities were found to be sensitive to the percentage of the IL additive. Shown in Figure 32 is the effect of the IL content of the electrolyte on the coulombic efficiencies (Ah. Eff. in the figure) of the graphite anode. The lowest value, the first Ah. Eff., was obtained with the pure IL electrolyte, and the highest was obtained with the pure organic electrolyte. The addition of IL to the organic electrolyte reduces the coulombic efficiency value. When the electrolyte is mixed 1 : 1, the first coulombic efficiency is 80%, the same value that we obtained with pure IL based on the FSI anion [19]. So we conclude that the mixture of IL–FSI with organic electrolyte should show an improvement in graphite anode performance.

In the second cycle, the coulombic efficiency reaches 97% with the 1 : 1 mixture. The other parameter of interest here is the reversible capacity; a low reversible capacity was obtained with pure IL, with a value of 94 mAh/g. This reversible capacity increases with the amount of organic solvent in the electrolyte. Passivation layer formation improved with increasing amounts of organic electrolyte in the mixture, allowing the subsequent formation of intercalation compounds and a higher reversible capacity. Results of efficiencies and capacities are summarized in Table 4. The highest reversible capacity is obtained with the pure organic solvent followed by the

TABLE 4 First Electrochemical Characteristics of the MCMB Graphite Anode in an EC–DEC–VC–LiPF₆/EMI–TFSI Mixed Electrolyte

	Dis. cap.	Char. cap.	Ah. eff.
% IL	First/Second	First/Second	First/Second
100	141/94	96/79	68/84
50	311/255	250/248	80/97
30	372/314	315/312	85/99
20	388/324	337/331	87/102
10	362/306	329/302	91/99
0	345/306	316/312	92/102

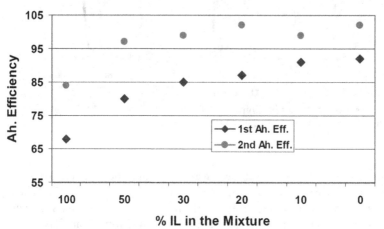

FIGURE 32 First and second coulombic efficiencies of Li/EC–DEC–VC–1 M LiPF₆/graphite cells with various EMI/TFSI ratios.

mixture having 10% IL, and thus decreasing proportionately as the percentage of IL is increased in the electrolyte.

Cathode The salient features of the LiFePO₄ cathode in these mixed electrolytes may now be described. The first and second cycles at the C/12 rate of the LiFePO₄ cathode are shown in Figure 33. The coulombic efficiencies and reversible capacities are given in Table 5. The first Ah. Eff. value for pure IL was found to be low, 82%. This behavior is probably related to the high viscosity of the IL and thus the low wettability of the carbon coating layer, both of which affect the reversible capacity, which is only 123 mAh/g. Nevertheless, the performance is reasonably good, independent of the IL content in the electrolyte mixture. The cells that have mixed electrolyte perform well and show data comparable to those for the pure organic electrolyte; hence, IL-based electrolytes exhibit excellent behavior for cathodes but less so for anodes.

The high rate performance of the LiFePO₄ was investigated as a function of the IL content and is represented in the Figure 34. The reversible capacity at high rates

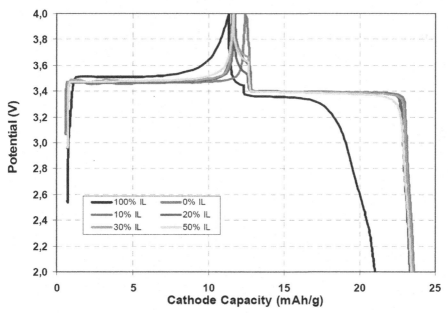

FIGURE 33   First-cycle behavior of Li/EC–DEC–VC–1 M $LiPF_6$/$LiFePO_4$ cells with various EMI/TFSI ratios.

decreases with increasing IL content, precipitously so for pure IL, perhaps due to its high viscosity and thus electrode wettability problem. With high viscous electrolytes such as in pure ionic liquids, the wettability of the carbon layer is more difficult to achieve, due to its large surface area arising from microroughness. Moreover, the high viscosity can also prohibit the wettability of all the electrodes in depth, both anodes and cathodes because, of the quasi-three-dimensional fractal nature of the electrodes, which contain not only mesopores (easily accessible by the electrolyte) but also micropores, which limit the electrolyte accessibility. With the electrolyte mixture (1 : 1), the high rate performance is sharply improved and resembles the pure organic electrolyte up to the 2C rate, but then starts decreasing slightly, although this

TABLE 5   First Electrochemical Characteristics of the LiFePO4 Cathode in an Organic–IL Mixed Electrolyte

	Capacity (mAh/g)		
% IL	Charge	Discharge	Ah. eff. (5)
100	150	123	82
50	154	153	99
30	154	154	100
20	154	152	99
10	154	154	100
0	155	156	100

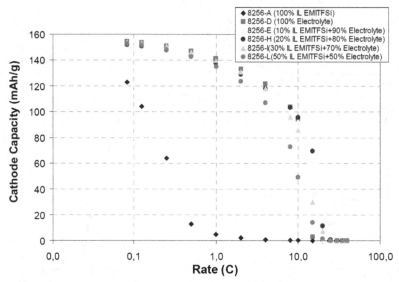

FIGURE 34   Rate capability of Li/EC–DEC–VC–1 M LiPF$_6$/LiFePO$_4$ cells with various EMI/TFSI ratios.

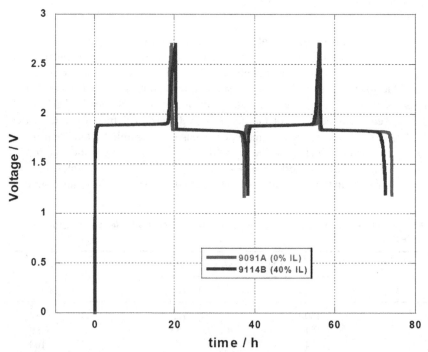

FIGURE 35   First-cycle behavior of a Li$_4$Ti$_5$O$_{12}$/LiFePO$_4$ cell at C/24 with 40% compared to 0% ionic liquid.

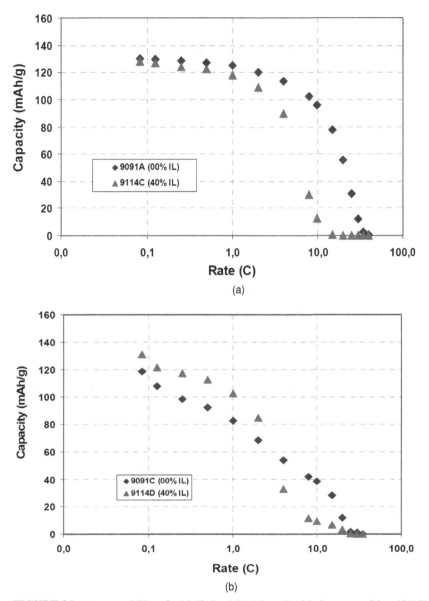

FIGURE 36   Rate capability of a $Li_4Ti_5O_{12}/LiFePO_4$ cell with the composition 40% IL.

trend continues up to the 10C rate. At rates higher than 10C, some departures can be noticed; cells having a mixture electrolyte with 10% IL and 20% IL showed a slightly higher capacity than that of the pure organic electrolyte, perhaps owing to their higher conductivity but yet pure organic electrolyte behavior: in other words, the higher quantity of salt available in the mixture electrolyte during peak power, arising from some IL in the electrolyte by increasing the salt concentration without yet creating high ion-pair interactions.

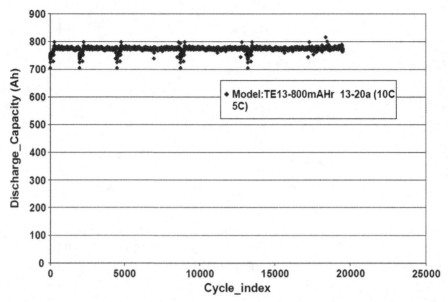

FIGURE 37   Cycling life of a $Li_4Ti_5O_{12}/LiFePO_4$ 18650 cell at Dis:10C–Ch:5C with an organic electrolyte.

Lithium-Ion Cell   Unfortunately, compositions having 10 to 30% IL showed flammability during flame tests (Fig. 27). To improve battery performance without losing any degree of safety, other mixed electrolyte compositions are used. Thus, Li-ion cells were assembled with mixture composition 40% IL as the electrolyte and $LiFePO_4$, and $Li_4Ti_5O_{12}$ as the cathode and anode, respectively. Figure 35 shows the first cycles at C/24 of Li-ion cells with this electrolyte (40% IL) and an electrolyte with no IL. At the C/24 rate, the cells yielded, respectively, 80% and 94% coulombic efficiency in the first cycle, and a reversible capacity of 115 and 128 mAh/g. The discharge and charge voltage plateaus are located at 1.84 and at 1.88 V for both cells. There is a very small polarization gap of around 40 mV between the charge and discharge curves, which reflects the high rate performance of the Li-ion cell.

To determine the rate capability performance, we carried out two types of evaluation: fast discharge and fast charge. A Li-ion cell having a mixture electrolyte (40% IL) was compared to one with a pure organic electrolyte (0% IL). The high-rate discharge capability of Li-ion cells is shown in Figure 36; the two types of cells deliver comparable capacity up to the 2C rate. A similar discharged capacity was obtained with both electrolytes. At 2C, the capacities are 119 and 109 mAh/g, respectively, for 0% and 40% IL. At rates higher than 2C, the discharge capacity for mixed electrolytes starts decreasing faster than for cells without IL, and the gap between capacities increases with increasing rate. In contrast to the high rate charging capability, a cell having a 40% IL electrolyte shows better performance up to 2C; then the data switch and a cell without IL improves (Fig. 36b). In both cases of high-rate discharge and charge tests, after a 2C rate the addition of 40% IL shows a lower performance than that of a pure organic electrolyte. This behavior can be associated with the availability of fewer organic solvent molecules to carry out solvation during charge and discharge at very high rates. However, this result shows an acceptable electrochemical performance and much improved safety. One of the safest material combinations for a Li-ion cell is $LiFePO_4/Li_4Ti_5O_{12}$. The evaluation of this cathode/anode configuration in the 18650 cell format using the organic standard electrolyte has already demonstrated a remarkable cycling life (Fig. 37). A very stable discharge capacity for a long cycle number discharging at 10C and charging at 5C was obtained for cells with a conventional electrolyte. The capacity is maintained constant at 800 mAh for more than 20,000 cycles. This type of accelerated aging is the equivalent of 50 years of the real life of the battery. For safe EV and HEV applications, however, using this technology, based on a $LiFePO_4$ cathode and a $Li_4Ti_5O_{12}$ anode in an optimal mixed electrolyte having around 40% IL could be of interest subject to further development and optimization.

5 CONCLUSIONS

The work at Hydro-Québec shows that IL is a good candidate for safer batteries; however, the performance should be improved. Mixed organic EC–DEC–1 M $LiPF_6$ and EMI–TFSI (IL) electrolytes can improve the performance of batteries, while maintaining the safety aspect of IL. An optimum range for the viscosity and ionic conductivity of a mixture electrolyte has been found between 30 and 40% IL in an organic electrolyte [19]. By adding a certain amount of IL, the performance of the graphite anode was also improved, and lithium intercalation occurred successfully, despite the utilization of IL based on TFSI anion. This mixed electrolyte gave results comparable to those of organic electrolytes for both a graphite anode and a $LiFePO_4$ cathode. From the point of view of a safer Li-ion battery, we have evaluated the configuration $LiFePO_4/Li_4Ti_5O_{12}$ in mixed electrolyte (40% IL) and compared it to the results for a standard organic electrolyte. The results were comparable up to the 2C rate for the discharge Ragone test, and then the capacity begins to decrease slightly with the mixed electrolyte. When the cells were evaluated using a high-rate charging test, cells with mixture electrolytes showed better performance up to 2C, and then the cells exhibited the same behavior as in the previous test. To make the best choice of electrolyte composition for use in a practical Li-ion battery, we should take into account the flammability, conductivity, and viscosity of the electrolyte. A battery with an electrolyte having an ionic liquid content between 30 and 70% should meet these requirements.

REFERENCES

1. A. Lewandowski and A. Swiderska-Mocek, *J. Power Sources*, **194**, 601 (2009).
2. A. Guerfi, S. Duchesne, Y. Kobayashi, A. Vijh, and K. Zaghib, *J. Power Sources*, **175**, 866 (2008).
3. (a) P. C. Howlett, D. R. MacFarlane, and A. F. Hollenkamp, *Electrochem. Solid-State Lett.*, 2004, **7**, A97. (b) J. H. Shin, W. A. Henderson, and S. Passerini, *Electrochem. Commun.*, 2003, **5**, 1016. (c) Y. Katayama, T. Morita, M. Yamagata, and T. Miura, *Eletrochemistry (Tokyo)*, 2003, **71**, 1033.

(d) H. Sakaebe and H. Matsumoto, *Electrochem. Commun.*, 2003, **5**, 594. (e) P. C. Howlett, D. R. MacFarlane, M. Forsyth, A. F. Hollenkamp, and S. A. Forsyth, World Patent WO2004082059-A1, 2004.

4. N. Bryne, P. C. Howlett, D. R. MacFarlane, and M. Forsyth, *Adv. Mater.*, **17**, 2497 (2005).

5. B. E. Conway, in *Modern Aspects of Electrochemistry*, Vol. 3, Butterworth, London, 1964, p. 43.

6. W. A. Van Schalkwijk and B. Scrosati, Eds., *Advances in Lithium-Ion Batteries*, Kluwer Academic/Plenum, New York, 2002, p. 3.

7. R. A. Huggins, *Advanced Batteries*, Springer-Verlag, New York, 2009, Chap. 7.

8. L. F. Nazar, B. Ellis, M. Makahnouk, and D. Ryan, in *Proceedings (Extended Abstracts) of the 48th Battery Symposium in Japan*, Electrochemical Society of Japan, Fukuoka, Japan, Nov. 13–15, 2007, p. 716.

9. A. Guerfi, S. Duchesne, P. Charest, Y. Kobayashi, A. Vijh, and K. Zaghib, in *Proceedings (Extended Abstracts) of the 48th Battery Symposium in Japan*, Electrochemical Society of Japan, Fukuoka, Japan, Nov. 13–15, 2007, p. 50.

10. S. Seki, Y. Kobayashi, Y. Ohno, H. Miyashiro, Y. Mita, N. Terada, P. Charest, A. Guerfi, and K. Zaghib, in *Proceedings (Extended Abstracts) of the 48th Battery Symposium in Japan*, Electrochemical Society of Japan, Fukuoka, Japan, Nov. 13–15, 2007, p. 174.

11. S. Seki, Y. Kobayashi, H. Miyashiro, Y. Ohno, Y. Mita, A. Usami, N. Terada, and M. Watanabe, *Electrochem. Solid-State Lett.*, **8**, A577 (2005).

12. M. Ishikawa, T. Sugimoto, M. Kikuta, E. Ishiko, and M. Kono, *J. Power Sources*, **162**, 658 (2006).

13. H. Matsumoto, H. Sakaebe, K. Tatsumi, M. Kikuta, E. Ishiko, and M. Kono, *J. Power Sources*, **160**, 1308 (2006).

14. Y. Wang, K. Zaghib, A. Guerfi, F. F. C. Bazito, R. M. Torresi, and J. R. Dahn, *Electrochim. Acta*, **52**, 6346 (2007).

15. A. Guerfi, M. Dontigny, Y. Kobayashi, A. Vijh, and K. Zaghib, *J. Solid State Electrochem.*, **13**, 1003 (2009).

16. M. Holzpfel, C. Josh, and P. Novak, *Chem. Commun.*, **4**, 2098 (2004).

17. M. Holzpfel, C. Jost, A. Prodi-Schaw, F. Krumeich, A. Wursig, H. Buqa, and P. Novak, *Carbon*, **43**, 1488 (2005).

18. T. Sato, T. Maruo, S. Marukane, and K. Takagi, *J. Power Sources*, **138**, 253 (2004).

19. A. Guerfi, M. Dontigny, P. Charest, M. Petitclerc, M. Legacé, A. Vijh, and K. Zaghib, *J. Power Sources*, **195**, 845 (2010).

CHAPTER 15

ELECTROLYTIC SOLUTIONS FOR RECHARGEABLE MAGNESIUM BATTERIES

Y. Gofer, N. Pour, and D. Aurbach

Lithium Batteries: Advanced Technologies and Applications, First Edition.
Edited by Bruno Scrosati, K. M. Abraham, Walter van Schalkwijk, and Jusef Hassoun.
© 2013 John Wiley & Sons, Inc. Published 2013 by John Wiley & Sons, Inc.

1 INTRODUCTION

Recent trends in energy security, concerns regarding the environmental impact of energy abuse, the ever-growing electrified world, and apprehension about resource depletion dictate global investment of funds and effort in the development of new, improved energy storage and conversion techniques in general and of rechargeable batteries in particular. There is no doubt that the past three decades showed remarkable progress in this field, owing mainly to the development of lithium-ion and NiMH batteries. Nonetheless, it is widely accepted that the search for alternative battery chemistries is crucial for future improvements. For one, current state-of-the-art battery systems fall short in the quest to set alternatives to the gas tank and to the internal combustion engine. Moreover, past lessons in energy and resource exploitation demonstrated unmistakably the necessity to diversify practices for energy conversion, transfer and storage, and resource utilization.

The attractive properties of magnesium, with respect to its utilization as an anode for high-energy-density batteries [1], were recognized early in the electrochemistry era. Magnesium is nontoxic and has high theoretical specific (charge) capacity, very low redox potential, and is light and abundant [2]. However, all attempts to develop workable rechargeable magnesium batteries over the past several decades come to nought [3–5]. There are two key issues in the quest to develop a feasible rechargeable magnesium battery system: (1) development of an electrolytic solution in which magnesium can be reversibly electroplated and stripped and in which a cathode can be reversibly charged and discharged; and (2) development of a cathode material that can be coupled with a Mg anode, presumably reacting reversibly with Mg cations.

In this chapter we summarize the basic electrochemical behavior of magnesium anodes in electrolytic solutions with respect to their possible utilization in rechargeable magnesium batteries. The basic hurdles in the development of feasible electrolytic solution systems are discussed, along with past and recent advancement in this field from our laboratory. Attention is also paid to the literature, both scientific and industrial. Due to space constraints, no specific account is given of cathodes for magnesium batteries, despite their marked importance. Potential future directions based on current knowledge are discussed in the concluding section.

2 ELECTROLYTES FOR RECHARGEABLE MAGNESIUM BATTERIES

Since magnesium is very reactive with water and oxygen, it is clear that for a rechargeable magnesium battery system, the electrolytic solution must be free of water and oxygen. As a result, the majority of past studies attempting to develop rechargeable magnesium systems concentrated on polar organic solvent–based solutions, with "simple" magnesium salts [6–8]. None of these attempts went far, as magnesium could never be electroplated or stripped reversibly from these solutions electrochemically. In the past two decades, following well-established lithium electrochemistry concepts, several additional attempts have been published [9,10]. Unfortunately, it was soon realized that, contrary to lithium, the range of possible solvents and salts is significantly narrower. Due to the divalent nature of its ion, few magnesium salts dissolve and dissociate sufficiently in polar organic solvents to yield ionic conductive media. In fact, of the commercial salts available, only magnesium perchlorate, triflate (ClO_4^- and $CF_3SO_3^-$) and, recently, TFSI [$(CF_3SO_2)_2N^-)$] show some, although relatively low, solubility in only a handful of organic solvents, such as propylene carbonate and acetonitrile. There were studies that claim to make use of tetrahydrofuran (THF) solutions with magnesium perchlorate [9]; however, our experience clearly demonstrated that such solutions are unattainable with thoroughly dried solvent and salt.

3 BASIC CONSIDERATIONS WITH RESPECT TO SOLUTION PROPERTIES FOR RECHARGEABLE MAGNESIUM BATTERIES

Besides being a magnesium ion–conducting medium, the electrolytic solution plays a fundamental role in the electrochemical reactions taking place at both the anode and the cathode. In the case of secondary batteries, in particular nonaqueous ones, the requirements regarding electrode–electrolyte interactions are exceedingly stringent. The basic ones are that the solutions must be stable with both electrodes and should support fully reversible electrochemical reactions to proceed. There are, however, several commercial secondary battery systems in which one or two of these requirements are not fully satisfied. In each of these cases there are compensating mechanisms that render the system functional.

Since any magnesium-based battery system is to be based on nonaqueous electrolytes, the mature models and mechanisms established in the last four decades for lithium electrochemistry are usually thought to be the guiding light for the development of rechargeable magnesium batteries. In our early studies, as well as those of others, it was very soon revealed that magnesium anodes behave in a manner radically different from that of lithium [1,11,12]. In fact, the very mechanism that allows the utilization of lithium (and other low-voltage lithium intercalation or alloying compounds, such as carbon and silicon) as an anode is the spontaneous buildup of "passivating" films [13]. This *passivation* layer protects the lithium–electrolyte couple from continuous reaction, which would lead to detrimental loss of active material and severe damage to the electrolyte. It is well established that the performance of lithium and lithium-ion batteries is dominated largely by the nature of these passivation interphases [14]. The fundamental characteristics of these surface films are that they are good electron insulators as well as good lithium-ion conductors. In this sense, the commonly used designation *passivation* is inappropriate [1]. Thus, whereas these films protect from an unmitigated reaction between the anode and the solution, they allow a reversible electrochemical deposition (or intercalation) of lithium by way of lithium-ion transport between the solution phase and the anode. Unfortunately, the case of magnesium is entirely different. Contrary to lithium, most magnesium salts are both electrons and ion insulators [10,15–17]. Hence, when magnesium anodes get in contact with most organic solvent–based solutions, a very tenacious passivation interphase forms, inactivating the anode totally. It is also reasonable to expect similar phenomenan to occur and block cathodes that are not thermodynamically stable with the solutions [18,19]. This is the fundamental reason that in the case of rechargeable magnesium batteries, the central requirement for the electrode–electrolyte system is to guarantee a surface-film-free condition [5,11,20].

To attain surface-film-free conditions, the electrolytic solution must possess several critical characteristics: First, the solvent must be thermodynamically compatible with magnesium metal; that is, it must be stable toward reduction by magnesium. Second, the salt must also be stable vs. magnesium. Third, the solutions must be totally free of water and oxygen, as well as free of any other reducible species (e.g., alcohols, CO_2, and the like).

The first solutions that fulfilled these three requirements were Grignard reagent solutions in ethers, such as THF [20–23]. Several studies demonstrated a very reversible electrochemical deposition and dissolution for Mg in several Grignard-based solutions. Nonetheless, this family of solutions is unrealistic for utilization in practical battery systems, for a number of reasons. First, the dissociation of Grignard reagents in ethers to ionic species is very minor, yielding solutions with extremely low ionic conductivity, in the range of tens of μS/cm [24]. Second, due to the reducing potential of these reagents, their stability vs. oxidation is very limited, forming solutions with a very narrow electrochemical stability window, in the range 0.5 to 1 V vs. Mg [22].

In 1990, a group from Dow Chemicals proposed a new class of electrolytes for magnesium batteries, superior to Grignard reagents [24]. These electrolytes were THF solutions of the reaction products of organoboron or organoaluminum reagents with diorganylmagnesium compounds. The complex salts thus produced can be looked upon as Lewis acid–Lewis base reaction products in which the aluminum or the boron core compounds serve as acids while the magnesium compounds are Lewis bases. Indeed, these families of solutions presented a considerable step up compared to Grignard solutions. They exhibited much higher ionic conductivity and a wider electrochemical stability window. The reversibility of the electrochemical deposition of magnesium was also very high, reaching values of close to 100%. In addition, these solutions had an extra intrinsic advantage. Since the complex salts are based on very reactive organometallic compounds, they react spontaneously with any residual contaminants in the solvent, such as water, oxygen and alcohols. This ensured the necessary conditions for reactive contaminant-free solutions, and the magnesium electrodes are held, thus are passivation-free. Unfortunately, this family of solutions still suffered from too narrow an electrochemical stability window, due to the reducing nature of the complex salts, rendering them unworkable for practical use.

4 KNOWLEDGE BASE CONCERNING THE MAJOR FACTORS THAT INFLUENCE THE CHARACTERISTICS OF SOLUTIONS FOR RECHARGEABLE MAGNESIUM BATTERIES

In the late 1990s we widened the scope of these families of electrolytes by exchanging one or more of the organic ligands for inorganic ligands. We tested a wide variety of

Lewis acid–Lewis base combinations and various solvents in the quest for an electrolytic solution with an optimal performance [1,4,5,26]. The performance parameters considered were the ionic conductivity, width of the electrochemical stability window, reversibility of the magnesium electrochemical redox process, exchange current density for magnesium deposition and dissolution, magnesium deposition overpotential, and magnesium deposition morphology, to name a few. After screening tests with several Lewis acid core elements (e.g., B, Al, P, Sb), we concluded that using Lewis acids based on aluminum compounds realized the best solution performance. Additionally, it was found that the chlorine atom serves as the best inorganic ligand [5,26]. (The oxidation stability of the Br ligands is inferior to that of Cl [27]).

Based on these conclusions, we synthesized an entire range of aluminum- and magnesium-based complex salts, with a variety of organic ligands, a varying Mg/Al atomic ratio, as well as the organic/inorganic ligand ratio, and total concentrations. Reactions and solutions were made in oxygen- and water-free Ar-filled glove boxes. All reaction procedures, conditions, and analysis conditions and data can be found in references cited 26, 28, and 29. These studies were carried out as a comprehensive electrochemical and analytical study, to determine the relationships between the nature of the reagents, the solution species formed, and their electrochemical properties. The analyses included multinuclear nuclear magnetic resonance (NMR), solid-state NMR, single-crystal x-ray diffraction Raman spectroscopy, Fourier transform infrared spectroscopy, inductively coupled plasma (ICP), x-ray photoelectron spectroscopy, and energy-dispersive x-ray analysis measurements, together with several electroanalytical techniques and theoretical density functional theory (DFT) calculations. Apart from a few examples, the measurements were carried out with THF as the solvent. Several conclusions had been drawn from these studies, as discussed next.

5 PRINCIPAL SOLUTION SPECIES OBTAINED FROM THE REACTION OF ORGANOCHLOROALUMINUM AND ORGANOCHLOROMAGNESIUM COMPOUNDS

The entire array of reagents, R_xMgCl_{2-x} and R_yAlCl_{3-y} (R = methyl, ethyl, butyl, and t-butyl, N-$(SiCH_3)_2$; $0 < x < 2$, $0 < y < 3$), react in any molecular ratio to yield an entire range of combinations of stable equilibrium species in the solution phase. The Lewis acid–base reactions between the magnesium and aluminum reagents result in transmetallation reaction, in which the chloride ligands reside primarily on the magnesium core and the organic ligands are bound to aluminum [26,29]. Naturally, this is true only if there is no relative excess of organic ligands. In such a case, when the aluminum molecules are saturated with organic ligands, the rest will also reside on Mg core species. The exact species existing in equilibrium depends on many parameters, such as the ratio between the solution components, Al/Mg/Cl/R, solvent, total concentration, temperature, and the identity of R. Apparently, the equilibrium species reflects a balance between a complete transmetallation reaction and a complete Lewis acid–base neutralization reaction. For example, the reaction products of 1 : 1 ($Et_2Mg + EtAlCl_2$) in THF gives the neutral $MgCl_2 + AlEt_3$, while the 1 : 2 reaction products are ionic $MgCl^+ + Et_2AlCl_2^- + Et_2AlCl$, and not the neutral species

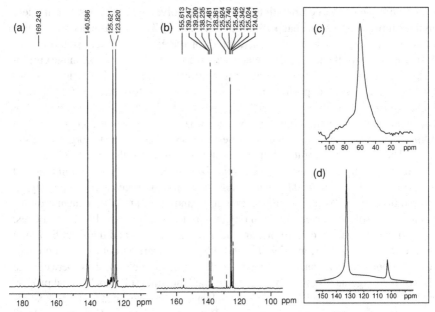

FIGURE 1 Transmetallation reaction between $AlCl_3$ + PhMgCl in THF. (a) ^{13}C NMR for PhMgCl in THF displaying Ph–Mg peaks at 169 and 140 ppm; (b) ^{13}C NMR for 0.25 M $AlCl_3$ + 0.5 M PhMgCl in THF, displaying transmetallated Ph–Al peaks at 155 and 139–137 ppm; (c) ^{27}Al NMR of $AlCl_3$ in THF; (d) ^{27}Al NMR of 0.25 M $AlCl_3$ + 0.5 M PhMgCl in THF.

$MgCl_2$ + $2Et_2AlCl$ (neglecting THF ligands) [26]. Figure 1 presents the basic ^{13}C and ^{27}Al NMR data showing the transmetallation products obtained from the reaction of PhMgCl with $AlCl_3$ (APC).

The aluminum species formed, anions or neutral molecules, always maintain a coordination number of 4, with either organic, chloride, or THF ligands. The magnesium species, either cations or neutral species, always maintain a coordination number of 6. In some cases, chloride bridging dimmers have been identified. In both cases the solvent molecule completes the coordination number of 4 or 6 for Al and Mg species, correspondingly.

LiCl and tetrabutylammonium chloride (TBACl) can be added to complex solutions in various proportions. The addition of these salts in appropriate proportions was found in some cases to impart improved characteristics to the solutions. Depending on the concentration added and the stoichiometric ratio, it may increase the ionic conductivity, improve the magnesium electrodeposition and stripping parameters, and widen slightly the electrochemical stability window of the solutions [28]. It is important to recognize that these two salts cannot by any means be considered as supporting electrolytes in the traditional sense. Both are introduced into the solution as reactants, with extensive impact on the equilibrium solution species. Note that LiCl is only slightly soluble in THF, and the solution thus produced possesses negligible ionic conductivity. From the behavior of the ionic conductivity vs. LiCl concentration in a variety of magnesium organochloroaluminum solutions, we concluded that LiCl reacts with the solution species in accordance with Lewis acid–base reaction schemes, where LiCl (or TBACl) acts as a Lewis base [30]. Based on ionic conductivity and NMR and Raman spectroscopy, it was established that LiCl reacts as a Lewis base with both aluminum and magnesium species. These reactions produce THF-solvated Li^+ (or TBA^+) ions in the solution, while Cl^- adjoins as a ligand to acidic species in solution, sequentially, according to the acidity strength order. As a rule of thumb, the Lewis acidity ranking goes as $AlL_3 > MgL_2 > LiCl > TBACl$. ($AlL_3$ and MgL_2 stand for any combination of the core atom with Cl and organic ligands. Naturally, the higher the Cl/R ratio for each such species, the stronger the Lewis acidity of the molecule.) Al-based anions and Mg-based cations are not included in this ranking, although they have significant involvement in the reactions and the resulting equilibrium species, since all molecules in the solution interact through dynamic, Schlenk-like equilibria [30]. In some cases, Al-based anions have been shown to exchange ligands and rearrange as a new set of equilibrium species [29]. Mg-based cations readily react with the two salts, after exhaustion of the stronger, Al-based Lewis acids in the solution, initially forming THF-coordinated $MgCl_2$ molecules [28]. It was found further that LiCl is such a strong base that in large excess it reacts even with $MgCl_2$ to yield the magnesiate anion $MgCl_4^{2-}$ [31,32]. Apart from changes in solution equilibrium species and their consequences on chemical and electrochemical behavior, an interesting result of LiCl addition is the ability to intercalate lithium ions from these solutions into a range of intercalation compounds. For example, lithium intercalation had been achieved from Li-containing APC solutions into such materials as V_2O_5 and $Fe_{0.2}Mn_{0.8}PO_4$, with which Mg shows no, or only very slow, intercalation. Figure 2 demonstrates the capability of Li intercalation into $Fe_{0.2}Mn_{0.8}PO_4$ from organometallic solution containing Li ions. This intercalation compound possesses two redox centers, pertaining to the transition metals Fe and Mn. The Fe^{+2}/Fe^{+3} redox lays within the electrochemical stability window of APC, whereas that of Mn^{+2}/Mn^{+3} lays beyond [33]. When the composite electrode cycled in Li-containing APC up to 3.0 vs. Mg, close to the theoretical intercalation level is obtained, with close to 100% reversibility. However, as seen in the insert, when the potential scan goes beyond the stability window, large positive currents flow, marking the irreversible oxidation of the electrolyte solution and the limit of the stability window. In fact, whenever intercalation cathodes are polarized in such complex salt solutions containing LiCl, the Li intercalation process constituted the preferred electrochemical reaction, owing to the rapid kinetics of lithium intercalation compared to that with magnesium.

6 CORRELATION BETWEEN SOLUTION IONIC CONDUCTIVITY AND THE EQUILIBRIUM SOLUTION SPECIES

The ionic conductivity of solutions depends primarily on the Al/Mg ratio, and to a lesser extent on the nature of the organic ligands and the total concentration of the complex salts [34] (Fig. 3). The Al/Mg ratio determines whether the solution species

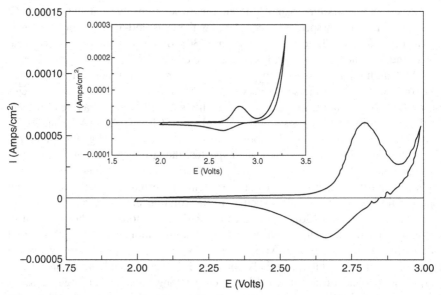

FIGURE 2 Cyclic voltammograms demonstrating lithium intercalation into $Fe_{0.2}Mn_{0.8}PO_4$ in 0.25 M APC + 0.5 M LiCl solution. The working electrode is $LiFe_{0.2}Mn_{0.8}PO_4$ (composite with carbon black and PVdF) on a Pt plate; Mg foil; scan rate = 0.1 mV/s. Insert: same electrode scanned up to 3.3 V.

would be ionic for the most part or mainly neutral. For example, with Et_2Mg and $EtAlCl_2$ as the reagents, at a Al/Mg ratio of 1 : 1, the solution contained chiefly the neutral species $MgCl_2$ and Et_3Al (neglecting THF ligands), yielding solutions with very low ionic conductivity. At any other mixing ratio, ionic species are generated, peaking in the vicinity of 1 : 2 and 2 : 1 ratio, the dominant species $MgCl_2 + EtMg^+ + Et_4Al^-$ and $MgCl^+ + Et_2AlCl_2^- + Et_2AlCl$, respectively [26]. The maximum ionic conductivity measured for a very wide array of complex solutions with different compositions ratios, ligands and total concentrations is around 0.8 to 2.4 mS/cm [26,31,34].

7 CORRELATION BETWEEN SOLUTION SPECIES AND THE METAL ELECTRODEPOSITED

With most complex solutions, anodic deposition yields a practically pure crystalline magnesium deposit [11] (Fig. 4). This has been verified numerous times, by energy dispersive spectroscopy (EDS), ICP, and atomic absorption. This point is not trivial, since the aluminum deposition potential (-1.66 V) is appreciably higher than that of magnesium, and in some of the solutions the total concentration of aluminum species can be twice or more that of magnesium. We have found however, that from a certain Al/Mg ratio in the complex solution [e.g., with DiChloroComplex (DCC), more than

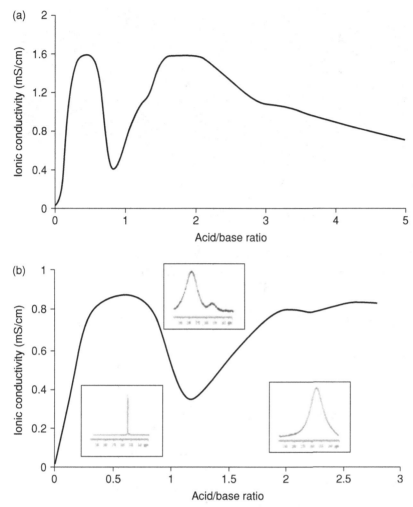

FIGURE 3   (a) Specific conductivity for 0.25 M solutions of the reaction products of Bu$_2$Mg ("base") with EtAlCl$_2$ ("acid") in THF, at various reactant ratios. (b) Specific conductivity for 0.25 M solutions of the reaction products of Et$_2$Mg (base) with EtAlCl$_2$ (acid) in THF, at various reactant ratios. The ^{27}Al NMR spectra for the reaction products of Et$_2$Mg with EtAlCl$_2$ at various ratios are presented in the inserts.

2 : 1 at 0.25 M in THF], codeposition of Al with Mg begins, with an increase in the Al content as the ratio increased [35]. This is in accordance with the well-established electrochemistry of nonaqueous aluminum chloride complexes used for aluminum electroplating. The critical parameter for aluminum electrodeposition is high solution acidity (i.e., excess of aluminum species). These results suggest that the utilization of ubiquitous Mg–Al alloys such as AZ31 and AZ91 may be used for the anode in a fully engineered battery: indeed, cycling efficiency tests with a few Mg alloys, and charge discharge measurements with complete cells (with Chevrel phase cathodes).

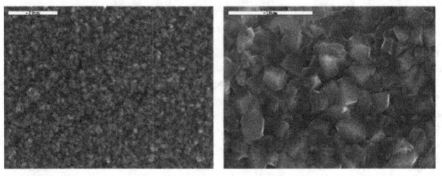

FIGURE 4   SEM micrographs of Mg deposits on copper substrates in 0.25 M DCC at two magnifications (a scale bar appears in each image). The deposition current density is 2 mA/cm^2 and the total charge density is 2.8 Ω/cm^2.

8   ELECTROCHEMICAL STABILITY WINDOW OF THE SOLUTION AND ITS CORRELATION WITH SOLUTION CHEMISTRIES

The electrochemical stability window of the complex salts solutions is dominated chiefly by the nature of the R ligand and by the Cl/R ligand [26] (Fig. 5). To a lesser degree it is influenced by the total concentration and by the Mg/Al content ratio. The measurement of the electrochemical stability window is also influenced strongly by

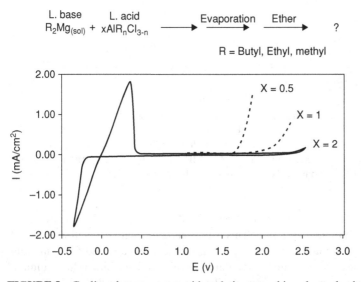

FIGURE 5   Cyclic voltammograms with a platinum working electrode, demonstrating the electrochemical stability windows for a THF solutions containing the complex electrolytes synthesized from Bu$_2$Mg + xEtAlCl$_2$ at different x values (Lewis acid–Lewis base ratios). Scan rate = 20 mV/s. The total electrolyte concentration was 0.25 M in all cases.

the electrode material used. Only a few materials, such as Pt and glassy carbon, are stable at the high-voltage side of the window.

Influence of the Cl/R Ratio   With a homolog series of complex solutions with aliphatic ligands it was found that for a constant Al/Mg ratio, R, and total concentration, the higher the Cl/R ratio the wider the electrochemical stability window, that is, the solutions are more electrooxidation resistant [26]. This behavior can be rationalized from two points of view.

1. For these organometallic salts, the most oxidation-susceptible bonds are the Al–C (and Mg–C bonds, if they exist). Hence, for the general case of the molecules R_xAlCl_{3-x}, $x = 0$ to 3 (or the anions, where $R_xAlCl_{4-x}^-$, $x = 0$ to 4), the higher the Cl/R ratio, the lower the electron density on the Al core. This, in turn, polarizes the Al–C, rendering it more oxidation resistant. Support for this perspective comes from ^{13}C and ^{1}H NMR measurements. Focusing at the peaks for carbon and hydrogen atoms α to the Al–C bond, with complexes containing various Cl/R ratios, showed that the higher the Cl/R ratio, the more the NMR peaks shifted lowfield.

2. Quantitatively, the higher the Cl/R ratio, the fewer the organic ligands attached to the metal core, and thus the lower the probability for an electrooxidation reaction to take place; accordingly, the more resistant is the compound to oxidation.

Influence of the Identity of R   NMR analysis of the electrolysis reaction products for a DCC complex suggested that the main route of the electrooxidation reaction is α elimination, the elimination of hydrogen on carbon α to the Al–C bond, and formation of an olefin [36]. Hence, organic ligands containing no C–H bond on α carbon, or ligands that cannot form C–C double bonds, such as methyl, *t*-butyl, phenyl and hexamethyldisilazide (HMDS), are expected to exhibit improved stability toward oxidation. The latter is also expected to be a weaker nucleophile, which may impart extra stability [37]. Indeed, measurements with complexes containing methyl, phenyl, and HMDS groups did show evidence of greater stability. Complex salts with *t*-butyl ligands were significantly inferior in this respect, even compared to the aliphatics. The complexes' solutions obtained with methyl ligands suffered from low solubility and ionic conductivity. To date, the widest electrochemical stability window measured in our laboratory involved the variety of complex salts containing phenyl ligands [29]. Optimized complex salts of this family obtained the best overall electrochemical characteristics so far. Replicating literature reports with HMDS ligands obtained solutions exhibiting somewhat inferior overall characteristics for optimized systems.

In general, the trend in the electrochemical stability window can also be a guiding rule as to the relative chemical stability of solutions toward oxidation. This point is critical when safety issues are considered, especially for potentially pyrophoric organometallic compounds solutions such as these. It must be emphasized, though, that an accurate and unambiguous electrochemical stability window is very difficult to determine [38]. Not only is determination of the exact breaking point frequently subjective, it is also exceedingly dependent on numerous factors, such as the ionic conductivity, the kinetics of the electrooxidation reaction, and the background currents. Also, the electrochemical window is always measured for a whole system, namely, the solution and the working electrode (and the reference electrode), usually neglecting the influence of the working electrode material, which is more than frequently unjustified.

9 KINETICS, REVERSIBILITY, AND MORPHOLOGY OF THE MAGNESIUM DEPOSITION AND STRIPPING PROCESS AND ITS CORRELATION WITH THE SOLUTION SPECIES

Undoubtedly, one of the most important electrochemical characteristics of solutions used in rechargeable magnesium batteries is the coulombic efficiency of the deposition–dissolution process. Several factors were found to have a profound influence on the cycling efficiency. Among them, the most important is the Cl/R ratio. For any series of complex salts with specific R, the lower the Cl/R ratio, the higher the cycling efficiency. In other words, the increasing content of organic ligands improves the electrochemical magnesium deposition redox characteristics. In solutions with a large R/Cl ratio—in particular those with complex salts containing organic ligands solely—the electrochemical deposition of magnesium frequently reached 100%. Other factors also have an influence, although less so, on the magnesium cycling efficiency: for example, the total concentration, the Mg/Al ratio, and the working electrode material (the highest cyclabilities were measured with a silver working electrode [22].)

In addition, another important parameter, the electrochemical deposition and dissolution kinetics, was found to vary in accordance with the same factors. Apart from the impact on the highest-attainable charge–discharge currents, the electrochemical kinetics also influences the deposition morphology. Smooth deposition and dissolution morphologies are crucial for long cycle life and for safety.

Influence of the Cl/R Ratio   Understanding the mechanisms and dynamics of the electrochemical processes taking place at the anode during these reactions is only in its infancy. Nonetheless, microelectrode electroanalytical measurements have led us to draw some coherent conclusions [39]. This study revealed that there is a consistent correlation between the exchange current density of the deposition reaction and the Cl/R ratio for DCC complexes. On the whole, the higher the Cl/R ratio, the lower the exchange current density. These findings match almost perfectly three other characteristics: the electroplating coulombic reversibility, the magnesium deposition and dissolution overpotential and as shown earlier, the electrochemical stability window. As a general rule, the conclusion from this study is that the richer are complexes with organic ligands, the larger are the deposition (and dissolution) exchange current densities, the smaller are the electrodeposition and electrodissolution overpotentials, and the better is the coulombic efficiency. Unfortunately, this comes at the expense of the width of the electrochemical stability, as presented above. A reasonable compromise is achieved, for example, with the normal DCC complex solution (Al/Mg/Cl/R 2 : 1 : 4 : 4 solution), yielding a cycling efficiency of close to 100%, a magnesium deposition overpotential of 50 to 150 mV, a solution conductivity of 1 to 1.5 mS, and an electrochemical stability window of about 2.2 V. The case of complexes made with phenyl ligand–bearing solutions is more complicated, as presented later.

Influence of the Identity of R   The identity of R, the organic ligand, also has some, albeit lesser influence on the electrodeposition characteristic. We did not find a coherent pattern or trend correlating R to these features. In fact, some of the complex salts tested, such as the one containing t-butyl ligands, showed significantly better electrochemical deposition characteristics. Unfortunately, with all but the four ligands mentioned above, the electrochemical stability window was unacceptably narrow. The origin of the relationship between the deposition overpotential and the exchange current density vs. the Cl/R ratio is not fully understood, yet we speculate that it is the result of strong interfacial interactions, such as specific adsorption of Cl-containing species onto the electrode surface.

The magnesium deposition morphology is dependent primarily on the current density, but also on the identity of the complex (Al/Mg and Cl/R ratio, nature of R), the total concentration, and the substrate. Comprehensive study of this subject was performed only for the DCC family of electrolytes [39]. Nevertheless, specifically interesting electrolytes had also been tested, with similar results, supporting the conclusions drawn from the comprehensive study [39]. In general, at relatively low current densities, in the range of several mA/cm^2, the deposit from DCC is characterized mainly by a homogeneous layer consisting of small crystallites about 0.5 to 4 μm in size. Each crystallite appears like a single crystal, with well-developed sharp edges. At higher current densities, different complexes, and high total deposition charge per area different morphologies have been observed, as elaborated in reference 39. Unquestionably, more important is the deposition and the electrodissolution morphology with pure and alloy magnesium foils as substrates, the true anodes in a fully engineered battery. Some work had been devoted to this issue, although a comprehensive study was never completed. In general, it was found that the most important factor in this respect is the initial anode surface chemical condition and the very first electrochemical process current densities used. The reason for this is that the native surface of even the cleanest magnesium anode is covered with tenacious stable surface oxides. For the anode to function effectively, these surface films must be permeated and expose homogeneous fresh metal surface. Some measures had been found to yield enhanced anode behavior, such as applying short spikes of large positive or negative currents as a "surface conditioning" treatment. Chemical surface conditioning of anodes with weak acids such as alcohols were also found to be beneficial. Further discussion on the subject is beyond the scope of this chapter.

10   SOLUTION CHEMICAL STABILITY AND SAFETY CONSIDERATIONS

Together with the electrochemical characteristics of electrolytic solutions, there are several other decisive parameters that could possibly render them commercially feasible or unfeasible; materials safety is unquestionably a central one. As mentioned above, all the complex salts solutions containing organometallic components are very susceptible to oxygen and water, let alone pyrophoric in some cases. THF and other ethers are naturally flammable, and many possess a very low flashpoint and high volatility. By contrast, the components of all these complex salt solutions, especially after degradation due to air exposure (after reaction with oxygen or water and CO_2), mainly aluminum and magnesium oxides and chlorides and various alcohols (from the reaction of the organic ligands with humidity), are not particularly poisonous or dangerous.

One of the immediate threats associated with any rechargeable battery system is related to electrochemical abuse during charging, particularly when a cell is overcharged. The primary concern relates to excessive pressure buildup that may cause battery bursting or venting. Other concerns relate to battery damage and the evolution of hazardous materials. In two experiments conducted with DCC and APC, we deliberately electrolyzed a large quantity of solutions at a constant potential well above the electrochemical stability window of the solutions. In both cases we passed a very large amount of charge, about one electron per anion in the solution. In both cases no gas evolution had been observed, and no pressure buildup was measured. The electrolysis products were identified with gas chromatograph–mass spectrometry and ^{1}H NMR and were found to be primarily benzene and butene, respectively. The lack of pressure buildup in the second case was presumably associated with the solubility of butene in the solution. These results testify that, at least in the absence of a cathode, even massive overcharging does not pose a hazard with these solutions.

Over the course of the exploration we developed two conceptions with respect of taming the reactivity of these solutions. The main hazard posed by these electrolytes is, undoubtedly, their pyrophoric nature. The hazard is increased dramatically when the complex salts dry off by solvent evaporation. The dry organometallic compound thus formed has a very large surface area and direct accessibility to water and oxygen. Replacing the highly volatile THF for ethereal solvents of the glyme family greatly reduced the tendency of the solutions to ignite spontaneously, even upon deliberately spilling relatively large quantities of the solutions on a flat surface under humid ambient conditions. This is achieved due to the low vapor pressure of these solvents and their slower evaporation. The effect was obtained with several members of the glyme family, such as diglyme, triglyme, and tetraglyme, as well as their mixtures with THF. Addition of as low as 20% v/v of the glymes had a strong positive influence. By the same token, utilization of these solvents, solely or as mixtures with THF, lowers the toxicity of the solutions, as the share of THF becomes smaller. Some of the electrochemical properties of the electrolytes are affected to some extent by this change in solvent. For example, due to their increased viscosity (and possibly their different polarity, donor number, and ability to form the desired species), the ionic conductivity is slightly decreased. However, for the most part, optimized solutions with glymes seem to be a feasible and safer alternative to THF. The second concept involves utilization of the appropriate gelling agents, so as to obtain semisolid gels or immobilized solutions. Poly(ethylene oxide) of various types (molecular weight, end groups, etc.) and poly(vinyl chloride) have been shown to be compatible with many complex salt solutions. It is conceivable that various other polymers or nanoparticle inorganic suspensions (such as nanosilica, nanoalumina, etc.) may serve the same purpose. In addition to many other advantages of using semisolid solutions for electrolytes, their beneficial effect on safety stems from the same reasoning as with the use of glymes. Proper selection of a gelling agent, polymeric or inorganic, is crucial for proper functioning of the full cell. Naturally, it is important to make sure that the electrochemical characteristics of the solutions will not be compromised. We believe that proper selection of such an additive and its formulation should be made only after completing full optimization of the electrolytic solution chemistry.

11   INSERTION OF MAGNESIUM IONS INTO INTERCALATION COMPOUNDS IN ORGANOMETALLIC COMPLEX SALT SOLUTIONS

To be practically feasible, apart from being chemically stable with the cathode, any electrolytic solution must also support cathodic electrochemical reactions. One of the most likely classes of cathode materials is, obviously, that of the magnesium intercalation compounds. Hence, to support the reversible intercalation reaction, complex salt solutions must carry liable magnesium species for intercalation under positive bias. This characteristic is by no means trivial, as no free magnesium ions are present in the solutions. Moreover, any of the magnesium-based species in these solutions carry at least one ligand, and in some cases it exists as a dimmer. For an Mg^{2+} ion to be freed of its ligands and solvation shell, it has to interact with Cl^- accepting species, such as $MgCl^+$, $Mg_2Cl_3{}^+$, or AlL_3. Thus, it is hypothesized that the electrochemical reaction requires the concerted interaction of at least two species on the electrode surface at the same location and time. This view is also true for the electrochemical deposition of metal on the anode. A few magnesium intercalation compounds, such as Mo_6T_8 (T = S, Se) Chevrel phases and TiS_2 (cubic and layered polymorphs), have been shown to readily undergo reversible electrochemical intercalation in many of the electrolytes described above [41]. Magnesium intercalation electrochemistry kinetics was found to be dominated by the solid-state diffusion of Mg ions within the intercalation compounds. That is, besides providing free magnesium ions liable for intercalation, the kinetics of the reaction is far faster than the rate determining step (RDS) for the intercalation process. Thus, with regard to constituting liable Mg ions for intercalation, these complex solutions have been proven to serve their purpose.

12 RECENT ADVANCEMENTS IN UNDERSTANDING OF THE STRUCTURE OF MAGNESIUM ORGANOHALOALUMINATE SOLUTIONS CONTAINING PHENYL LIGANDS

Lacking α-hydrogen, it was hypothesized that the homolog complex salts encompassing phenyl as the organic ligand will act just as the aliphatics, apart from being more oxidation resilient. Indeed, these solutions, denoted as the APC family, exhibit the widest electrochemical window achieved thus far (Fig. 6), Interestingly, comprehensive analysis of these solution structures revealed that some of their characteristics deviate from the picture established for the aliphatic ligand–bearing complexes [29]. For example, the coexistence of $AlCl_4^-$ and Ph_4Al^- had been identified, while with the aliphatics, uniform ligand distribution to yield $R_2AlCl_2^-$ was established persistently. These solutions, correspondingly, exhibited unique electrochemical behavior, noncompliant with the trends discovered with aliphatic ligands bearing complex salts. The most intriguing solution is the uncorrelated electrochemical stability window vs. Cl/Ph ratio. In agreement with all previous conclusions, DFT calculations have shown that for the homolog series of complex salts with phenyl ligands, the higher the Cl/Ph ratio, the higher the HOMO energy level (Table 1).

Since the electrooxidation reaction involves electron capture from the molecular HOMO level, it is conceivable that the higher the HOMO energy level, the higher will be the oxidation potential for the related species [41]. Measurements of the electrochemical stability window for a wide range of solutions with varying Cl/Ph ratios did not show a consistent trend. Surprisingly, the electrochemical window measured with a Pt working electrode for the exceedingly phenyl-rich complex salt,

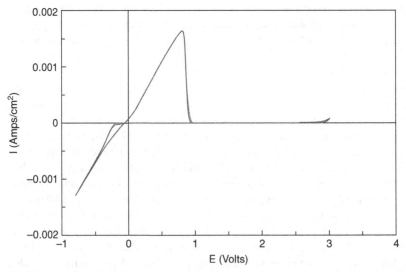

FIGURE 6 Typical cyclic voltammogram demonstrating the electrochemical windows of 0.25 M APC with Pt as working electrode and Mg as counter and reference electrodes. Scan rate = 0.25 mV/s.

TABLE 1 Summary of the HOMO and LUMO Energy Levels Calculated and Their Energy Difference in Electron Volts for Haloorgano Aluminum and Magnesium Species in THF

Molecule	Energy level (eV)		
	HOMO	LUMO	ΔE
Ph_4Al^-	-5.33	0.64	5.97
Ph_3AlCl^-	-5.76	0.38	6.15
$Ph_2AlCl_2^-$	-5.98	0.34	6.32
$PhAlCl_3^-$	-6.20	0.23	6.43
$Ph_3Al \cdot THF$	-6.55	-0.62	5.93
$Ph_2AlCl \cdot THF$	-6.71	-0.71	6.00
$PhAlCl_2 \cdot THF$	-6.99	-0.77	6.22
$MgCl^+ \cdot THF_5$	-7.46	-0.79	6.67
$Mg_2Cl_3^+ \cdot THF_6$	-7.51	-0.69	6.82
$AlCl_4^-$	-7.53	1.01	8.54

comprising about a 1 : 9 Cl/Ph ratio, showed no distinctive positive ascending currents in excess of 5.5 V. This result obviously reflects unique behavior: not only that the organometallics species expected to electrooxidize well below this voltage, but both THF oxidation and chlorine evolution occur well below this potential. These and additional results substantiated the notion that this behavior is related to interfacial phenomena rather than thermodynamic stability. To date, the mechanism that leads to the apparent inordinate stability is still not well understood. We hypothesize that a very effective specific adsorption of certain solution species promotes this extraordinary stability. As the only difference between these and the other complex salt solutions is the existence of phenyl ligands, it is rational to hypothesize that strong interactions involving the phenyl's π electrons is responsible for this interesting behavior.

Preliminary electroanalytical studies suggested that this surface phenomenon is potential dependent, and it may also be strongly dependent on the working electode's material nature. It was shown, for example, that with carbon and gold electrodes, the electrochemical behavior is somewhat different. In any case, thorough studies with a variety of working electrode materials, including composite cathode materials such as V_2O_5 and $Fe_{0.2}Mn_{0.8}PO_4$ (Fig. 2), have proven that the APC solutions indeed possess a "true" very wide electrochemical stability window, well better than any of the other family of complexes, approaching 2.9 to 3.1 V. DFT calculations, multinuclear NMR, and Raman spectroscopy studies have shown several interesting chemical insights related to this family of solutions. For example, it has been discovered that for the homolog $Ph_{4-x}AlCl_x$, $x = 1$ to 3, series of molecules, excess in the electron-withdrawing Cl ligands causes changes in the electronic structure of the aromatic organic ligands, and it acquires a quinoid structure. Additional interesting features have been revealed from electroanalytical measurements, such as the predeposition of aluminum atoms, on the order of one to several monolayers, prior to magnesium bulk deposition. Such a phenomenon may have very important consequences related to the choice of anode material, the anode current collector, and the origins of the magnesium deposition overpotential.

13 SURVEY OF RECENT PUBLICATIONS ON ELECTROLYTIC SOLUTIONS FOR RECHARGEABLE MAGNESIUM BATTERIES

With the increased interest in room-temperature ionic liquids (RTILs) in recent years, several publications have reported on reversible magnesium deposition from such solutions [42–44]. It claimed to possess advantageous properties, such as greater stability, lower vapor pressure, and better safety features. Examining selected RTILs in our laboratories has revealed that developing ILs for practical purposes in the field of magnesium batteries is not a trivial venture. One reason for this is that very few Mg salts are soluble in ILs. Another is that just like any other selection of electrolyte, it must be absolutely stable with Mg metal. To achieve this, proper selection of the cation and the anion, as well as meticulous purification of the RTIL and the added Mg salt, are critical. Many common anions used with RTILs (e.g., BF_4^-, $AlCl_4^-$, PF_6^-) are inadequate, due to the passivating surface films formed on the metal in contact with them or due to their limited oxidation stability. Some of the common quaternary ammonium cations are also instable with Mg. For example, ethylmethyl imidazolium [45] was found to be reduced directly by Mg. Therefore, while a wide variety of Mg-stable RTILs cations exist, the greatest effort will be to hit upon the proper anion, which will encompass all the required properties.

An interesting Grignard reagent was proposed in 2000 by Liebenow et al., comprising hexamethyl disalazide [37] and magnesium chloride (HMDSMgCl) in THF, from which magnesium was shown to be reversibly deposited. This solution also showed the widest electrochemical window achieved with Grignard reagent solutions alone. Recently, Kim et al. [46] proposed an improved electrolyte based on the reaction product of HMDSMgCl with $AlCl_3$. The authors state that in this way they have obtained an electrolyte with an excellent reversible magnesium deposition process, an electrochemical window of about 3.2 V, and good ionic conductivity. Additionally, the Al-based anion generated is nonnucleophilic, thus allowing the use of elemental sulfur to construct a high-energy-density rechargeable battery. The core of the researchers' innovation is purification by recrystallization of the solution so-obtained, yielding a superior electrolyte due to the removal of HMDSMgCl surplus, which oxidizes at 0.8 V below that of the purified solution. Studies with this electrolytic system in our lab showed similar results obtained by using the correct reactant ratio from the beginning ($2:1$ HMDSMgCl + $AlCl_3$ for Mg_2Cl_3 $HMDS_2AlCl_2$ or HMDSMgCl + $MgCl_2$ + $AlCl_3$ for Mg_2Cl_3 $HMDSAlCl_3$), thus omitting the purification stage. In any event, this electrolytic solution is interesting and certainly constitutes a promising candidate for practical rechargeable magnesium batteries.

The literature is also peppered with papers presenting various conceptions to mitigate one or more of the deficiencies present in current solutions. For example, the utilization of Grignard reagents, represented by the formula RMgBr, with fluorine-derivatized organic ligands R, was studied by Youngsheng et al. [47]. Derivatizing the organic ligand with strong electron-withdrawing fluorine atoms provided Grignard solutions with an improved electrochemical window. It will be interesting to learn whether the addition of Lewis acids such as $AlCl_3$ will further improve the properties of these solutions. Certainly, one of the important actions will be to eliminate the Br ligands, as bromine evolution beginning below the potential for chlorine evolution will lower the highest electrochemical stability window attainable. In a paper by Padany et al. [48], the authors present ion-conducting gel-polymer electrolytes with dispersed silica. The dispersed silica improved both the mechanical and the electrochemical properties of the electrolytes. Unfortunately, the use of $Mg(ClO_4)_2$ precludes the applicability of these gel polymer electrolytes for rechargeable Mg cells. In yet another paper, Liang et al. [49] suggest the use of Mg nanoparticles as the anode in a cell consisting of a DCC-type solution. According to the authors, the metallic nanoparticulate Mg-based anode, coupled with the solution, afford, better cycling efficiency and higher working potentials. Allegedly, the improved performance is achieved due to thinner passivation layers on the Mg particles. There are still, occasionally, reports on the development of rechargeable magnesium cells, or electrolytes for such cells, that make use of common highly polar organic solvents such as propylene carbonate and acetonitrile, with "simple" magnesium salts such as $Mg(ClO_4)_2$, $Mg(CF_3SO_3)_2$, and $Mg((CF_3SO_2)_2N)_2$. Most of these reports appear in patents, quite frequently presenting electrochemical results obtained with complete cells, that is, with intercalation cathodes. To date, none of the results could be repeated in our lab, both for the anode side and for the intercalation cathode [50–53].

In recent publications Matsui et al. [54,55] propose the utilization of fast $Mg =$ conducting alloys [56] such as Bi and Sb as an alternative to a pure Mg anode. The authors demonstrated fast and reversible Mg alloying into some of these alloys, both from organometallic-based solutions and from simple Mg salts in organic solvents, such as MgTFSI in acetonitrile. The electrochemical alloying of Mg at some 200 to 350 mV above the magnesium deposition potential may result in the suppression or elimination of the notorious passivation encountered with pure Mg anodes at a lower potential. Although the alloying compounds proposed are of very high density, calculations have shown that their volumetric specific capacity is still sufficient to considered them to be potential anode alternatives. Since the alloys formed have potentials appreciably higher than that for Mg bulk deposition, it is obvious that they are thermodynamically more stable than Mg, which contributes to their lower reducing power and their improved stability with organic solvent–based solutions. It is also possible that volume changes during electrochemical alloying and de-alloying results in instability of passivation films in case they form, rendering the anode electrochemically active. It is evident, though, that due to the appreciable lower working potentials of complete cells utilizing such alloys, high-specific-capacity high-voltage cathodes will be crucial in overcoming the shortcomings of such anodes.

14 CONCLUDING REMARKS AND FUTURE PROSPECTS

It has been shown that despite many similarities, the electrochemistry of magnesium is very different from that of lithium. Different factors dominate anode behavior, and different strategies are applicable to the development of electrolytic solutions.

Despite the complexity associated with the strict conditions crucial for the functioning of workable electrolytic solutions for magnesium batteries, it seems that the goal is not farfetched. With solutions based on complex salts containing phenyl and Hexamethyldisilazide (HMDS), electrochemical stability windows as high as about 3 V are attainable. Development of appropriate cathode materials that will make use of most of this voltage span may result in commercially feasible magnesium-based rechargeable batteries that have competitive properties.

REFERENCES

1. D. Aurbach, Y. Gofer, Z. Lu, A. Schechter, O. Chusid, H. Gizbar, Y. Cohen, V. Ashkenazi, M. Moshkovich, and R. Turgeman, *J. Power Sources*, **97–98**, 28–32 (2001).
2. P. Novak, R. Imhof, and O. Haas, *Electrochimi. Acta*, **45**(1–2), 351–367 (1999).
3. J. O. Besenhard and M. Winter, *ChemPhysChem*, **3**(2), 155–159 (2002).
4. D. Aurbach, Z. Lu, A. Schechter, Y. Gofer, H. Gizbar, R. Turgeman, Y. Cohen, M. Moshkovich, and E. Levi, *Nature (London)* **407**(6805), 724–727 (2000).
5. D. Aurbach, I. Weissman, Y. Gofer, and E. Levi, *Chem. Record*, **3**(1), 61–73 (2003).
6. Z. Wenzhi, X. Dafeng, and W. Xizun, *WuliHuaxueXuebao* (Chinese) **5**(1), 103–106 (1989).
7. H. J. Bittrich, R. Landsberg, and W. Gaube, *Wiss. Z. Tech. Hochsch. Chem. Leuna-Merseburg*, **2**, 449–451 (1960).
8. J. H. Connor, W.E Reid, and G. B. Wood, *J. Electrochem. Soc.*, **104**, 38–41 (1957).
9. N. Yoshimoto, M. Ishikawa, and M. Morita, *Hyomen Gijutsu*, **50**(9), 839–840 (1999).
10. N. Yoshimoto, S. Yakushiji, M. Ishikawa, and Masayuki Morita, *Electrochim. Acta*, **48** (14–16) 2317–2322 (2003).
11. H. M. Walborsky, *Acc. Chem. Res.*, **23** (1990).
12. E. Levi, E. Lancry, Y. Gofer, and D. Aurbach, *J. Solid State Electrochem.*, **10**, 176–184 (2006).
13. Z. Ogumi and M. Indaba, in *Advances in Lithium-Ion Batteries*, W. A. van Schalkwijk and B. Scrosati, Eds., Kluwer Academic/Plenum, New York, 2002, Chap. 2.
14. D. Aurbach in *Advances in Lithium-Ion Batteries*, W. A. van Schalkwijk and B. Scrosati, Eds., Kluwer Academic/Plenum, New York, 2002, Chap. 1.
15. P. Novak, R. Imhof, and O. Haas, *Electrochim. Acta*, **45**, 351 (1999).
16. Z. Lu, A. Schechter, M. Moshkovich, and D. Aurbach, *J. Electroanal. Chem.*, **466**, 203 (1999).
17. E. Levi, Y. Gofer, and D. Aurbach, *Chem. Mater.*, **22**, 860–868 (2010).
18. M. Wakihara, H. Ikuta, and Y. Uchimoto, in *Advances in Lithium-Ion Batteries*, W. A. van Schalkwijk and B. Scrosati, Eds., Kluwer Academic/Plenum, New York, 2002, Chap. 3.
19. Y. Nishi, in *Advances in Lithium-Ion Batteries*, W. A. van Schalkwijk and B. Scrosati, Eds., Kluwer Academic/Plenum, New York, 2002, Chap. 7.
20. J. H. Connor, W. E. Reid, G. B. Wood, *J. Electrochem. Soc.*, **104** 38-41 (1957).
21. J. D. Genders and D. Pletcher, *J. Electroanal. Chem.*, **199**, 92 (1986).
22. C. Liebenow, *J. Appl. Chem.*, **27**, 221 (1997).
23. W. Vonau and F. Berthold, *J. Prakt. Chem.*, **336**, 140 (1994).
24. W. V. Evans and R. Pearson, *J. Am. Chem. Soc.*, **64**(12), 2865–2871 (1942).
25. T. D. Gregory, R. J. Hoffman, and R. C Winterton, *J. Electrochem. Soc.*, **137**(3), 775–780 (1990).
26. D. Aurbach, H. Gizbar, A. Schechter, O. Chusid, H. E. Gottlieb, Y. Gofer, and I. Goldberg, *J. Electrochem. Soc.*, **149**(2), A115–A121 (2002).
27. http://en.wikipedia.org/wiki/Standard_electrode_potential_(data_page).
28. Y. Vestfried, O. Chusid, Y. Goffer, P. Aped, and D. Aurbach, *Organometallics*, **26**, 3130–3137 (2007).
29. N. Pour, Y. Gofer, D. T. Major, and D. Aurbach, *J. Am. Chem. Soc.*, **133**(16), 6270–6278 (2011).
30. Y. Gofer, O. Chusid, H. Gizbar, Y. Viestfrid, H. E. Gottlieb, V. Marks, and D. Aurbach, *Electrochem. Solid-State Lett.*, **9**(5), A257 A260 (2006).
31. P. Sobota, *J. Organomet. Chem.*, **290**(1), C1–C3 (1985).

32. P. Sobota, T. Pluzinski, and T. Lis, *Bull. Polish Acad. Sci.*, **33**(11–12), 491–496 (1986).
33. S. K. Martha, J. Grinblat, O. Haik, E. Zinigrad, T. Drezen, J. H. Miners, I. Exnar, A. Kay, B. Markovsky, and D. Aurbach, *Angew. Chem. Int. Ed.*, **48**, 8559–8563 (2009).
34. H. Gizbar, Y. Vestfrid, O. Chusid, Y. Gofer, H. E. Gottlieb, V. Marks, and D. Aurbach, *Organometallics*, **23**(16), 3826–3831 (2004).
35. H. Gizbar, The Study of Non Aqueous Electrochemical Systems of Magnesium, Ph.D. dissertation, Department of Chemistry, Bar Ilan University, 2005.
36. Y. Gofer, O. Chusid, H. Gizbar, Y. Vestfrid, H. E. Gottlieb, V. Marks, and D. Aurbach, *Electrochem. Solid-State Lett.*, **9**, A257 (2006).
37. C. Liebenow, Z. Yang, and P. Lobitz, *Electrochem. Commun.*, **2**, 641–645 (2000).
38. A. J. Bard, G. Inzelt, and F. Scholz, in *Electrochemical Dictionary*, Springer-Verlag, New York, 2008, p. 195.
39. Y. Viestfrid, M. D. Levi, Y. Gofer, and D. Aurbach, *J. Electroanal. Chem.*, **576**, 183–195 (2005).
40. E. Lancry, E. Levi, Y. Gofer, M. D. Levi, and D. Aurbach, *J. Solid State Electrochem.*, **9**, 259–266 (2005).
41. M. Murakami, A. Ue, and S. Nakamura, *J. Electrochem. Soc.*, **149**, A1572 (2002).
42. S. Osamu, Y. Nobuko, M. Mami, E. Minato, and M. Masayuki, *J. Power Sources*, **196**, 1586–1588 (2011).
43. K. Takeshi, Y. Nobuko, E. Minato, and M. Masayuki, *Electrochem. Commun.*, **12**, 1630–1633 (2010).
44. N.S. Venkata Narayanan, B.V. Ashok Raj, and S. Sampath, *Electrochem. Communi.*, **11**, 2027–2031 (2009).
45. N. Amir, Y. Vestfrid, O. Chusid, Y. Gofer, and D. Aurbach, *J. Power Sources*, **174**, 1234–1240 (2007).
46. H. S. Kim, T. S. Arthur, G. D. Allred, J. Zajicek, J. G. Newman, A. E. Rodnyansky, A. G. Oliver, W. C. Boggess, and J. Muldoon, *Nat. Communi.*, **2**(Aug.), 1435/1–1435/6 (2011).
47. G. Yongsheng, Y. Jun, N. Yanna, and W. Jiulin, *Electrochem. Communi.*, **12**, 1671–1673 (2010).
48. G. P. Pandey, R. C. Agrawal, and S. A. Hashmi, *J. Solid State Electrochem.*, **15**, 2253–2264 (2011).
49. Y. Liang, R. Feng, S. Yang, H. Ma, J. Liang and J. Chen, *Adv. Mat.*, **23**, 640–643 (2011).
50. J. Zhiping, S. Rimma, I. N. Krastev, Magnesium cell with improved electrolyte, Patents PCT/US2010/036581, WO/2010/144268, 2010.
51. V. Aravindan, G. Karthikaselvi, P. Vickraman, and S. P. Naganandhini, *J. App. Polym. Sci.*, **112**(5), 3024–3029 (2009); D. J. Kim, H. S. Ryu, I. P. Kim, K. K. Cho, T. H. Nam, K. W. Kim, J. H. Ahn, and H. J. Ahn, *Physi. Scr.*, **T129**, 70–73 (2007).
52. O. Atsushi, Magnesium secondary battery, Japanese Patent 2007280627, 2008.
53. I. Masaharu, M. Masahide, F. Masahisa, K. Hideyuki, and D. Kazunori, Electrolyte for nanaqueous battery, method for producing the same, and electrolytic solution for nonaqueous battery, U.S. Patent 20040137324, 2004.
54. T. S. Arthur, N. Singh, and M. Matsui, *Electrochem. Communi.*, **16**, 103–106 (2012).
55. M. Matsui, T. Arthur, and N. Singh, 220th ECS Meeting, 2011, Abstr. 616.
56. A. C. Barnes, C. Guo, and W. S. Howells, *J. Phys.*, **32**, L467 (1994).

CHAPTER 16

RECHARGEABLE SODIUM AND SODIUM-ION BATTERIES

K. M. Abraham

1 INTRODUCTION

The concept of high-energy-density rechargeable batteries with sodium negative electrodes (anodes) predates the development of rechargeable lithium and lithium-ion batteries. A sodium–sulfur (Na–S) battery consisting of a molten sulfur cathode and a sodium-metal anode, contained in a Na^+-conducting solid electrolyte ceramic tube which also serves as the separator in the battery, has been under development since the 1960s. This system, first described by Weber and Kummer [1], has the cell configuration: liquid Na/solid electrolyte/molten S, current collector. Here, the solid electrolyte is a Na-β-Al_2O_3 ceramic [2–4] which allows the transport of Na^+ while completely preventing permeation by anionic and neutral chemical species. The fully charged cathode in this battery is comprised of a conductive graphite–felt matrix immersed in a pool of nonconductive molten S. The battery operates at about 300°C, to keep the molten S and its discharge products in a liquid state during discharge and charge.

Lithium Batteries: Advanced Technologies and Applications, First Edition.
Edited by Bruno Scrosati, K. M. Abraham, Walter van Schalkwijk, and Jusef Hassoun.
© 2013 John Wiley & Sons, Inc. Published 2013 by John Wiley & Sons, Inc.

When a Na–S battery is discharged, Na ions migrate through the solid electrolyte to the cathode compartment, where they react with molten sulfur to form various sodium polysulfides (e.g., Na_2S_5 and Na_2S_3) in the porous graphite cathode. The initial discharge reaction is

$$2Na + 5S \rightarrow Na_2S_5 \tag{1}$$

The Na_2S_5 is reduced further to lower sodium polysulfides as discharge proceeds. The reverse reactions occur on charge. The Na–S battery has certain unique characteristics, including high discharge rates, a relatively high voltage of about 2 V, and a low self-discharge rate. However, it has a few drawbacks, stemming from the high temperature required for its operation. Thus, cell failure occurs because of the accumulation of corrosive materials on the positive electrode, deterioration of the solid electrolyte or of seals, or more general corrosion [1] of the cell container. Another drawback of this battery is that at its high operating temperatures, Na sulfides precipitate out when the overall composition of the discharge product reaches Na_2S_3. Consequently, the energy density of the battery is considerably less than that corresponding to the theoretical reaction involving the two-electron reduction of sulfur to S^{2-} and is close to the discharge process involving the reduction of sulfur to S_3^{2-}, corresponding to about 750 Wh/kg.

$$2Na + 3S \rightleftharpoons Na_2S_3 \tag{2}$$

When the amount of S present in this battery during discharge is greater than that equivalent to an overall composition of Na_2S_3, the two immiscible liquid phases of Na_2S_3 and S, coexist. The sulfur-rich phase is nonconductive and can cause cell cycling difficulties. Despite the significant interest in and development of the Na–S battery, the high temperatures required for its use have prevented it from being adapted for powering portable consumer products and electric automobiles. The main focus of its current development and implementation is for power storage and load-leveling applications in electric utilities.

The dream of developing practical sodium batteries for ambient-temperature applications has persisted, with research efforts on two other types of systems: (1) batteries capable of operation at moderately high temperatures of 100 to 200°C, and (2) batteries able to function at room temperature. The cell configuration of the former is similar to that of a high-temperature battery with a liquid Na anode, separated from the cathode by a Na^+-conducting solid electrolyte separator, typically Na-β-Al_2O_3 or its more conductive variant, Na-β″-Al_2O_3 [2–4]. The identification of rechargeable cathodes has been a major aspect of the development of these moderately high-temperature batteries.

Their room-temperature counterpart has a cell configuration similar to that of ambient-temperature rechargeable Li batteries. In these the Na-metal anode and the rechargeable cathode, which are separated from each other electronically by means of a porous polymer membrane, are immersed in a nonaqueous Na^+-conducting electrolyte. Since metallic Na is in direct contact with the electrolyte, identification of Na compatible electrolytes is a major task in the development and implementation of these ambient-temperature batteries. Alternatively, ambient-temperature Na-ion batteries in which two Na insertion electrodes serve as cathode and anode active materials are also sought.

In a moderately high-temperature battery operating at 100 to 200°C, cathode materials exhibiting both Na intercalation and deintercalation reactions (more generally called Na insertion and extraction reactions) and displacement-type electrode processes can be used, whereas room-temperature systems are exclusively in the realm of Na insertion and extraction electrodes. These various room- and moderately elevated-temperature Na batteries are the focus of discussion in this chapter.

2 MODERATE-TEMPERATURE RECHARGEABLE SODIUM BATTERIES

Rechargeable sodium batteries with the ability to perform in the range 100 to 200°C range are very similar to the aforementioned high-temperature batteries in which a liquid Na anode is used in conjunction with a Na-β''-Al$_2$O$_3$ separator. The difference is in the cathode active materials which are chosen to function at the lower operation temperature. The first effort to develop such a battery was made by Abraham et al. in the late 1970s [5]. They reasoned that many of the problems encountered in high-temperature Na–S batteries could be overcome when lowering the operating temperature by using a soluble sulfur cathode consisting of sodium polysulfides such as Na$_2$S$_4$ dissolved in a solvent. In this way, the operating temperature of the battery could be slightly above the melting point of Na, about 100°C. By using a solvent that dissolves Na$_2$S appreciably, the final reduction product of sulfur in an Na–S battery, it might also be possible to achieve an end-of-discharge cathode composition closer to the theoretical two-electron reduction of S. Moreover, the use of a solvent that dissolves higher Na-polysulfides (Na$_2$S$_n$) could eliminate the undesirable phase separation encountered in a molten S battery. The Na-soluble S rechargeable battery described by Abraham et al. has the general configuration

liquid Na|Na-β''-Al$_2$O$_3$|Na$_2$S$_n$, organic solvent, conducting Na salt

A major challenge in this work was identification of a solvent that had high solubility for sodium sulfides Na$_2$S$_n$, where $n = 1$ to 8, that formed highly Na$^+$-conducting electrolytes and was thermally stable in the operational temperature range of the battery, 100 to 150°C. The solvent found to be useful was dimethylacetamide, CH$_3$CON(CH$_3$)$_2$ (DMAC). Physical properties of this solvent relevant to the Na-soluble S battery are listed in Table 1. It dissolves 1.6 M Na$_2$S$_4$ and 1.2 M Na$_2$S

TABLE 1   Physical Properties of Dimethylacetamide

Boiling point (°C)	165
Melting point (°C)	−20
Density (g/mL)	0.94
Viscosity (mp at 25°C)	1.44
Dielectric constant	37
Specific conductance (S/cm at 25°C)	20

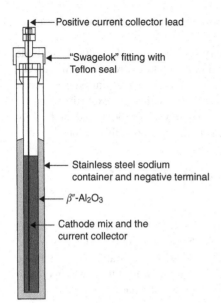

Positive current collector lead

"Swagelok" fitting with Teflon seal

Stainless steel sodium container and negative terminal

β''-Al_2O_3

Cathode mix and the current collector

FIGURE 1   Rechargeable sodium cell. (From [6], with permission of the *Journal of the Electrochemical Society.*)

at 150°C. A solution containing approximately 5 M S as Na_2S_4 and 0.4 M $NaBF_4$ exhibited a conductivity of 1.89×10^{-2} S/cm at 150°C.

The liquid Na-β''-Al_2O_3 /Na_2S_4, DMAC (0.4 M $LiBF_4$) laboratory cell used to study the discharge/charge cycling performance is shown in Figure 1. This cell exhibited an open-circuit voltage of 1.83 V at 130°C where it was operated. It exhibited several charge–discharge cycles (Fig. 2), albeit at moderate current densities,

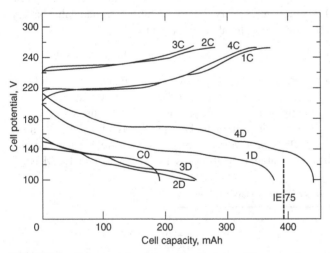

FIGURE 2   Galvanostatic charge (C)–discharge (D) curves for a cell of configuration molten Na–Na_2S_4, DMAC at 130°C. The cathode current collector is carbon felt with an area of 12.6 cm^2. 4C and 4D at 1.59 mA/cm^2; 1C and 1D at 2.4 mA/cm^2; 2C, 2D, 3C, and 3D at 3.2 mA/cm^2. (From [5], with permission of *Electrochim. Acta.*)

due primarily to the low conductivity of the ceramic Na^+ conductor. Assuming full consumption of Na_2S_4, the total charge passed in the first discharge of this cell corresponded to the reduction of $S_4{}^{2-}$ to $S_{1.3}{}^{2-}$, with an overall cathode utilization of 0.8 electron per mole of S. Material balance calculations revealed that subsequent discharges at lower rates also followed more or less the same pattern, with very little capacity showing below a polysulfide composition of approximately $S_2{}^{2-}$. It *was* found that polysulfides of composition as high as Na_2S_8 could be formed during cell charging without precipitation of elemental S. Some cells were charged and discharged for up to one month at 130°C, although chemical reactions between the solvent and the sodium polysulfides were evident.

It became clear from the early studies that to build a long-lived Na-soluble sulfur battery, solvents that were thermally and chemically more stable than DMAC were needed. However, very little work was carried out subsequently to identify better organic electrolytes. Instead, efforts in our laboratory and in others focused on sodium batteries utilizing molten $NaAlCl_4$-based electrolytes in combination with either transition metal sulfide or transition metal chloride cathode active materials.

2.1   Rechargeable Sodium Batteries with Transition Metal Sulfide Cathodes in Molten NaAlCl₄

In an attempt to circumvent the problems associated with organic electrolytes at moderately elevated temperatures, we used molten $NaAlCl_4$ as an alternative electrolyte [6–9]. Some physical properties of $NaAlCl_4$ presented in Table 2 show why it is an attractive electrolyte for use in moderately elevated–temperature Na batteries. These Na cells were operated at 160 to 190°C . The cell construction details were the same as those described above for organic electrolyte cells. A series of transition metal sulfides such as NiS, NiS_2, CuS, FeS_2, and VS_2 were investigated as the cathode active materials. Typically, the metal sulfide powder was sprinkled into a graphite–felt current collector, which was placed on one side of the Na-β''-Al_2O_3 tube. Metallic sodium was added to the other side of the tube (Fig. 1). The cell was then sealed and heated to 165°C, where it was typically operated. The general cell configuration is

$$\text{liquid Na/}\beta''\text{-Al}_2O_3|\text{molten NaAlCl}_4, \text{ metal sulfide}$$

In molten $NaAlCl_4$, the discharge and charge reactions of a metal sulfide cathode (MS_n) can be described by the general reaction

$$MS_n + 2yNa^+ + 2ye^- + yNaAlCl_4 \rightleftharpoons M + yNaAlSCl_2 + 2yNaCl \qquad (3)$$

TABLE 2   Some Properties of NaAlCl₄ and Na-β″-Al₂O₃

Electrlyte	Temp. (K)	Viscosity (cP)	Density (g/mL)	Conductivity (S/cm)
NaAlCl₄	460	2.65	1.43	0.46
Na-β″-Al₂O₃	573	solid	3.2	0.16

Source: From [6].

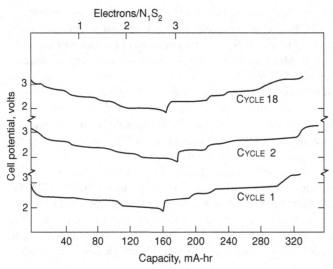

FIGURE 3 Discharge cycles of a Na–NiS$_2$ cell at 170°C between 1.7 and 3.5 V. Current density, discharge = charge = 1 mA/cm^2. (From [6], with permission of the *Journal of the Electrochemical Society*.)

The cathode reaction involves the participation of NaAlCl$_4$, since it is converted into NaAlSCl$_2$ during discharge. The metal sulfide electrode discharge reaction is believed to proceed first to Na$_2$S and the corresponding transition metal, followed by the reaction of the Na$_2$S with NaAlCl$_4$ to form NaAlSCl$_2$ and NaCl. The theoretical cathode capacities of NiS, NiS$_2$, FeS$_2$, and CuS correspond to reduction reactions involving two electrons per NiS, four electrons per NiS$_2$, four electrons per FeS$_2$, and two electrons per CuS. The actual cathode utilization in cells depended on cathode loading, graphite–felt electrode porosity, discharge current density, and the metal sulfide/NaAlCl$_4$ ratio.

The Na/NiS$_2$ cell was the cell studied most extensively, as it showed promise for practical use. It has an open-circuit voltage of 3.0 V and a mid-discharge voltage of 2.3 V at 165°C. Cells were cycled at current densities of 2 to 10 mA/cm^2 between voltage limits of 1.7 to 3.5 V. Typical charge–discharge cycles for a cell are depicted in Figure 3, and the capacity vs. cycle number plot is shown in Figure 4. The cell exhibited excellent rechargeability. Some cells were cycled for more than 650 cycles at temperatures between 170 and 190°C . Cells built initially with NiS as the cathode active material also exhibited good cycling behavior. In addition, we showed that cells could be built in the discharged state using a mixture of Na$_2$S and Ni and activated by charging first to oxidize the cathode mixture to NiS or NiS$_2$.

To gain insights into the reaction mechanism of Na–NiS$_2$ and Na–NiS cells, we calculated their equilibrium potentials at 500 K using thermodynamic data for the reactions NiS$_2$ + 2Na = NiS + Na$_2$S and NiS + 2Na = Na$_2$S and Ni. The values were 1.6 and 1.3 V, respectively, significantly lower than the cell open-circuit voltages observed. In addition, we identified NiS, Ni$_3$S$_4$, and Ni$_3$S$_2$ as intermediate phases during various stages of the charge and discharge of Na–NiS$_2$ cells [6]. From

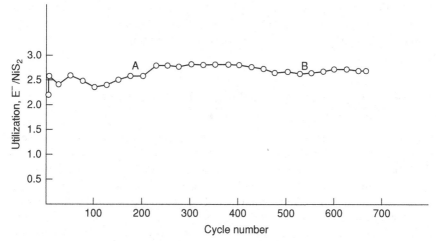

FIGURE 4   Cathode utilization vs. cycle number for a Na–NiS$_2$ cell. At point A the temperature was raised from 170°C to 190°C. At point B the cell was left on open-circuit voltage at room temperature for two months prior to resuming cycling. (From [6], with permission of the *Journal of the Electrochemical Society*.)

this analytical and thermodynamic information, we proposed the following reversible reactions in the cycling of a Na–NiS$_2$ cell:

$$3NiS + 4Na^+ + 4e^- + 2NaAlCl_4 \rightleftharpoons Ni_3S_4 + 2NaAlSCl_2 + 4NaCl \qquad (4)$$

$$Ni_3S_4 + 2Na^+ + 2e^- + NaAlCl_4 \rightleftharpoons 3NiS + NaAlSCl_2 + 2NaCl \qquad (5)$$

$$3NiS + 2Na^+ + 2e^- + NaAlCl_4 \rightleftharpoons Ni_3S_2 + NaAlSCl_2 + 2NaCl \qquad (6)$$

$$Ni_3S_2 + 4Na^+ + 4e^- + 2NaAlCl_4 \rightleftharpoons 3Ni + 2NaAlSCl_2 + 4NaCl \qquad (7)$$

Reactions [6] and [7] are common to both NiS and NiS$_2$ electrodes. In extensively cycled cathodes we found small amounts of NiCl$_2$, probably formed from the reaction of NaAlCl$_4$ and Ni sulfide. In separate experiments we found that the reduction of NiCl$_2$ to Ni in a Na cell is very reversible and occurs at about 2.7 V. To our regret, we did not investigate this Na/NiCl$_2$ cell further. This rechargeable Na battery later became known as the *zebra cell*, developed by South African researchers (see Section 2.2).

We also studied rechargeable Na batteries with other metal sulfides, including FeS$_2$, CuS, VS$_2$, NbS$_2$Cl$_2$, and MoS$_3$, in combination with a molten NaAlCl$_4$ electrolyte [7–9]. The Na–FeS$_2$ cell had an open-circuit voltage of 3.1 V and a mid-discharge voltage of 2.4 V at 165°C. In laboratory cells, reversible cathode capacities exceeded three electrons per FeS$_2$ at 1 mA/cm^2 [8]. The Na–CuS cell had a higher mid-discharge voltage of 2.9 V. Reversible capacities of 1.5 electrons per CuS were achieved in laboratory cells cycling at 1 mA/cm^2 [8].

We made the general observation that metal sulfides reacted with NaAlCl$_4$ in varying degrees to form metal sulfur chlorides or metal chlorides in situ, and that the resulting cathode, consisting of mixtures of metal sulfides, chlorides, and sulfur chlorides, exhibited good rechargeability [6–9]. Interest in metal sulfide cathode materials for moderate-temperature Na batteries has continued with recent investigations utilizing cathode mixtures of metal sulfides and chlorides in molten $NaAlCl_4$ electrolytes [10].

2.2 Rechargeable Sodium Batteries with Transition Metal Chloride Cathodes in NaAlCl₄

Rechargeable Na–metal chloride batteries were developed by South African researchers, who called them zebra cells [11,12]. They utilized either $FeCl_2$ or $NiCl_2$ cathode active material in conjunction with molten $NaAlCl_4$ as the electrolyte. The cell has the same general features as the cell in Figure 1. The Na–$FeCl_2$ cell was constructed with an initial cathode mixture of Fe and $FeCl_2$ combined with $NaAlCl_4$. This cell exhibited a constant open-circuit voltage of 2.35 V over the entire charge–discharge cycle and operated in the range 175 to 300°C. The preferred operating temperature was 250°C. The reversible cell discharge reaction is

$$FeCl_2 + 2Na \rightleftharpoons 2NaCl + Fe \qquad (8)$$

Figure 5 shows the cycles selected for an 8-Ah Na–$FeCl_2$ cell operating at 250°C. All of the cathode reactants, Fe, $FeCl_2$, and NaCl, are insoluble in the $NaAlCl_4$ electrolyte and therefore remain in intimate contact with each other and the current collector.

Flat discharge curves were observed at all rates up to the l-h rate, and rechargeability of the Na–$FeCl_2$ cell system was demonstrated with more than 1000 charge–discharge cycles at about 250°C . Great strides were made to optimize the system,

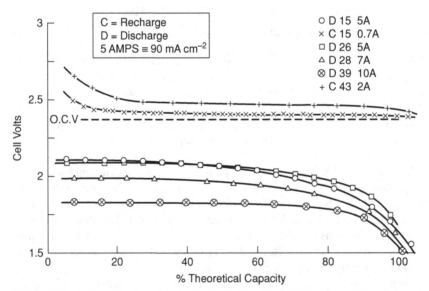

FIGURE 5 Selected cycles of an 8-Ah Na–$FeCl_2$ cell at 250°C. (From [11], with permission of the *Journal of the Electrochemical Society*.)

giving a practical cell-specific energy of 130 Wh/kg at the 5-h rate. The same research group demonstrated the usefulness of $NiCl_2$ as a rechargeable cathode for the zebra cell either alone or as a mixture with $FeCl_2$ [11,12]. The Na–$NiCl_2$ zebra cell had an open-circuit potential of 2.59 V at 250°C and was cycled more than 2000 times, demonstrating the good rechargeability of the battery.

We also demonstrated a moderate-temperature Na cell that utilized FeOCl as the cathode material in conjunction with $NaAlCl_4$ in a cell operating at a more moderate 180°C [13]. We found that the Na–FeOCl cell exhibited discharge–charge behavior very similar to that of a cell made initially with a cathode mixture of $NaFeO_2$ and $NaAlCl_4$. We proposed that both of these cathode materials reacted with $NaAlCl_4$ to generate the same product, $NaFeOCl_2$, in situ, which then cycled in an identical manner. The reactions proposed in the Na–FeOCl and Na–$NaFeO_2$ cells are shown by

$$NaAlCl_4 + FeOCl \rightarrow NaFeOCl_2 + AlCl_3 \tag{9}$$

$$NaFeO_2 + NaAlCl_4 \rightarrow NaFeOCl_2 + NaAlOCl_2 \tag{10}$$

The $NaFeOCl_2$ is then believed to cycle as depicted in the reactions

$$NaFeOCl_2 + Na \rightleftharpoons Na_2O + FeCl_2 \tag{11}$$

$$FeCl_2 + 2Na \rightleftharpoons 2NaCl + Fe \tag{12}$$

The Na–$NaFeO_2$ cell exhibited an average cell potential of 2.63 V and a theoretical specific energy [including the weight of $NaAlCl_4$ from reaction (9)] of 628 Wh/kg and the Na–FeOCl exhibited a cell potential of 2.61 V and a specific energy of 603 Wh/kg.

In conclusion, Na–metal sulfide and metal chloride batteries based on molten $NaAlCl_4$ electrolyte have proven to be highly rechargeable. Interest in these moderately elevated–temperature batteries for practical applications persists, with current development in several laboratories around the world focusing on scale-up and cell and battery packaging .

3   RECHARGEABLE SODIUM BATTERIES WITH Na INSERTION CATHODES

Rechargeable sodium batteries are attractive compared to their lithium counterparts because of the lower cost of sodium, but the development of such batteries with insertion cathodes has lagged behind that of their rechargeable lithium cousins. The main reason has been the lack of suitable rechargeable cathode materials. Sodium insertion cathode materials are often characterized by crystallographic phase changes when Na is introduced into their crystal structures, with the result that the amount of Na that can be cycled reversibly at ambient temperatures is often limited.

An intercalation (or insertion) reaction involves the interstitial introduction of a guest species (Na^+ in the present context) into a host crystal lattice:

$$nA^+ + ne^- + MX_y \rightleftharpoons A_nMX_y \tag{13}$$

An ideal intercalation reaction by definition involves little or no structural change of the host when the alkali metal atom is inserted to its maximum capacity. In such cases the reaction is highly reversible because similar transition states are readily achieved for both the forward and backward reactions, leading to close compliance with the thermodynamic principle of microscopic reversibility. This is the behavior of many Li intercalation electrodes. In actual reactions, the bonding within the host lattice may be slightly perturbed, a slight expansion of the host lattice may occur, and crystallographic phase changes of the host compound may take place. Depending on the extent of the structural perturbations, the reaction may be reversible, partially reversible or irreversible. The requirements of intercalation-positive electrode materials include a large free energy of reaction, ΔG, affording a high cell voltage; a wide compositional range, n in reaction (13), resulting in high cell capacities; high diffusivity for Na^+ in the host, allowing high power densities; good electronic conductivity over a wide range of x; and minimal structural changes with the degree of intercalation, resulting in a reversible reaction and long cycle life. Sodium atoms, because they are larger ($rNa^+ = 0.98$ Å) than Li atoms ($rLi^+ = 0.71$ Å), usually form multiple crystallographic phases, as a larger amount of Na is inserted into the host electrode material lattice, with the result that the reactions are often only partially reversible. Table 3 provides examples of phase changes that accompany the insertion of Na into some transition metal sulfides and selenides [14].

In the past 30 years a number of transition metals—sulfides, selenides, and oxides—have been investigated as Na insertion cathodes. Their suitability as Na battery cathodes has been studied both at room temperature and at moderately elevated temperatures up to 130°C.

TABLE 3   Phase Changes Accompanying Na Insertion into Metal Sulfides and Selenides[a]

Chalcogenide	Transition metal site symmetry in the chalcogenide	Na-intercalate, Na_xMY_2, compositions	Phase
TiS_2	0	$0.17 < x \leq 0.3$	Second-stage compound
		$0.38 < x \leq 0.72$	First-stage TP
		$0.79 < x \leq 1$	First-stage TAP
$TiSe_2$	0	$x < 0.32$	Second-stage compound
		$0.68 < x \leq 0.74$	First-stage TP
		$0.82 < x \leq 0.91$	First-stage TAP
ZrS_2	0	$x < 0.32$	Second-stage compound
		$0.64 < x \leq 1$	First-stage TP
VS_2	0	$x < 0.3$	Second-stage compound
		$0.3 < x \leq 0.8$	First-stage TP
		$0.8 < x \leq 1$	First-stage TAP

Source: From [14].
[a]0, octahedral; TP, trigonal prismatic; TAP, trigonal antiprismatic or elongated 0.

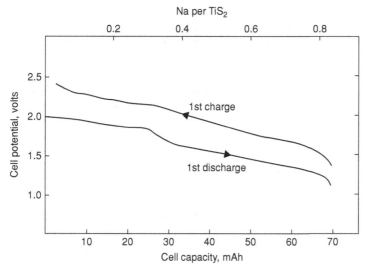

FIGURE 6   Discharge and charge cycles of a Na–TiS$_2$ cell at 130°C. (From [15], with permission of the *Journal of the Electrochemical Society.*)

3.1 Transition Metal Sulfide and Selenide Cathodes

Much of the early work on the use of transition metal disulfides and selenides (generally called metal chalcogenides) as Na battery cathodes investigated by this author and others has been reviewed [14]. The Na test cells used at the moderately elevated temperature of 130°C were similar in configuration to that depicted in Figure 1 and had the configuration

liquid Na|Na-β″-Al$_2$O$_3$/Na$^+$-conducting organic electrolyte, metal chalcogenide

The Na$^+$-conducting organic electrolyte was a 1 M solution of NaI in 1,2-bis(2-methoxyethoxy)ethane (triglyme) [15–17]. Figure 6 displays the charge–discharge cycle for a Na/TiS$_2$ cell at 130°C [15].

The first discharge shows a two-plateau voltage-capacity, curve with a limiting capacity, equivalent to the insertion of 0.8 Na per TiS$_2$. The upper voltage plateau encompassed the Na composition in Na$_x$TiS$_2$, $0 < x \leq 0.3$, and the lower, $0.3 < x \leq 0.85$. The capacity of the first phase gradually decreased with cycling, although even after eight cycles about half of the capacity in the upper plateau was utilized reversibly, along with all of the capacity in the lower plateau. The data indicated an average rechargeable capacity of 0.65 Na per TiS$_2$ for the Na cell with a 1.7-V mid-discharge, corresponding to a cathode specific energy of 215 Wh/kg [15].

The cycling behavior of this elevated temperature Na/TiS$_2$ cell is similar to that observed by Klemann and Newman [18] for a Na cell cycled at room temperature with both the Na anode and TiS$_2$ cathode immersed in an organic electrolyte in a configuration similar to that of a Li anode cell. The electrolyte consisted of sodium triethyl(N-pyrrolyl)borate in 1,3-dioxolane. At a discharge current density of

2.5 mA/cm^2, a capacity of 0.8 Na per TiS$_2$ was obtained at room temperature with a limiting cathode composition of Na$_{0.8}$TiS$_2$. The discharge is characterized by two voltage plateaus, one at about 1.9 V and the other at about 1.5 V. The former plateau spans the Na composition range x, in Na$_x$TiS$_2$, at 0 < x 0.4, and the latter at 0.4 < x 0.8, very similar to the behavior for the high-temperature cell in Figure 6. The capacity in the first region in this cell diminished with cycling and only about 5% could be recovered after eight cycles. The capacity of the lower plateau was maintained after 16 cycles.

The voltage plateaus seen in both the room-temperature and 130°C cells during discharge encompass the Na compositions x in Na$_x$TiS$_2$, where crystallographic structural changes have been identified as shown in Table 3. At less than 0.3 Na insertion, a second-stage Na$_x$TiS$_2$, compound is formed; at intermediate Na levels between 0.35 and 0.72, a first-stage trigonal prismatic phase is formed; and at Na compositions larger than 0.8, a first-stage trigonal antiprismatic Na$_x$TiS$_2$ phase is formed. Sodium can be intercalated electrochemically up to the end of the first two phases at both room temperature and 130°C, but full reversibility is found only for the Na in the intermediate trigonal prismatic phase. This makes the rechargeable capacity of the TiS$_2$ cathode low.

The cycling of a Na–TiS$_2$ cell with a β-Al$_2$O$_3$ solid electrolyte separator was investigated at temperatures between 230 and 280°C by Zanini and co-workers [20]. In this cell, solid TiS$_2$ was used without any liquid electrolyte. In contrast to the results at 25 and 130°C, this cell was found to be rechargeable for x between 0.0 and 0.95 in Na$_x$TiS$_2$. The difference could be attributed to factors such as higher Na$^+$ diffusivities at higher temperatures, the extremely low TiS$_2$ loading capacity employed in that study, or possibly the different phase regions observed at these higher temperatures. The Na composition ranges of the three voltage plateaus in the Na–TiS$_2$ cell discharged at 280°C were observed at 0 < x ≤ 0.2, 0.2 < x ≤ 0.69, and 0.69 < x ≤ 0.95. It appears that at temperatures near 300°C, a higher specific energy of about 337 Wh/kg is possible with a TiS$_2$ electrode.

Other transition metal sulfides investigated as cathodes for sodium batteries include VS$_2$, TiS$_3$, TaS$_2$, MoS$_3$, Cr$_{0.5}$V$_{0.5}$S$_2$, and NiPS$_3$, all in conjunction with triglyme/NaI electrolytes at 130°C [14]. All these sulfides showed multiple phase formations and varying degrees of reversibility. VS$_2$ showed better capacity and rechargeability than TiS$_2$ and it cycled more than 20 times at 130°C [15]. Interestingly, two of the transitional metal was diselenides that we studied, VSe$_2$ and TiSe$_2$, showed much better rechargeability at 130°C [16]. The cycling data for the Na–VeS$_2$ cell with an organic electrolyte in contact with the selenide cathode shown in Figure 7 illustrate the superior behavior of metal selenide cathodes in rechargeable Na batteries. Ten cycles were demonstrated with little loss in capacity. The larger crystallographic sites available in the metal selenides may allow the insertion of Na without significant sliding of its selenide slabs in the crystal layers and with minimal structural changes in the Na$_x$VeSe$_2$ phases formed during discharge.

Very little further work on Na cells utilizing transition metal sulfide and selenide cathodes was carried out subsequent to these pioneering investigations. Most recent effort on Na intercalation cathodes has been focused on transition metal oxides.

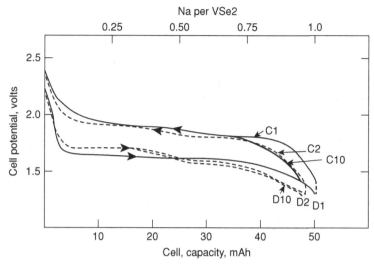

FIGURE 7   Cycles of a Na–VSe$_2$ cell at 130°C. (From [16], with permission of the *Journal of the Electrochemical Society*.)

3.2   Na Insertion Transition Metal Oxide Cathodes and Na-Ion Batteries

A number of transition metal oxides have been investigated recently as cathodes for room-temperature rechargeable Na batteries. All of these cells were tested at room temperature in a cell configuration in which the Na anode and the metal oxide cathode, which were separated from each other electronically by a porous polymer membrane separator, were in direct contact with an organic liquid electrolyte as in ambient-temperature rechargeable Li anode cells. Many of these oxides were synthesized and tested in the discharged state as sodium intercalated metal oxides. The oxide electrode for test cells was fabricated in the same fashion as for testing electrodes for Li-ion cells. The emphasis of the investigations to date has been on understanding the rechargeability of metal oxide cathode materials at room temperature, with little effort devoted to the study of either anode rechargeability or the performance of full cells.

The transition metal oxides that exhibited good reversibility have been materials with layered crystal structures. Recent electrochemical results indicate that layered oxides structures such as NaCrO$_2$ [20,21], NaNi$_{0.5}$Mn$_{0.5}$O$_2$ [20], and NaMnO$_2$ [22] have high Na insertion and extraction capacities and reversibility at room temperature. Komba et al. [20] studied NaCrO$_2$ and NaNi$_{0.5}$Mn$_{0.5}$O$_2$ in a cell using a 1 M NaClO$_4$ solution in propylene carbonate as the electrolyte. Cell cycling was started by electrochemical extraction of Na from the cathode by charging. The removal of Na was accompanied by several phase transitions, as depicted in Figures 8 and 9. Almost all Na ions were extracted from the NaNi$_{0.5}$Mn$_{0.5}$O$_2$ and NaCrO$_2$ electrodes by galvanostatic charging to 4.5 V.

The NaNi$_{0.5}$Mn$_{0.5}$O$_2$ electrode showed a highly reversible capacity of 185 mAh/g in the potential region between 2.5 and 4.5 V vs. Na/Na$^+$ (Fig. 8).

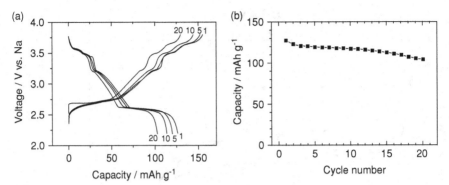

FIGURE 8   (a) Galvanostatic charge and discharge curves and (b) variation in discharge capacity of Na–NaNi$_{0.5}$Mn$_{0.5}$O$_2$ cells at a rate of 4.8 mA/g. (From [20], with permission of the *Journal of the Electrochemical Society.*)

When the discharge was carried out after fully charging to 4.5 V, several voltage plateaus appeared at the same potentials as in the charge curve, indicating the same phase transitions during sodium extraction and insertion. The discharge capacity of 185 mAh/g is one of the highest values reported for layered sodium transition metal oxides. There were irreversible reactions above the 4-V region, probably due to electrolyte decomposition. The irreversible capacity was suppressed when the potential range was limited to 3.8 V, and at this potential limit reversible capacity was 125 mAh/g. Typical cycles are shown in Figure 8.

Figure 9 displays the cycling performance of a Na–NaCrO$_2$ cell. The repeatable voltage plateaus in both charge and discharge curves of the NaCrO$_2$ electrode were interpreted to mean reversible Na intercalation, similar to NaNi$_{0.5}$Mn$_{0.5}$O$_2$. The NaCrO$_2$ electrode was electrochemically inactive after oxidation to 4.5 V. When galvanostatic cycling was carried out between 2 and 3.5 V, discharge capacities of 100 to 120 mAh/g, with satisfactory capacity retention, were observed. It is to be noted that

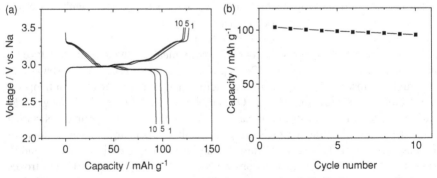

FIGURE 9   (a) Galvanostatic charge and discharge curves and (b) discharge capacity vs. cycle number of NaCrO$_2$ in Na cells at 4.8 mA/g. (From [20], with permission of the *Journal of the Electrochemical Society.*)

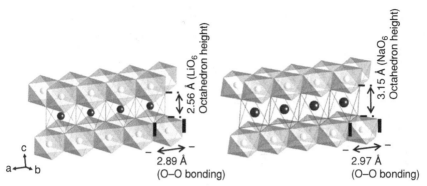

FIGURE 10 Comparison of crystal models of LiCrO$_2$ and NaCrO$_2$. (From [20], with permission of the *Journal of the Electrochemical Society.*)

although both layered LiCrO$_2$ and NaCrO$_2$ possess quite similar crystal structures (*R3m*) the former is inactive in Li cells while the latter is active in rechargeable Na cells. This is attributed to the crystallographic structural differences of the two materials (Fig. 10).

Reviewing the structures of LiCrO$_2$ and NaCrO$_2$ in Figure 10, it is seen that the interstitial tetrahedron (O–O bond length and tetrahedron height are 2.89 and 2.56 Å, respectively) in LiCrO$_2$, which is face-sharing with CrO$_6$ octahedra, matches well with the tetrahedron Cr^{VI}O$_4{}^{2-}$. On the other hand, the interstitial tetrahedron (O–O bond length and tetrahedron height are 2.97 and 3.15 Å, respectively) in an NaO$_2$ slab is considered to match with the tetrahedron Cr^{VI}O$_4{}^{2-}$. Consequently, irreversible lithium intercalation in LiCrO$_2$ is most probably due to irreversible migration of Cr^{VI} into the tetrahedral site by electrochemical oxidation. However, such migration could not be possible for NaCrO$_2$ because of the mismatch of the bond length.

Recently, Xia and Dahn [21] combined electrochemical studies of NaCrO$_2$ with thermal stability measurements using accelerating rate calorimetery and thermal gravimetry. They confirmed the Na insertion and extraction behavior reported by Komba et al. and showed that NaCrO$_2$ is very stable in contact with organic carbonate electrolytes up to 350°C. According to Dahn, NaCrO$_2$ is a safe cathode for use in Na-ion batteries.

Ma et al. studied [22] the usefulness of α-NaMnO$_2$ as a positive electrode for a Na-ion battery. They found that 0.85 Na could be extracted and 0.8 Na inserted back reversibly, corresponding to a 210-mAh/g charge capacity and 197-mAh/g discharge capacity, respectively (Fig. 11). The charge and discharge profiles show a different series of plateaus, indicating different reaction paths with different intermediate phases. Current relaxation in potentiostatic intermittent titration indicated a phase transformation barrier at the 2.63-V plateau, which was confirmed to be a two-phase reaction from x-ray diffraction data. While some capacity fade was observed after 20 cycles, no significant structural change was found after cycling.

Uebou et al. [23] carried out a study of sodium insertion in (MoO$_2$)$_2$P$_2$O$_7$ with liquid organic electrolytes at room temperature. Specific capacities of up to 250 mAh/g were obtained in the first discharge at room temperature. At constant

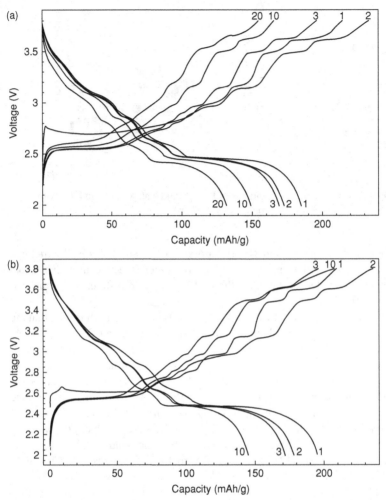

FIGURE 11 Voltage profile of NaMnO$_2$ after multiple cycles at rates of C/10 (a) and C/30 (b), respectively. The cell is cycled galvanostatically between 2.0 and 3.8 V. (From [22], with permission of the *Journal of the Electrochemical Society.*)

current densities of 0.2 mA/cm^2, the reactions were reversible within the range of composition $0.85 < x < 4.0$ for Li and $0.5 < x < 3.1$ for Na in Na$_x$(MoO$_2$)$_2$P$_2$O$_7$ (A: Li, Na), respectively. Structural changes induced by lithium or sodium intercalation followed by ex situ x-ray diffraction measurements indicated crystal-to-amorphous phase transitions in both cases.

Doeff et al. studied [24] the Na insertion extraction behavior of several Mn oxides with tunnel structures, for example Na$_{0.44}$MnO$_2$. This oxide has an unusual double tunnel structure which can accommodate intercalation and de-intercalation of either Na or Li ions without undue stresses, rendering it extremely robust. They concluded that these materials might be useful as cathodes for rechargeable Na batteries.

Berthelot et al. studied [25] the Na insertion and extraction properties of layered $NaCoO_2$, with special attention paid to elucidating the relationship between Na cyclability and crystallographic phase changes with respect to the Na content in the oxide. They showed that a $NaCoO_2$ structure termed two-layer-type phase (P2-Na_xCoO_2) reversibly cycled about 0.6Na per $NaCoO_2$, although several potential plateau regions were observed. Further studies are needed to fully assess the practical potential of this material.

Su et al. [26] studied the Na insertion capacities of thin films of V_2O_5 prepared by a sol–gel technique, and the Na chemical diffusion coefficient was measured using the potentiostatic intermittent-titration technique in a room-temperature ionic liquid prepared from the reaction of methanesulfonyl chloride (MSC) with $AlCl_3$. The diffusion coefficient (DNa^+) in the V_2O_5 film ranges from a minimum of 5×10^{-14} cm^2/s to a maximum of 9×10^{-12} cm^2/s. They found V_2O_5 films to have an insertion, capacity of about 1.6 Na per V_2O_5. Upon sodium insertion, V_2O_5 undergoes a phase change, resulting in a new structure with stable and reproducible cycling properties. Sodium insertion into V_2O_5 in the MSC molten salt shows high coulombic efficiency, up to 98%. They concluded that with further development, V_2O_5 might be suitable as a cathode material in sodium batteries.

Researchers from Argonne National Lab [27,28] recently reported two innovations for rechargeable sodium ion batteries, one a TiO_2 nanotube electrode material for use as an anode and the other a Na intercalating cathode material of the composition $Na_{0.85}Li_{0.17}Ni_{0.21}Mn_{0.64}O_2$. These electrode materials were first studied for their reversible Na capacities at room temperature. The layered structure of $Na_{0.85}Li_{0.17}Ni_{0.21}Mn_{0.64}O_2$ allows for a single-phase reversible Na intercalation reaction demonstrating 100 mAh/g. The cathode can be discharged with high rate (e.g., 65 mAh/gram at a 25C rate). The authors claim that the presence of Li and the electronic ordering of Ni^{II}/Mn^{IV} in the transition metal layer is responsible for material stability in this electrode.

Reversible sodium insertion into TiO_2 proceeds only in amorphous large-diameter nanotubes (>80 nm inside diameter). These electrodes maximized their capacity in continued cycling of cells to reach a reversible capacity of 150 mAh/g in 15 cycles.

The Argonne group also demonstrated a full Na-ion cell with the TiO_2 anode and the layered $Na_{0.85}Li_{0.17}Ni_{0.21}Mn_{0.64}O_2$ cathode in a test cell in which the cell capacity was limited by the mass of the anode. The cell showed an operation voltage of about 1.8 V and a discharge capacity of about 80 mAh/g. The cell voltage was smaller than with a Li-ion battery, but still higher than 1.2 V, the operation voltage of the NiMH battery used in most hybrid batteries the authors noted.) Some cycles for this Na-ion cell are given in Figure 12.

Na-ion batteries exhibiting higher energy densities could be obtained by coupling any of the high-capacity cathodes discussed above with a high-capacity anode that inserts Na reversibly at low voltages. In this respect, high-capacity Na insertion carbon anodes would be attractive. Stevens and Dahn studied [29] the Na insertion capacities and cyclability of graphite and an amorphous hydrogen-containing carbon. They concluded that graphite is not useful as an anode for Na batteries, as it inserted reversibly only small amounts of Na, whereas a nanoporous carbon that contains

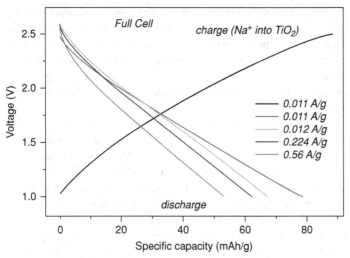

FIGURE 12 Charge–discharge voltage profiles of a Na-ion battery at ambient temperature cycled between 2.6 and 1 V at various rates. (From [26,27], with permission of the *Journal of Physical Chemistry Letters.*)

residual hydrogen in the carbon framework is more attractive. The latter reversibly inserted between 250 and 300 mAh/g of carbon with a sloping potential profile spanning between 1.2 and 0.0 V. Detailed cycling studies of the latter electrode have not yet been carried out. The usefulness of carbon electrodes for Na-ion batteries would be worth investigating further with attention paid to the selection of carbon structures, and electrolytes and additives to from stable electrode–electrolyte interface structures for long-term Na cycling stability.

4 PERSPECTIVES

The development of rechargeable sodium batteries followed two paths: moderate- to high-temperature systems in which the Na anode is separated from the rechargeable cathode by a Na^+-conducting ceramic separator. This cell configuration largely eliminated the rechargeability issue of the Na anode, and the chemical instability between Na anode and the electrolyte or the catholyte. Issues such as materials corrosion, stability of the ceramic Na-ion conductors, relatively high cost, and packaging have prevented their full practical implementation, with only some of these systems currently being considered for applications, primarily in power storage and load leveling. They will continue to be developed, with some of these, especially the moderate-temperature Na batteries utilizing metal chlorides, metal sulfides, or mixtures of metal chlorides and sulfide cathodes, finding future applications.

Room-temperature Na-ion batteries are receiving increased attention following the advent of several high-capacity rechargeable metal oxide cathodes. Suitable anode materials with Na high insertion capacities at low potentials are still lacking for the development of practical Na-ion batteries. The identification of such anode materials is an area for future investigations. Also needed are advanced nonaqueous Na^+-conducting electrolytes and additives to form the stable electrode–electrolyte interface structures necessary for long-lived Na-ion batteries. Despite many technical challenges, interest in the development of rechargeable Na batteries will continue in view of the abundance and low cost of metallic sodium and its insertion compounds for reversible electrodes.

Acknowledgment

Financial support to write this chapter was provided by E-KEM Sciences, Needham, Massachusetts.

REFERENCES

1. N. Weber and J. T. Kummer, *Ann. Power Sources Conf.*, **21**, 37 (1967). See also Automotive Engineering Congress, Detroit, MI, S.A.E. 670179 (1976).
2. S. Gratch, J. V. Petrocelii, R. P. Tischer, R. W. Minck, and T. J. Whalen, *Proceedings of the 7th IECEC Meeting*, 1972, p. 38.
3. C. Levine, *Proceedings of the 10th IECEC Meeting*, 1975, p. 721.
4. S. P. Mitoff and J. B. Bush, Jr., *Proceedings of the 9th IECEC meeting*, 1974, p. 916.
5. K. M. Abraham, R. D. Rauh, and S. B. Brummer, *Electrochim. Acta*, **23**, 501 (1978).
6. K. M. Abraham and J. E. Elliott, *J. Electrochem. Soc.*, **131**, 2211 (1984).
7. K. M. Abraham, M. W. Rupich, and L. Pitts, *J. Electrochem. Soc.*, **128**, 2700 (1981).
8. K. M. Abraham, M. W. Rupuich, and L. Pitts, Fall Meeting of the Electrochemical Society, Detroit, MI, 1982, Extended Abst. 351.
9. K. M. Abraham, M. W. Rupuich, and L. Pitts, *Electrochim. Acta*, **30**, 1635 (1985).
10. D. C. Bogdan and M. Vallance, Fall ECS Meeting, Las Vegas, NV, Oct. 2010, Extended Abstr. 216.
11. R. J. Bones, J. Coetzer, R. C. Galloway, and D. A. Teagle, *J. Electrochem. Soc.*, **134**, 2379 (1987).
12. R. J. Bones, D. A. Teagle, S. D. Brooker, and C. F. Cullen, *J. Electrochem. Soc.*, **136**, 1274 (1989).
13. K. M. Abraham and D. M. Pasquariello, *J. Electrochem. Soc.*, **137**, 1189 (1990).
14. K. M. Abraham, *Solid State Ionics* 7, 199–212 (1982) and references therein.
15. K. M. Abraham, L. Pitts, and R. Schiff, *J. Electrochem. Soc.*, **127**, 2545 (1980).
16. K. M. Abraham and L. Pitts, *J. Electrochem. Soc.*, **128**, 1060 (1981).
17. K. M. Abraham and L. Pitts, *J. Electrochem. Soc.*, **128**, 2575 (1981).
18. G. H. Neman and L. P. Klemann, *J. Electrochem. Soc.*, **127**, 2097 (1980).
19. M. Zanini, J. L. Shaw and G. J. Tennen house, *J. Electrochem. Soc.* **128**, 1647 (1981).
20. S. Komaba, T. Nakayama, A. Ogata, T. Shimizu, C. Takei, S. Takada, A. Hokura, and Y. Naoaki, *ECS Trans.*, **16**, 43 (2009).
21. X. Xia and J. R. Dahn, *Electrochem. Solid-State Lett.*, **15**(1), A1 (2012).
22. X. Ma, H. Chen, and G. Ceder, *J. Electrochem. Soc.*, **158**(12), A1307 (2011).
23. Y. Uebou, S. Okada, and J. Yamaki. *J. Power Sources*, **115**, 119 (2003).
24. M. M. Doeff, T. J. Richardson, and L. Kepley, *J. Electrochem. Soc.*, **143**, 2507 (1996).
25. R. Berthelot, D. Carlier, and C. Delmas. *Nat. Mater.*, **10**, 75 (2011).
26. L. Su, J. Winnick, and P. Kohl, *J. Power Sources*, **101**, 226 (2001).
27. D. Kim, S. H. Kang, M. Slater, S. Rood, J. T. Vaughey, N. Karan, M. Balasubramanian, and C. S. Johnson, *Adv. Energy Mater.*, **1**, 333 (2011).
28. H. Xiong, M. D. Slater, M. Balasubramanian, C. S. Johnson, and T. Rajh, *J. Phys. Chem. Lett.*, **2**(20), 2560–2565 (2011).
29. D. A. Stevens and J. R. Dahn, *J. Electrochem. Soc.*, **148**, A803 (2001).

THE ELECTROCHEMICAL SOCIETY SERIES

Corrosion Handbook
Edited by Herbert H. Uhlig

Modern Electroplating, Third Edition
Edited by Frederick A. Lowenheim

Modern Electroplating, Fifth Edition
Edited by Mordechay Schlesinger and Milan Paunovic

The Electron Microprobe
Edited by T. D. McKinley, K. F. J. Heinrich, and D. B. Wittry

Chemical Physics of Ionic Solutions
Edited by B. E. Conway and R. G. Barradas

High-Temperature Materials and Technology
Edited by Ivor E. Campbell and Edwin M. Sherwood

Alkaline Storage Batteries
S. Uno Falk and Alvin J. Salkind

The Primary Battery (in Two Volumes)
Volume I *Edited by* George W. Heise and N. Corey Cahoon
Volume II *Edited by* N. Corey Cahoon and George W. Heise

Zinc-Silver Oxide Batteries
Edited by Arthur Fleischer and J. J. Lander

Lead-Acid Batteries
Hans Bode
Translated by R. J. Brodd and Karl V. Kordesch

Thin Films-Interdiffusion and Reactions
Edited by J. M. Poate, M. N. Tu, and J. W. Mayer

Lithium Battery Technology
Edited by H. V. Venkatasetty

Quality and Reliability Methods for Primary Batteries
P. Bro and S. C. Levy

Techniques for Characterization of Electrodes and Electrochemical Processes
Edited by Ravi Varma and J. R. Selman

Electrochemical Oxygen Technology
Kim Kinoshita

Synthetic Diamond: Emerging CVD Science and Technology
Edited by Karl E. Spear and John P. Dismukes

Corrosion of Stainless Steels
A. John Sedriks

Semiconductor Wafer Bonding: Science and Technology
Q.-Y. Tong and U. Göscle

Uhlig's Corrosion Handbook, Second Edition
Edited by R. Winston Revie

Atmospheric Corrosion
Christofer Leygraf and Thomas Graedel

Electrochemical Systems, Third Edition
John Newman and Karen E. Thomas-Alyea

Fundamentals of Electrochemistry, Second Edition
V. S. Bagotsky

Fundamentals of Electrochemical Deposition, Second Edition
Milan Paunovic and Mordechay Schlesinger

Electrochemical Impedance Spectroscopy
Mark E. Orazem and Bernard Tribollet

Fuel Cells: Problems and Solutions
Vladimir S. Bagotsky

Lithium Batteries: Advanced Technologies and Applications
Edited by Bruno Scrosati, K. M. Abraham, Walter van Schalkwijk, and Jusef Hassoun

INDEX

Lithium Batteries: Advanced Technologies and Applications, First Edition.
Edited by Bruno Scrosati, K. M. Abraham, Walter van Schalkwijk, and Jusef Hassoun.
© 2013 John Wiley & Sons, Inc. Published 2013 by John Wiley & Sons, Inc.

369

Printed in the United States
By Bookmasters